METHODS IN MOLECULAR BIOLOGY™

Series Editor
John M. Walker
School of Life Sciences
University of Hertfordshire
Hatfield, Hertfordshire, AL10 9AB, UK

For other titles published in this series, go to
www.springer.com/series/7651

Cancer Stem Cells

Methods and Protocols

Edited by

John S. Yu

Department of Neurosurgery, Maxine Dunitz Neurosurgical Institute, Cedars-Sinai Medical Center, Los Angeles, CA, USA

Humana Press

Editor
John S. Yu
Department of Neurosurgery
Maxine Dunitz Neurosurgical Institute
Cedars-Sinai Medical Center
Los Angeles, CA
USA

ISSN 1064-3745 e-ISSN 1940-6029
ISBN 978-1-61779-753-8 e-ISBN 978-1-59745-280-9
DOI 10.1007/978-1-59745-280-9
Springer Dordrecht Heidelberg London New York

Printed on acid-free paper

Springer is part of Springer Science+Business Media (www.springer.com)

This book is dedicated to Helena, Jeffrey, and Lauren whose love and devotion has illuminated my path and made each step joyful.

Foreword

Cancer is a devastating disease that affects millions of people in the world. Our therapies for tumors have mostly been based on classical chemotherapy and antiproliferative treatments. More recently, directed therapies against a causative oncogene have led to prominent reduction in tumor rates for some cancers. For instance, chronic myelogenous leukemia that is due to a BCR–ABL translocation can be targeted with Gleevec.

Despite these advances in individual tumors, a number of patients are treated for their primary tumors and ultimately relapse. Relapsing tumors can be due to resistance to chemotherapy or to antioncogene drugs. Another hypothesis is that heterogeneity in the tumor leads to an inability of classical chemotherapy to completely eradicate all cells of the tumor. This concept of heterogeneity has led some investigators to propose a cancer stem cell model.

As patients are treated with chemotherapy, most of the dividing cells are killed, but this leaves a small subset of cells that have the ability to remake the entire tumor. These are cancer stem cells. They possess the signals of self-renewal and yet can also differentiate. If one could understand more about the cells that remain after classical chemotherapy or the cells that can remake the tumor among the heterogeneous population, this information would have a huge impact on our treatment of cancer.

Over the past 3 years, a number of investigators have developed assays for stem cell populations within tumors. A small subset of cells within a tumor have the ability to be transplantable in mice. The work was initiated in the leukemia field mostly because the cell surface markers exist to purify progenitor and stem cell populations. Cancer stem cells have now been isolated from a number of solid tumors including tumors of the nervous system as well as breast cancer. Signaling events within this stem cell population is distinct from the other cells in the tumor. Pathways that would prevent self-renewal and perhaps further differentiate the stem cells are actively being investigated.

At the core of this field of cancer stem cells is a major question of the cell of origin of the cancer. It is possible that the oncogene transforms a particular stage of differentiated cell in the body and this leads to a cell that can be of acquired self-renewal, thereby allowing the tumor to always recur. Such is the case for the transformation of the hematopoietic stem cell with an oncogene leading to leukemia. It is also possible that a progenitor is transformed and acquires self-renewal in an aberrant fashion, dedifferentiates, or maintains itself in the presence of specific signals. Thus, the transplantable population of cells within a tumor or the self-renewing cell population of cells in a tumor may not equate to the cell that ultimately was transformed at the very beginning of the cancer.

There is much to be done for this field, including the understanding of signaling pathways that affect self-renewal in normal stem cell populations as well as cancer stem cell populations. In addition, each cancer stem cell should be purified to homogeneity and evaluated for such signaling pathways. This will involve the genomics and proteomics fields for finding surface molecules that can be used to purify these populations. Lastly, the field of cancer stem cells in human tumors still remains to be developed. Given that there are very few ways of actually propagating human tumors, this becomes a very difficult task. A number of studies have been done in immunodeficient mice in which tumors can be maintained and cell populations purified. This seems to be a promising approach, yet it still leaves us with placing tumors in an abnormal environment, and the tumors can react and change their particular fate. It is also possible that culture conditions can alter the ability of cells to take on self-renewal programs and/or behave like cancer stem cells. Therefore, we need to have a better understanding of the tumor itself and its heterogeneity and what makes individual cells within a tumor undergo symmetric cell division, metastasize, or invade.

This is a very bright field, and this book has a substantial impact on our understanding of how tumors self-renew or differentiate.

Leonard Zon, Ph.D.

Preface

The concept of cancer stem cells has reinvigorated cancer research with a novel paradigm to study the cause and treatment of cancers. The role of putative cancer stem cells in initiating and supporting cancer has been described by several groups for a growing number of cancers. The characterization of the virulent cancer stem cell has focused the attention of clinician-scientists to this therapeutic target. With the interest of aiding cancer investigators to enter into the cancer stem cell field or to work with scientists who study cancer as a stem cell disease, we have compiled cancer stem cell research techniques and protocols from preeminent researchers in their respective fields. The methods involved in cancer stem cell research have all too often been shrouded in mystery, inhibiting healthy competition, discourse, and collaboration. It is my hope that this volume not only helps to aid a new investigator enter this fascinating field but also "levels the playing field" so that investigators may point to the same or comparable methods when discussing experiments and results.

To be considered as cancer stem cells, clonally derived cells from a tumor specimen must display the following characteristics: (1) they must self-renew and proliferate, (2) they must be able to differentiate and express markers typical of the end terminal cells of that organ (i.e. markers for astrocytes, oligodendrocytes, and neurons in the case of brain tumor stem cells), and (3) they must be able to generate tumors after in vivo transplantation in animal models that resemble the original tumor from patients. With this simplified definition as a point of departure, this volume opens with two animal models of cancer stem cells. Major categories of cancers from which cancer stem cells are derived are represented: leukemia stem cells, brain cancer stem cells, prostate cancer stem cells, pancreatic cancer stem cells, head and neck cancer stem cells, and pituitary adenoma stem cells. Methylation profiling to study cancer stem cells and the contribution of the niche in the regulation of cancer stem cells are explored. The presence and high expression of ABC transporter proteins, antiapoptosis protein, and DNA repair checkpoint protein in cancer stem cells could explain why common therapies, in particular chemotherapy and radiation, are not sufficient to eradicate the tumor. Immunologic targeting or specific targeting of self-renewal mechanisms of cancer stem cells may be an optimal means of targeting and clearing chemo- and radioresistant cancer stem cells. A method for immunologic targeting of cancers stem cells is described as a novel therapeutic strategy and, finally, the use of normal stem cells to treat its evil twin is proposed.

Prospective isolation of cancer stem cells through cell surface marker expression and isolation of cancer stem cells through the formation of spherical cell culture phenotypes are described. These methods are not without controversy. Neurosphere-derived cells are able to differentiate after the withdrawal of growth factors and give rise to a progeny of cells expressing markers of end terminal cells. However, concerns have been raised about the true reflection of "stemness" with the generation of neurospheres. The culturing of tumor cells until the formation of neurospheres precludes the direct comparison of cancer stem cells with the nonstem cell populations. However the use of cell surface markers is not without controversy, as cancer stem cells identified by cell surface markers inevitably have set the stage for the identification of a group of cancer stem cells that do not express the markers of interest. Methods to characterize self-renewal and differentiation are described as well as orthotopic transplantation in immunodeficient animal models. A balanced introduction to the various methods of isolating, characterizing, and propagating cancer stem cells is made with the understanding that each method has its advantages and pitfalls.

Cancer stem cells are likely to share many of the properties of normal stem cells that provide for a long lifespan, including: relative quiescence, resistance to drugs and toxins through the expression of several ABC transporters, an active DNA-repair capacity, and resistance to apoptosis. Several groups have demonstrated that cells expressing stem cell markers from multiple cancer types exhibit resistance to conventional cancer therapies. Recent studies have indicated that the presence of a cancer stem cell population may be highly correlated with prognosis. Hence, therapeutic strategies to target cancer stem cells will be an important goal for clinician-scientists over the years to come.

The field of cancer stem cell research will create the foundation for the understanding of cancer initiation and propagation and the development of novel targets for cancer therapy over the next 10 years. It is our desire that this volume will aid in generating interest and support for this nascent field during this seminal time.

Los Angeles, CA *John S. Yu*

Contents

Contributors

FARID AHMED • *GSF National Research Center for Environment and Health and the Clinical Cooperative Group – Leukemia, Department of Medicine III, Klinikum Grosshadern, Munich, Germany*

LAURIE AILLES • *Division of Stem Cell and Developmental Biology, University Health Network, Ontario Cancer Institute, Toronto, ON, Canada*

SURINDER K. BATRA • *Department of Biochemistry and Molecular Biology, Eppley Institute for Research in Cancer and Allied Diseases, University of Nebraska Medical Center, Omaha, NE, USA*

CHRISTIAN BUSKE • *Helmholtz Center Munich and the Clinical Cooperative Group - Leukemia, Department of Medicine III, Klinikum Grosshadern, Munich, Germany*

TAMMY CALHOUN-DAVIS • *Department of Carcinogenesis, the University of Texas M.D. Anderson Cancer Center, Smithville, TX, USA*

KAISORN L. CHAICHANA • *Brain Tumor Stem Cell Laboratory, Department of Neurosurgery, Johns Hopkins School of Medicine, Baltimore, MD, USA*

ANIRUDDHA J. DESHPANDE • *GSF National Research Center for Environment and Health and the Clinical Cooperative Group – Leukemia, Department of Medicine III, Klinikum Grosshadern, Munich, Germany*

ALBERT D. DONNENBERG • *Hillman Cancer Center, Pittsburgh, PA, USA*
Division of Hematology and Oncology, Department of Medicine, University of Pittsburgh School of Medicine, Pittsburgh, PA, USA

VERA S. DONNENBERG • *Hillman Cancer Center, Pittsburgh, PA, USA*
Department of Surgery, Heart, Lung and Esophageal Surgery Institute, University of Pittsburgh School of Medicine, Pittsburgh, PA, USA

MICHAEL C. DOVEY • *Harvard Medical School, Boston, MA, USA*

GUOPING FAN • *Department of Human Genetics and Institute of Stem Cell Biology and Medicine, David Geffen School of Medicine, UCLA, Los Angeles, CA, USA*

BRUNHILDE FELDING-HABERMANN • *Burham Institute, La Jolla, CA, USA*

GAETANO FINOCCHIARO • *Unit of Experimental Neuro-Oncology, Fondazione IRCCS Istituto Neurologico Besta, Milano, Italy*

SHAUN D. FOUSE • *Department of Human Genetics and Institute of Stem Cell Biology and Medicine, David Geffen School of Medicine, UCLA, Los Angeles, CA, USA*

HUGO GUERRERO-CÁZARES • *Brain Tumor Stem Cell Laboratory, Department of Neurosurgery, Johns Hopkins School of Medicine, Baltimore, MD, USA*

SIMON W. HAYWARD • *Department of Urologic Surgery, Vanderbilt University Medical Center, Nashville, TN, USA*

SOFIA HONORIO • *Department of Carcinogenesis, the University of Texas M.D. Anderson Cancer Center, Smithville, TX, USA*

COLLENE R. JETER • *Department of Carcinogenesis, the University of Texas M.D. Anderson Cancer Center, Smithville, TX, USA*

MING JIANG • *Department of Urologic Surgery, Vanderbilt University Medical Center, Nashville, TN, USA*

KYEUNG MIN JOO • *Department of Neurosurgery, Samsung Medical Center, Cancer Stem Cell Research Center, Sungkyunkwan University School of Medicine, Seoul, Korea*

YIBIN KANG • *Department of Molecular Biology, Princeton University, Princeton, NJ, USA*

JENNIFER KATZ • *Burham Institute, La Jolla, CA, USA*

HARLEY I. KORNBLUM • *Pediatrics, Pharmacology, and Psychiatry, David Geffen School of Medicine at UCLA, Los Angeles, CA, USA*

JOSEPH KRUEGER • *Burham Institute, La Jolla, CA, USA*

C.J. LEE • *Department of Surgery, University of Michigan, Ann Arbor, MI, USA*

CHENWEI LI • *Department of Surgery, University of Michigan, Ann Arbor, MI, USA*

HANGWEN LI • *Department of Carcinogenesis, the University of Texas M.D. Anderson Cancer Center, Smithville, TX, USA*

Division of Nutritional Science, Department of Human Ecology, The University of Texas at Austin, Austin, TX, USA

E. MICHAEL MEYER • *Hillman Cancer Center, Pittsburgh, PA, USA*

MURIELLE MIMEAULT • *Department of Biochemistry and Molecular Biology, Eppley Institute for Research in Cancer and Allied Diseases, University of Nebraska Medical Center, Omaha, NE, USA*

ICHIRO NAKANO • *Departments of Neurosurgery and Pediatrics, David Geffen School of Medicine at UCLA, Los Angeles, CA, USA*

DO-HYUN NAM • *Department of Neurosurgery, Samsung Medical Center, Cancer Stem Cell Research Center, Sungkyunkwan University School of Medicine, Seoul, Korea*

LUBNA PATRAWALA • *Department of Carcinogenesis, the University of Texas M.D. Anderson Cancer Center, Smithville, TX, USA*

SERENA PELLEGATTA • *Unit of Experimental Neuro-Oncology, Fondazione IRCCS Istituto Neurologico Besta, Milano, Italy*

MARK PRINCE • *Department of Otolaryngology-HNS, University of Michigan, Ann Arbor, MI,* USA

ALFREDO QUIÑONES-HINOJOSA • *Brain Tumor Stem Cell Laboratory, Department of Neurosurgery, Johns Hopkins School of Medicine, Baltimore, MD, USA*

YIN SHEN • *Department of Human Genetics and Institute of Stem Cell Biology and Medicine, David Geffen School of Medicine, UCLA, Los Angeles, CA, USA*

DIANE M. SIMEONE • *Department of Surgery, University of Michigan, Ann Arbor, MI, USA*

JULIE B. SNEDDON • *Howard Hughes Medical Institute, Harvard University, Cambridge, MA, USA*

Evan Y. Snyder • *Burham Institute, La Jolla, CA, USA*

Dean G. Tang • *Department of Carcinogenesis, the University of Texas M.D. Anderson Cancer Center, Smithville, TX, USA*

Program in Molecular Carcinogenesis, Graduate School of Biomedical Sciences (GSBS), Houston, TX, USA

Patrizia Tunici • *Maxine Dunitz Neurosurgical Institute, Cedars-Sinai Medical Center, Los Angeles, CA, USA*

Shunyou Wang • *Department of Pharmacology, UCLA, Los Angeles, CA, USA*

John S. Yu • *Department of Neurosurgery, Maxine Dunitz Neurosurgical Institute, Cedars-Sinai Medical Center, Los Angeles, CA, USA*

Leonard I. Zon • *Stem Cell Program and Division Hematology/Oncology Children's Hospital and Dana Farber Cancer Institute, Howard Hughes Medical Institute, Harvard Stem Cell Institute, Harvard Medical School, Boston, MA, USA*

Chapter 1

Defining Cancer Stem Cells by Xenotransplantation in Zebrafish

Michael C. Dovey and Leonard I. Zon

Summary

The zebrafish (*Danio rerio*) has become an increasingly utilized and relevant model organism in the study of cancer. The use of transgenic and reverse genetic approaches has yielded several strains that model a variety of human neoplasms. In addition to modeling human disease, these strains provide a platform for the analysis of tumor stem cells. Here we describe the basic technique for the isolation and transplantation of tumor tissue in the zebrafish. This technique was designed to study metastasis and invasive potential of zebrafish tumor cells. Additionally, the basic protocol can be modified in order to describe cancer stem cell characteristics, including proliferative capacity, self-renewal, and the minimum number of tumor cells required for engraftment.

Key words: Zebrafish, Cancer stem cells, Transplantation, Metastasis

1. Introduction

As increasing numbers of cancer studies utilize zebrafish *(1–5)*, an array of standardized assays must be developed in order to characterize the tumors developed. One qualitative phenotype to characterize includes invasiveness. In mice, many tumors engraft when transplanted into immunocompromised recipients. Similarly, the injection of tumor tissue into sublethally irradiated zebrafish can be used to describe the invasive potential of a tumor *(4, 5)*.

Transplantation can also provide information about the cancer stem cells of a tumor (*see* **Note 1**). The theory of tumor stem cells is predicated on the observation that some tumor cells behave similar to somatic stem cells *(6)*. Somatic stem cells are defined by their unlimited proliferative capacity and the ability

John S. Yu (ed.), *Cancer Stem Cells*, Methods in Molecular Biology, vol. 568
DOI 10.1007/978-1-59745-280-9_1, © Humana Press, a part of Springer Science+Business Media, LLC 2009

to differentiate into other cell types. Similarly, some cancer cells have been found that exhibit an expanded proliferative potential and the ability, when transplanted into immunodeficient recipients, to give rise to heterogeneous tumors *(7–10)*. Tumor stem cells, then, are a subset of cells within a tumor that contain these two fundamental somatic stem cell properties *(11)*.

Successful serial transplantation of tumor cells provides evidence for both expanded proliferative potential and self-renewal capacity. The genesis of a tumor from serially transplanted tissue is indicative of the unlimited proliferative capacity because it shows that at least some tumor cells are capable of unlimited proliferation. Additionally, the repeated generation of heterogeneous tumors is proof of self-renewal activity by a tumor stem cell. Finally, in order to estimate the approximate number of tumor stem cells within a tumor, limiting dilution transplantations can be done.

2. Materials

2.1. Irradiation of Recipients

1. 60 × 15 mm Petri dish (Falcon)
2. Tricaine-S, 2 g/l (Western Chemical, Inc.)

2.2. Preparation of Tumor Tissue

1. 4 mg/ml tricaine stock (Fisher)
2. 0.9× PBS
3. Heat-inactivated fetal bovine serum
4. Razor blades
5. 100 × 20 mm Petri dish (Fisher)
6. 10-ml pipettes
7. 40-μm cell strainer (BD Falcon)
8. 50-ml conical tube
9. 0.4% trypan blue (Gibco)
10. Liberase 3 Blendzyme, 9.5 mg/ml stock (Roche)

2.3. Transplantation

1. 701 N 10 μl Syringe (Hamilton)

3. Methods

3.1. Irradiation of Recipients: Day 0

1. Irradiation of recipient fish must occur 24–48 h prior to transplantation (*see* **Note 2**).

2. Anesthetize five recipient fish and transfer to a 60 × 15 mm Petri dish containing 0.02% tricaine in fish water (*see* **Note 3**).
3. Expose fish to a final dose of 23 Gy using a ^{137}Cesium source irradiator (*see* **Note 4**).
4. Let fish recover in fish water.

3.2. Preparation of Tumor Tissue: Day 2

1. Kill tumor-bearing fish in an overdose of tricaine. Dissect out the tumor.
2. Transfer tumor tissue to a Petri dish with 1 ml 0.9× PBS and mince tumor with a razor blade.
3. Add 8.5 ml 0.9× PBS and 100 μl Liberase stock solution to Petri dish and pipette up and down to promote disaggregation. Working concentration of liberase should be 0.1 mg/ml.
4. Incubate at room temperature for 30 min, pipetting vigorously every 10 min to promote dissaggregation.
5. Add 500 μl fetal bovine serum (FBS) to stop enzymatic digestion and filter over a 40-um cell strainer into a 50-ml falcon tube. Rinse Petri dish with 5 ml 0.9× PBS supplemented with 5% FBS and filter into the same 50-ml tube.
6. Centrifuge filtered cells at 1,000×*g* for 5 min. Resuspend tumor cells in 1 ml 0.9× PBS supplemented with 5% FBS.
7. Add 10 μl of the tumor cell solution to 90 μl of trypan blue.
8. Count the number of live cells on a hemacytometer.
9. Spin sample and resuspend cells to the desired concentration in 0.9× PBS.

3.3. Transplantation: Day 2

1. Clean syringe by withdrawing and extruding the following solutions into the syringe: 10% bleach followed by two separate solutions of 100% ethanol followed by 0.9× PBS.
2. Anesthetize one previously irradiated recipient fish in 0.02% tricaine (*see* **Note 5**).
3. Withdraw 10 μl of tumor cell suspension into the syringe and inject cells into the intraperitoneal cavity of a lightly anesthetized recipient fish (*see* **Note 6** and 7). Repeat **steps 2** and **3** for all remaining recipient animals.
4. Clean syringe by withdrawing and extruding following solutions into the syringe: 10% bleach followed by two separate solutions of 100% ethanol.
5. Analyze transplant recipients for engraftment 7, 14, and 21 days post injection (*see* **Note 8**).

4. Notes

1. It is important to note that transplantation alone is incapable of identifying tumor stem cells. Instead, transplantation as it is described here is a technique that probes the stem cell-like properties in a tumor. The isolation of cancer stem cells requires the use of cell-surface antigens or some other genetic marker that, when coupled with fluorescence activated cell sorting, identifies a subpopulation in a tumor that is uniquely capable of engraftment.
2. Irradiation of recipients should be done in the late afternoon of Day 0 to ensure that transplantation on Day 2 occurs within the 24–48 h window.
3. Anesthetizing recipient animals is optional but may make it easier to transfer them into Petri dishes.
4. Irradiation doses from 20 to 25 Gy have been used successfully for irradiation.
5. Anesthetic is not required but does make handling fish easier.
6. The volume of cells to be transplanted is dependent on the size of the recipient fish and the site of injection. Large adult fish have greater intraperitoneal space and can therefore easily handle 10 μl of injected fluids. Young adults or fish kept at high density may be limited to 5 μl of injected fluid.
7. Tumors can also be transplanted subdermally or intramuscularly in the side of a recipient animal. Although this is more technically difficult, visualizing engraftment, especially when tumor cells are not fluorescently labeled, is often easier.
8. It is not uncommon for engrafted tumors to regress starting 2 weeks post irradiation. This is consistent with the kinetics of hematopoetic recovery and return of the immune system *(12)*. To minimize this effect, transplant recipients should be of a similar genetic background as the tumor-generating fish. The genetic similarities of common zebrafish strains have been described *(13)*.

References

1. Amatruda, J.F., J.L. Shepard, H.M. Stern, and L.I. Zon, *Zebrafish as a cancer model system.* Cancer Cell, 2002. 1(3): p. 229–31.
2. Berghmans, S., R.D. Murphey, E. Wienholds, D. Neuberg, J.L. Kutok, C.D. Fletcher, J.P. Morris, T.X. Liu, S. Schulte-Merker, J.P. Kanki, R. Plasterk, L.I. Zon, and A.T. Look, *tp53 mutant zebrafish develop malignant peripheral nerve sheath tumors.* Proc Natl Acad Sci U S A, 2005. 102(2): p. 407–12.
3. Haramis, A.P., A. Hurlstone, Y. van der Velden, H. Begthel, M. van den Born, G.J. Offerhaus, and H.C. Clevers, *Adenomatous polyposis coli-deficient zebrafish are susceptible to digestive tract neoplasia.* EMBO Rep, 2006. 7(4): p. 444–9.
4. Langenau, D.M., D. Traver, A.A. Ferrando, J.L. Kutok, J.C. Aster, J.P. Kanki, S. Lin, E. Prochownik, N.S. Trede, L.I. Zon, and A.T. Look, *Myc-induced T cell leukemia*

in transgenic zebrafish. Science, 2003. 299 (5608): p. 887–90.

5. Patton, E.E., H.R. Widlund, J.L. Kutok, K.R. Kopani, J.F. Amatruda, R.D. Murphey, S. Berghmans, E.A. Mayhall, D. Traver, C.D. Fletcher, J.C. Aster, S.R. Granter, A.T. Look, C. Lee, D.E. Fisher, and L.I. Zon, *BRAF mutations are sufficient to promote nevi formation and cooperate with p53 in the genesis of melanoma.* Curr Biol, 2005. 15(3): p. 249–54.
6. Pardal, R., M.F. Clarke, and S.J. Morrison, *Applying the principles of stem-cell biology to cancer.* Nat Rev Cancer, 2003. 3(12): p. 895–902.
7. Singh, S.K., I.D. Clarke, M. Terasaki, V.E. Bonn, C. Hawkins, J. Squire, and P.B. Dirks, *Identification of a cancer stem cell in human brain tumors.* Cancer Res, 2003. 63(18): p. 5821–8.
8. Al-Hajj, M., M.S. Wicha, A. Benito-Hernandez, S.J. Morrison, and M.F. Clarke, *Prospective identification of tumorigenic breast cancer cells.* Proc Natl Acad Sci U S A, 2003. 100(7): p. 3983–8.
9. Bonnet, D. and J.E. Dick, *Human acute myeloid leukemia is organized as a hierarchy that originates from a primitive hematopoietic cell.* Nat Med, 1997. 3(7): p. 730–7.
10. Lapidot, T., C. Sirard, J. Vormoor, B. Murdoch, T. Hoang, J. Caceres-Cortes, M. Minden, B. Paterson, M.A. Caligiuri, and J.E. Dick, *A cell initiating human acute myeloid leukaemia after transplantation into SCID mice.* Nature, 1994. 367(6464): p. 645–8.
11. Tan, B.T., C.Y. Park, L.E. Ailles, and I.L. Weissman, (2006) *The cancer stem cell hypothesis: a work in progress.* Lab Invest, 86(12): p. 1203–7.
12. Traver, D., A. Winzeler, H.M. Stern, E.A. Mayhall, D.M. Langenau, J.L. Kutok, A.T. Look, and L.I. Zon, (2004) *Effects of lethal irradiation in zebrafish and rescue by hematopoietic cell transplantation.* Blood, 104(5): p. 1298–305.
13. Guryev, V., M.J. Koudijs, E. Berezikov, S.L. Johnson, R.H. Plasterk, F.J. van Eeden, and E. Cuppen,(2006) *Genetic variation in the zebrafish.* Genome Res, 16(4): p. 491–7.

Chapter 2

Analysis of Cancer Stem Cell Metastasis in Xenograft Animal Models

Yibin Kang

Summary

Metastatic spread of cancer cells from the primary tumors to distant vital organs, such as lung, liver, brain, and bone, is responsible for the majority of cancer-related deaths. Cancer stem cells are likely to play essential roles in the metastatic spread of primary tumors because of their self-renewal capability and their potential to give rise to differentiated progenies that can adapt to different target organ microenvironments. Investigating the metastatic behavior of cancer stem cells (CSCs) is critical for the development of more effective therapies to prevent or delay the progression of malignant diseases. Animal models have been developed to mimic the multistep process of metastasis to various target organs. In this chapter, I will describe several xenograft methods to introduce human breast cancer cells into nude mice in order to generate spontaneous and experimental metastases. Similar experimental approach can be applied to analyze the metastatic behavior of CSCs derived from other tumor types.

Key words: Metastasis, Xenograft, Mammary fat pad injection, Intravenous injection, Intracardiac injection, Animal model, In vivo imaging

1. Introduction

Metastasis represents the most devastating stage of cancer progression and has attracted intense research efforts in recent years *(1–3)*. In order to successfully colonize a distant target organ, metastatic carcinoma cells need to overcome several rate-limiting hurdles, including the degeneration of epithelial polarity, the invasion and migration through extracellular matrix and basement membrane, intravasation, and survival in blood circulation, and finally, extravasation and adaptation to a foreign microenvironment *(4)*. It is believed that the survival and outgrowth of tumor

John S. Yu (ed.), *Cancer Stem Cells,* Methods in Molecular Biology, vol. 568
DOI 10.1007/978-1-59745-280-9_2, © Humana Press, a part of Springer Science+Business Media, LLC 2009

cells in the target organ is the most challenging step in the entire metastatic cascade *(5)*. Due to their intrinsic self-renewal and differentiation capabilities, cancer stem cells are likely to play essential roles in the seeding and outgrowth of a heterogeneous population in a macrometastasis. Understanding the cellular origin of cancer stem cells and the role of cancer stem cell in metastasis has great potential implications in the prognosis, detection, prevention, and treatment of metastatic cancer *(6)*.

Many important aspects of cancer biology, such as cell cycle control or apoptosis, have been routinely modeled and studied in cell cultures or even in test tubes. However, the multistep process of metastasis, which involves intricate interactions between tumor cells and the neighboring host environment, can hardly be faithfully recapitulated in vitro. Therefore, animal models have been indispensable for studying the basic mechanisms of cancer metastasis *(7, 8)*. Depending on the method of tumor introduction and the relationship between the tumor and the host, animal metastasis models can be classified into several different categories. Spontaneous metastases is relatively rare in mouse tumors and often do not reflect the full spectrum of organ tropism typically seen in human cancers. Therefore, the study of tissue-specific metastasis has relied heavily on transplantable models of metastasis *(2, 8–13)*.

Transplantable models of metastasis can be classified into syngeneic models and xenograft models. In syngeneic models, the transplanted tumors and the host animal are from the same species and same genetic background so that the tumor cells will not be rejected by the host. However, syngeneic models may lack the genetic complexity of human tumors. In addition, mechanisms of human cancer metastasis may not be fully conserved in animal models. Xenograft models often involve the introduction of human tumor cell lines or tumor tissue fragments into immunocompromised animals, such as the nude mice (deficient in T-cell function), the SCID mice (deficient in T-cell and B-cell function), the beige mice (deficient in NK cells), or mouse strains with combinations of different deficiencies. The disadvantage of xenograft models is that the species difference may limit tumor-stromal interactions. In addition, the use of immunocompromised animals limits the opportunity to examine the key role of the immune system in metastasis.

Among transplantable models, injection of tumor cells into anatomically relevant sites (orthotopic injections) has the advantage of generating physiological relevant "primary tumors" that lead to spontaneous metastases to different distant sites. For example, breast cancer cells are often injected into the mammary fat pad (MFP) to generate primary tumors in the mammary gland. Alternatively tumor cells can be introduced directly into blood circulation, a method that is often termed "experimental metastasis." This method bypasses the early steps of the metastasis cascade and can generate metastasis much more efficiently to facilitate meaningful statistical analysis. Additionally it has the advantage of the shorter incubation time for metastasis development and the higher possibility of generating metastases in organs that may be difficult or even impossible to produce using orthotopic injections. However, experimental metastasis models do not encompass the initial stages of primary tumor growth, invasion, and metastasis from an orthotopic site. The route of inoculation greatly influences the metastasis pattern. For example, lateral tail vein injection results primarily in pulmonary metastases because the lung is the first capillary bed that tumor cells encounter following tail vein injections. Intracardiac injection of cells allows widespread distribution of tumor cells throughout the body of the animals and may produce metastases in multiple sites. In summary, each animal metastasis model has its advantages as well as shortcomings and should be evaluated and chosen carefully depending on the specific questions that need to be addressed in the study.

In this chapter, I will describe several commonly used xenograft methods for the analysis of breast cancer metastasis, including MFP injection to generate spontaneous metastasis of breast cancer from the mammary gland, and intravenous as well as intracardiac injections to generate experimental metastases in lung, bone, and other organs. These procedures can be applied to analyze bulk tumor population or purified CSC populations. Similar experimental metastasis models can also be applied to the analysis of tumor cells from other tumor types.

2. Materials

2.1. Mice

For analysis of breast cancer cells, we use female nude mice of about 5–6 weeks old at the beginning of the experiment. We commonly use NCr-nu/nu strain for intracardiac injections and BALB/c-nu/nu strain for tail vein and MFP injections. Mice can be ordered from the NCI animal production program, The Jackson Laboratory, the Charles River Laboratories, and other vendors (*see* **Note 1**).

2.2. Tumor Cells

Human tumor cell lines, free of mycoplasma and murine pathogenic viruses (reovirus type 3, pneumonia virus, K virus, Theiler's encephalitis virus, Sendai virus, minute virus, mouse adenovirus, mouse hepatitis virus, lymphocytic choriomeningitis virus, ectromelia virus, lactate dehydrogenase virus). Checking the cell lines for these viruses will reduce the risk of introducing pathogens into the animal facility. In addition, contamination of cell lines will dramatically influence their metastatic behaviors in vivo.

2.3. Media and Solutions

1. Cell culture medium with serum and antibiotics (penicillin, streptomycin, and fungizone).
2. PBS without Ca^{2+} and Mg^{2+}.
3. Trypsin–EDTA: 0.25% w/v trypsin and 0.02% w/v EDTA in PBS without Ca^{2+} and Mg^{2+}.
4. Anesthetic solution: make 10 mg/ml ketamine and1 mg/ml xylazine from stock solutions (Sigma, K-113 or K4138). Store the solution at 4°C. Use 10 μl of anesthetic solutions per 1 g weight of mice.
5. D-Luciferin substrate solution for bioluminescence imaging: dissolve D-Luciferin (from Xenogen, Promega, or Nanolight) in PBS to make 15 mg/ml solutions. Aliquots can be stored with protection from light at –80°C for several months. Use 100 μl/mouse for bioluminescence imaging.

2.4. Surgical Equipment

1. Sterile instruments and surgical supplies for surgery, cloth tape, alcohol wipes, scissvors, pins, 12-mm wound clips, and wound clip applier (Harvard Apparatus).
2. Mouse-restraint device for intravenous injections.
3. Sterile 1-ml tuberculin syringes, insulin needles (for retro-orbital injection of luciferin solution), 26 gauge (G) (for intracardiac injections) and 27G (for intravenous injections) × 1/2in. (13 mm) needles. 10 μl Hamilton syringe with a 26G needle for MFP injections.
4. Dissecting microscope.
5. Warming lamp and warming pad.
6. Calipers for tumor measurements.

3. Methods

3.1. Preparation of Tumor Cell Suspensions

The following procedure is used to prepare single-cell suspensions from bulk (unsorted) tumor cell culture. CSCs are often enriched through fluorescence-activated cell sorting (FACS) from clinical tumor specimens or cancer cell cultures using methods provided in other chapters. CSC-enriched cell suspensions (with appropriate cell concentration) obtained from FACS experiments can be used directly in transplantation experiments.

1. Aspirate culture medium from rapidly growing cancer cell culture that is 75–90% confluent. Wash with 10 ml of PBS per 10 cm dish; add 1–2 ml of the trypsin–EDTA solution. Shake the dish to cover the cells with trypsin and incubate until the

cells begin to round up under the microscope. Shake or tap the dish to detach the cells (*see* **Note 2**).

2. Resuspend the cells in 10 ml of culture medium and transfer to a centrifuge tube. Determine total cell number by using a hemacytometer. Centrifuge at 200 × *g* for 5 min. at 4°C, and resuspend the cell pellet in 10 ml PBS.
3. Centrifuge again at 200 × *g* for 5 min. at 4°C and resuspend the cells in an appropriate volume of PBS to reach the desired concentration of cells for each injection method (see later). Make sure to have more than twice the amount of cells that are actually needed for the injection into mice.
4. Place the suspension in ice and proceed immediately to inject the cells.

3.2. Mammary Fat Pad Injection for Spontaneous Metastasis Assay

1. Make a 10^7/ml cell suspension for MFP injection as in **Subheading 3.1**. 10 μl of the cell suspension will be injected into each mouse (*see* **Note 3**).
2. Anesthetize mice by using intraperitoneal injection of ketamine at 100 mg/kg and xylazine 10 mg/kg.
3. Invert tube or vortex the cell suspension gently to mix settled cells. Be certain that the cells are free of aggregates to prevent embolic obstruction (*see* **Note 2**). Gently mix the cells periodically and prior to each inoculation.
4. Use scissors to make an incision along the abdominal midline and laterally between the fourth and fifth nipples midway down to right hind leg (ventral side up). Carefully separate the skin flap from the body wall. The separated skin flap should be pinned to the surgery board thus exposing the #4 MFP.
5. Locate MFP under a dissection microscope. Use a 10-μl Hamilton syringe with a 26-G needle to 10 μl of cell suspension into the intact fat pad (below the draining lymph node). Close the wound by using 4–5 surgical staples.
6. Perform the same procedure for the other #4 mammary gland, if necessary.
7. Transplanted mice will then be imaged immediately after surgery using bioluminescence imaging as time 0 point (*see* **Subheading 3.5**).
8. The primary tumor outgrowth should be monitored weekly by taking measurements of the tumor length (L) and width (W). Tumor volume was calculated as $\pi LW^2/6$. For metastasis assays, tumors should be surgically resected when they reached a volume greater than 300 mm^3. After resection, the mice should continue to be monitored by bioluminescent imaging for the development of metastases.

3.3. Intravenous (Tail Vein) Injection for Experimental Lung Metastasis Assay

1. Make a 2×10^6/ml cell suspension for intravenous (I.V.) injection as in **Subheading 3.1**. 100 μl of the cell suspension will be injected into each mouse.
2. Use a heating lamp to warm up the mice in order to dilate the tail veins (*see* **Note 4**)
3. Immobilize mice by using a mouse-restraint device (**Fig. 1**) and without anesthesia.
4. Invert tube or vortex the cell suspension gently to mix settled cells (*see* **Note 2**). Be certain that the cells are free of aggregates to prevent embolic obstruction. Gently mix the cells periodically and prior to each inoculation.

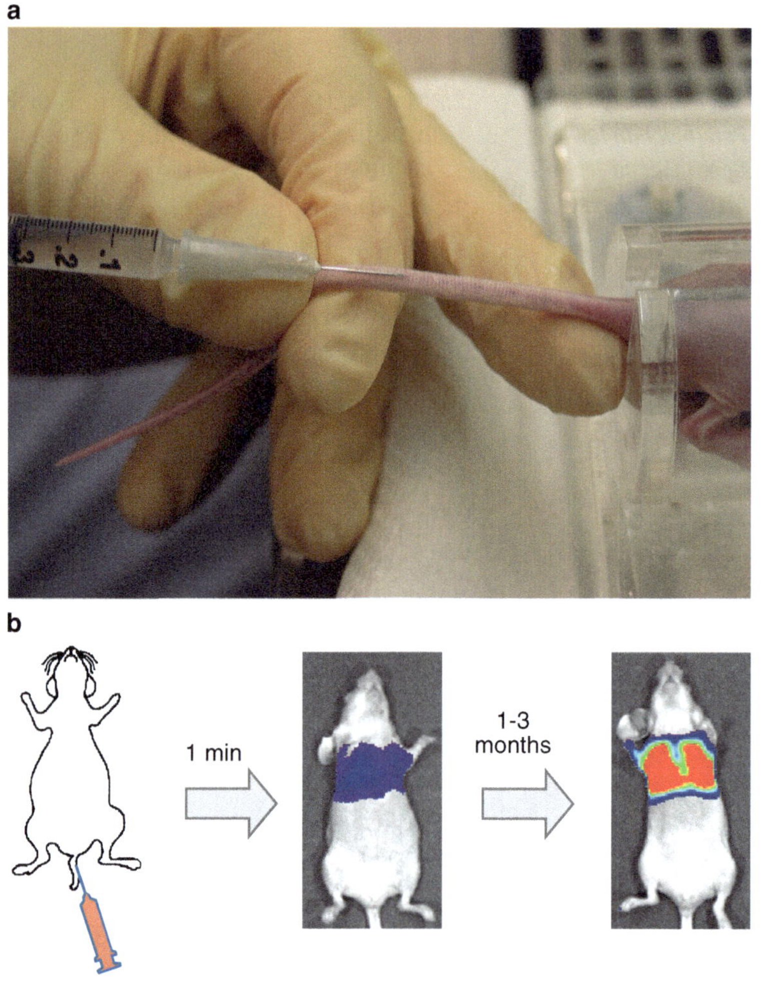

Fig. 1. The intravenous injection method and the development of lung metastasis after injection. (**a**) A mouse was stabilized by a mouse-restraint device and tumor cells were injected via the lateral tail vein. Note the positioning of needle and the angle of injection. (**b**) Bioluminescence imaging was performed immediately after injection to visualize the localization of luciferase-labeled tumor cells in the lung. The increase of the lung metastasis burden over time can be quantified by bioluminescence imaging.

5. Inject 2×10^5 cells (100 μl) into the lateral tail vein through a 27-G needle attached to a 1.0-ml syringe. The needle is inserted with its bevel pointing downward, at an angle of that is almost parallel to the vein (**Fig. 1**). Apply a pressure on the needle downward toward the vein while moving the needle straight forward (*see* **Note 5**). Injections should begin close to the tip of the tail in order to preserve injection sites further up the tail if necessary.
6. Following injections, pressure should be applied by gently holding cotton or gauze over the injection site for approximately 30 s to stop bleeding and prevent hematoma formation.
7. Transplanted mice should then be imaged immediately after surgery using bioluminescence imaging as time 0 point (*see* **Subheading 3.5**). These recipient mice should be imaged weekly to track the development of lung metastasis.

3.4. Intracardiac Injection for Experimental Bone and Brain Metastasis Assay

1. Make a 10^6/ml cell suspension for intracardiac injection as in **Subheading 3.1**. 100 μl of the cell suspension will be injected into each mouse.
2. Anesthetize mice by using intraperitoneal injection of ketamine at 100 mg/kg and xylazine 10 mg/kg.
3. Invert tube or vortex the cell suspension gently to mix settled cells (*see* **Note 2**). Be certain that the cells are free of aggregates to prevent embolic obstruction. Gently mix the cells periodically and prior to each inoculation.
4. Once the mouse is fully anesthetized, place it on its back with arms and legs extended (**Fig. 2**).
5. Place a piece of tape (cloth tape is better as it does not irritate the skin of nude mice and is less sticky) across the abdomen to secure the animal firmly on the working surface. Then tape the upper extremities in the extended position (**Fig. 2**). Do not place tape across the abdomen too tightly or the mouse will not be able to breathe properly. It is imperative that the mouse is symmetrically positioned.
6. When the mouse is firmly secured and not moving, wipe the chest with 70% ethanol. This will enable you to visualize the landmarks as well as cleaning and sterilizing the inoculation site.
7. Gently mix the cell suspension. Leave an air space or bubble of about 200 μl near to the plunger and draw up the cell suspension into a syringe and attach a 26-G needle. Tap the syringe with the needle pointing down to make sure all air bubbles are removed from the hub of the needle while maintaining the air bubble near the plunger. This is an important step as it allows the spontaneous entrance of blood into the hub of the needle when the left ventricle is entered. The arterial blood cannot enter the syringe if no air space is left or if the needle

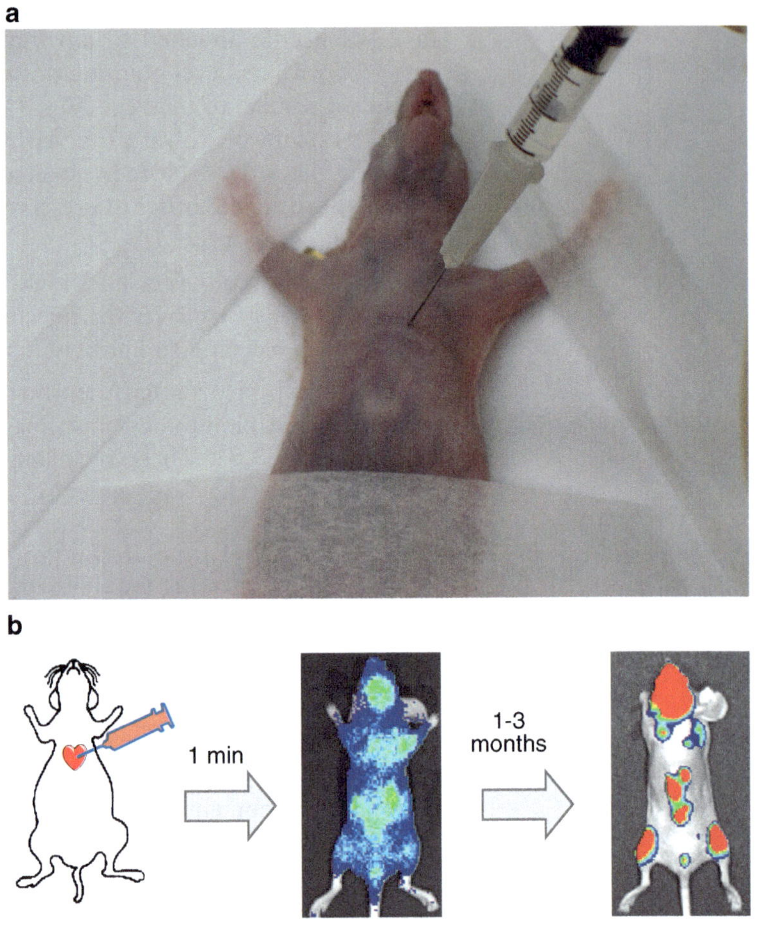

Fig. 2. The intracardiac injection method and the development of bone and brain metastasis after injection. (**a**) A mouse was anesthetized and stabilized with cloth tapes. Tumor cells were injected via the lateral tail vein. Note the positioning of needle and the angle of injection. (**b**) Bioluminescence imaging was performed immediately after injection to visualize the diffusive distribution of luciferase-labeled tumor cells throughout the mouse. The development of metastases in the brain, spine, and limbs can be detected and quantified by noninvasive bioluminescence imaging.

is blocked. It is imperative that a new needle is used for each inoculation as blood can coagulate and clog the needle; however, the same syringe can be used.

8. Insert the needle, at an angle that is 45° to the right, 45° to the horizontal plane and pointing toward you. Insert the needle into the second intercostal space 3 mm to the left of the sternum, directing the tip into the center of the chest, to a depth of 6 mm. Pulsatile flow of red blood into the hub of the needle will indicate correct placement of the needle in the left ventricle, and gentle turning of the needle may be needed if red blood does not appear immediately (*see* **Note 6**). Steady the syringe and slowly inject 0.1 ml of suspension with one hand over 20–30 s. Do not inject the air bubble.

9. Quickly withdraw the needle from the chest to prevent seeding of tumor cells into the heart and the lung. Place pressure on the chest with alcohol wipes for about thirty seconds to stop bleeding. Place the mouse on a heating pad to accelerate recovery and observe for bleeding or other unusual behavior. Return the mouse to a clean cage after it has completely recovered from the anesthetic.
10. Transplanted mice should then be imaged immediately after surgery using bioluminescence imaging as time 0 point (*see* **Subheading 3.5**). These recipient mice should be imaged weekly to track the development of bone, brain, and other types of metastases.
11. Monitor the mice 1 h after injection to make sure they fully recover from injection. You can anesthetize one cage *(5–6)* of mice at a time and perform their injections together. Occasionally, administer warm sterile saline intraperitoneally (about 1–3 ml) to help the recovery if the mouse appears to be hypotensive (cold and pale) after getting anesthesia or an injection.
12. Observe the mice daily for signs of tumor burden, including paralysis, hunched posture, or weight loss. Euthanize when moribund or at predetermined time points, and necropsy and preserve tissue for histology if required. Examination of the skeleton by X-ray radiography can detect skeletal lesions. Soft tissue metastases can be detected and monitored non-invasively by bioluminescence imaging (*see* **Notes 7 and 8**).

3.5. Bioluminescence Imaging of Metastasis with the Caliper (Xenogen) IVIS 200 System

One major hurdle in experimental metastasis research is the difficulty to follow the progression of metastasis in living animals. Early stages of metastasis are often asymptomatic in experimental animals. Mice can display no sign of distress even when they are already under considerable metastasis burdens. In the past, detection of metastatic lesions relied on the endpoint assays that involve dissection of animals and histological analysis of tissue sections. This approach is tedious and inefficient at best, and important metastasis events could be missed due to incomplete sectioning of tissue blocks. In addition, the animals are killed at each data collection time point, making it impossible to follow the progression of metastasis chronically in each animal. This has resulted in high cost, large data variations, and poor reproducibility that have frustrated metastasis research for years. The application of small animal imaging techniques, such as bioluminescence imaging, has allowed highly sensitive localization and quantitation of metastatic growth in living animals *(14, 15)*. Tumor cells are typically engineered to express the firefly luciferase reporter gene that upon exposure to luciferin substrate can produce signals detect-able by specialized sensors such as the Xenogen IVIS imaging systems *(16)*. Noninvasive imaging ensures the quality of injection. In addition, the progression of metastasis can be monitored and quantified by bioluminescence imaging.

1. Initialize IVIS system and wait for the status light to turn green.
2. Anesthetize mice by using intraperitoneal injection of ketamine at 100 mg/kg and xylazine 10 mg/kg.
3. Inject 100 µl of luciferin solution through the orbital plexus using an insulin needle.
4. Image the mice using the Xenogen IVIS 200 imaging system with appropriate field of view (E for imaging five mice at the same time) and binning (small for high resolution). The default imaging time is 1 min. Shorten the exposure time if the image is saturated.

4. Notes

1. The genetic background and the age of the nude mice may influence the outcome of xenograft metastasis assays. It is important to keep these parameters consistent in different sets of experiments in order to obtain reproducible results. When placing animal orders, specify the desired age (usually 4 weeks old) of mice at the time of delivery. Allow 1 week for the mice to adjust to the environment before the experiment. Place ear tags on mice and weigh them a day before the xenograft experiment. Coordinate the cell culture with the delivery of the mice to achieve the optimal time of the animal experiment. Do not use mice more than 6 weeks old for xenograft experiments as they are more resistant to the development of metastasis.
2. Using a consistent method to prepare cell suspensions is another important consideration for reducing the variability in metastasis experiments. Use cells that are recently thawed from the frozen stock or freshly obtained from cancer patients. The cells should be in subconfluent, actively growing cultures. The cells from confluent cultures are more likely to form clumps or aggregates, which can lead to dramatic changes in their metastatic behavior in vivo. In addition, the degree of confluence in vitro has been reported to regulate gene expression, which might also influence the metastasis phenotype. Using the Ca^{2+} and Mg^{2+}-free buffer will reduce the formation of clumps, and gentle vortexing before injection may help to break up aggregates. However, vigorous pipetting or vortexing should not be used in order to avoid the damage to the cells. Avoid overtrypsinizing the cells. Keep the suspension on ice and proceed

with inject experiment as soon as possible. Depending on the setup of animal facility, it may be necessary to arrange the experiment to be performed through cooperation of a team of two or three researchers.

3. MFP injection has also been done through direct injection of tumor cells through nipples. However, such method does not guarantee the delivery of tumor cells into the MFP and in fact often create subcutaneous tumors. Injection of tumor cell suspension in a volume bigger than 10 μl will likely lead to the leakage of cells out of the MFP.
4. Dilation of lateral tail vein by heating the mice greatly facilitates intravenous injections. Dilation of the blood vessel allows the insertion of the needle which has a diameter that is often bigger than normal lateral tail vein vessels. Mice can be heated up by a heating lamp. Alternatively, the vein can be heated by immersing the tail in hot water. Such method is less effective and more time consuming than the heating lamp method. When mice are heated by a heating lamp, they need to be monitored closely. Overheating of the animals can lead to heat shock and death.
5. It is very easy to insert the needle "through" the tail vein (instead of "into" the vein) and inject the tumor cells intramuscularly instead of intravenously. To avoid such mistake, make sure that the needle can move up and down the vein without much resistance after the insertion of the needle into the blood vessel. Injection of the tumor cell suspension should meet little resistance if the needle is inserted correctly. If the blood vessel is clogged during the injection, remove the needle immediately and attempt the injection again at a different location up the tail.
6. It is important to inject the tumor cells into the left cardiac ventricle instead of other compartment of the heart to ensure the diffusive distribution of the cancer cells throughout the body of the mice. The appearance of dark color blood is often an indication of misplaced needles. In such circumstance, the needle needs to be retrieved immediately and the injection can be attempted again. However, it is not recommended to attempt intracardiac injection on the same mouse more than 3 times. Bioluminescence imaging immediately after the injection can be used to ensure the success of each injection.
7. All the animal procedures need to be approved by the Institutional Animal Care and Use Committee (IACUC). Experimental design should take into account the well being of the mice and use appropriate procedures to reduce pain and suffering. In the context of this chapter this means careful monitoring of mice for development of tumor and metastasis burden, appropriate animal handling and efficient execution

of surgical procedures, and the humane and timely use of euthanasia. Usage of bioluminescence imaging helps reducing the number of mice required for each experiment. Using a moribund endpoint, rather than a death endpoint for a study, is more practical if the point of the study is to assess the extent of tumor spread. The following endpoint criterion is routinely used in animal metastasis assays.

8. The animals should be euthanized as soon as possible and not longer than 24 h following detection of any the following symptoms (a) severe cachexia (weight loss approaching 25%), (b) inability to obtain food or water, general lack of moving activities, (c) pale appearance with body temperature below 99?F, (d) breathing problem, (e) pathological bone fracture, (f) bite wound or ulcer, (g) infection, and (h) apparent neurological disorder as judged from the pattern of movement.

Acknowledgments

I thank members of my laboratory for the critical reading of this manuscript and Dr. Teresa Guise for her advice in various xenograft techniques. Research in our laboratory is supported by American Cancer Society (RSG MGO-110765), Department of Defense (BC051647), Susan G. Komen Foundation (BCTR0503765), and the NJ Commission on Cancer Research (05-2408-CCR-E0).

References

1. Gupta, G. P., and Massague, J. (2006) Cancer metastasis: building a framework, *Cell* **127**, 679–95.
2. Kang, Y. (2005) Functional genomic analysis of cancer metastasis: biologic insights and clinical implications, *Expert Rev Mol Diagn* **5**, 385–95.
3. Steeg, P. S. (2006) Tumor metastasis: mechanistic insights and clinical challenges, *Nat Med* **12**, 895–904.
4. Chambers, A. F., Groom, A. C., and MacDonald, I. C. (2002) Dissemination and growth of cancer cells in metastatic sites, *Nat Rev Cancer* **2**, 563–72.
5. Fidler, I. J. (2003) The pathogenesis of cancer metastasis: the 'seed and soil' hypothesis revisited, *Nat Rev Cancer* **3**, 453–8.
6. Li, F., Tiede, B., Massague, J., and Kang, Y. (2007) Beyond tumorigenesis: cancer stem cells in metastasis, *Cell Res* **17**, 3–14.
7. Khanna, C., and Hunter, K. (2004) Modeling metastasis in vivo, *Carcinogenesis* **25**, 2285–2292.
8. Welch, D. R. (1997) Technical considerations for studying cancer metastasis in vivo, *Clin Exp Metastasis* **15**, 272–306.
9. Khanna, C., Khan, J., Nguyen, P., Prehn, J., Caylor, J., Yeung, C., Trepel, J., Meltzer, P., and Helman, L. (2001) Metastasis-associated differences in gene expression in a murine model of osteosarcoma, *Cancer Res* **61**, 3750–9.
10. Minn, A. J., Gupta, G. P., Siegel, P. M., Bos, P. D., Shu, W., Giri, D. D., Viale, A., Olshen, A. B.,

Gerald, W. L., and Massague, J. (2005) Genes that mediate breast cancer metastasis to lung, *Nature* **436**, 518–24.

11. Kang, Y., Siegel, P. M., Shu, W., Drobnjak, M., Kakonen, S. M., Cordon-Cardo, C., Guise, T. A., and Massague, J. (2003) A multigenic program mediating breast cancer metastasis to bone, *Cancer Cell* **3**, 537–49.
12. Kang, Y., He, W., Tulley, S., Gupta, G. P., Serganova, I., Chen, C. R., Manova-Todorova, K., Blasberg, R., Gerald, W. L., and Massague, J. (2005) Breast cancer bone metastasis mediated by the Smad tumor suppressor pathway, *Proc Natl Acad Sci U S A* **102**, 13909–14.
13. Yin, J. J. (1999) Osteoblastic bone metastases: tumor-produced endothelin-1 mediates new bone formation via the endothelin A receptor, *J Back Musculoskeletal Rehabil* **14**, F400.
14. Gross, S., and Piwnica-Worms, D. (2005) Spying on cancer: molecular imaging in vivo with genetically encoded reporters, *Cancer Cell* 7, 5–15.
15. Weissleder, R. (2002) Scaling down imaging: molecular mapping of cancer in mice, *Nat Rev Cancer* **2**, 11–8.
16. Contag, P. R., Olomu, I. N., Stevenson, D. K., and Contag, C. H. (1998) Bioluminescent indicators in living mammals, *Nat Med* **4**, 245–7.

Chapter 3

Identification of Murine and Human Acute Myeloid Leukemia Stem Cells

Aniruddha J. Deshpande, Farid Ahmed, and Christian Buske

Summary

There is now compelling evidence to show that tumors, once believed to be a homogeneous mass of abnormally proliferative cells, comprise a heterogeneous population of transformed cells resembling the hierarchically organized populations in the corresponding tissue. At the top of this tumor hierarchy are the cancer stem cells (CSCs) which are critical for tumor growth and maintenance as they constantly replenish the tumor bulk and are believed to be resistant to most conventional chemotherapies. The first evidence for both malignant hierarchy and CSCs came from studies on acute myeloid leukemia (AML) *(1, 2)* followed by mounting evidence in other types of cancer *(3–5)*. Murine models of leukemia have been the proving ground for the elucidation of key novel concepts in this nascent field of CSC research. A spate of recent studies highlighting the importance of CSCs in cancer has accentuated the need for identifying and characterizing these cells. The frequency of CSCs in the leukemic bulk (LSCs) is usually estimated by in vivo limiting-dilution transplantation assays of leukemic cells into recipient animal hosts in which identical leukemias can be regenerated. Each cell that is capable of propagating the leukemia in secondary recipients is termed a leukemia-propagating cell (LPC).

LSC candidates have been typically identified by the systematic dissection of compartments within a heterogeneous tumor to determine the subpopulation which shows maximal enrichment for LPCs. This is performed by determining the frequency of LPCs in each of those purified subpopulations, as has been described in both human and murine models of leukemia *(1, 2, 6)*. It is important to mention that even though leukemia repopulation is often used as a surrogate for LSC estimation, the term LPCs can be equated with LSCs only if the phenotypic heterogeneity of the original tumor mass is regenerated upon tumor propagation into secondary recipients.

The first part of the chapter deals with the estimation of LPCs in syngenic or congenic murine–murine models. In the second part, murine xenograft models for the estimation of LSCs in human leukemias are described.

Key words: AML, Cancer stem cells, Leukemia, Limiting dilution assays, Leukemic stem cells, Mouse bone marrow transplantation

John S. Yu (ed.), *Cancer Stem Cells,* Methods in Molecular Biology, vol. 568
DOI 10.1007/978-1-59745-280-9_3, © Humana Press, a part of Springer Science+Business Media, LLC 2009

1. Introduction

Leukemia stem cells (LSCs) are operationally defined as cells that can initiate identical tumors in recipient animals. These cells have been characterized in human AML samples as well as in several murine models of leukemia. The details of LSC estimation from both kinds of samples are discussed in this chapter. The identification of these cells by the transfer of titrated leukemia cell numbers into recipient mice is similar to the quantification of normal stem cells by the classical limiting dilution competitive repopulating unit (CRU) assay *(7)*.

Several in vitro and in vivo methods have been developed for the identification and characterization of candidate leukemic stem cells (LSCs) in leukemia *(11–13)*. Although in vitro assays have been useful in providing information about the presence of clonogenic AML cells or AML long-term culture initiating cells (AML LTC-IC), these methods have to be examined and ultimately validated in vivo to prove their usefulness in identifying the LSC. The identification of LSCs requires addressing two important features of cancer stem cells (CSCs), namely tumor propagation and self-renewal, which can currently only be demonstrated by in vivo models. In human AML, LSCs have been defined as those cells that are capable of regenerating identical AMLs in irradiated nonobese diabetic/severe combined immunodeficient (NOD/SCID) mice and are therefore termed NOD/SCID leukemia initiating cells (NOD/SL-IC or SL-IC) *(1, 2)*. The NOD/SL-IC defined so far generate the phenotypic heterogeneity of the AMLs from which they were originally isolated and can therefore be equated with LSCs.

Although the frequency of NOD/SL-IC is much lower than the AML LTC-IC and AML CFU, the ability to transplant secondary recipients is the only measure of self-renewal capacity of LSCs in vivo. Moreover, the development and spread of leukemia in NOD/SCID mice closely resembles the clinical disease in patients and reflects the outcome of AML *(14, 15)*. Due to its ability to sensitively detect and quantitate LSCs from a majority of AML samples, this model is useful for preclinical evaluation of candidate drugs and other therapeutic strategies *(16, 17)*. Leukemic cells from mouse models or mononuclear cells (MNC) enriched from fresh or frozen AML samples are injected into irradiated syngenic or NOD/SCID mice, respectively, with or without enrichment for potential LSC markers *(1, 18)*. Intravenous injections are performed on preconditioned mice and leukemic engraftment is detected either by BM aspiration or by killing the recipient animals. Limiting dilution analysis of different isolable subpopulations can be performed in order to detect the frequency of LSCs in each leukemia population or sorted subpopulations. Using Poisson statistics, the frequency of LSC is calculated based on the frequency of mice engrafted with the leukemia.

Recent data suggest that the frequency of LSCs could be underestimated by the use of conditioned recipients. Still, the use of this in vivo assay to identify the leukemia-propagating populations would throw up valuable insights into CSC biology in general and leukemia in particular.

2. Materials

2.1. Prerequisites

1. A fully equipped experimental facility for mouse experiments with individually ventilated cage (IVC) systems.
2. Syngenic mice with discernible antigen expression or a marker for specifically identifying leukemic cells (for murine leukemia).

Or

NOD/SCID mice for engraftment of human AML samples. Some related strains of NOD/SCID such as NOD/SCID-bnx and NOD/LtSz-*Prkdc*^{scid} *Prkdc*^{scid}–β2 microglobulin$^{-/-}$ (NOD/SCID-β2m$^{-/-}$) have also been successfully used to transplant human AML cells (*see* **Note 1**).

2.2. Murine Bone Marrow Preparation and Injection (for Mouse Leukemia Samples)

1. *Sterile syringes and needles.* BD Plastipak™ 1-ml syringe (BD Biosciences, Palo Alto, CA) with 0.5 × 25 mm needles for tail vein injection of cells in mice and Kendall Monoject™ 3-ml syringes (Tyco Healthcare, UK) with 0.55 × 25 mm needles (BD Microlance™, Drogheda, Ireland) for flushing (harvesting) of bone marrow from extracted bones.
2. *RBC lysis buffer.* 0.8% NH_4Cl with 0.1 mM EDTA, pH adjusted with $KHCO_3$ to a final pH of 7.2–7.6; (Stem Cell Technologies, Vancouver, Canada).
3. *Cell culture apparatus.* BD Falcon 40-μm nylon strainer for filtering the harvested bone marrow suspension (BD Biosciences, Palo Alto, CA), sterile disposable pipettes (Corning Inc., Corning, NY), sterile FACS tubes (BD Biosciences, Palo Alto, CA).
4. *Medium.* Dulbecco's Modified Eagle's Medium (DMEM) 4.5 g/l glucose, L-glutamine, sodium pyruvate, and 3.7 g/l $NaHCO_3$ (PAN biotech GmbH, Aidenbach, Germany) Dulbecco's phosphate-buffered saline (DPBS) without magnesium and calcium, sterile filtered (PAN biotech GmbH, Aidenbach, Germany).

2.3. AML Mononuclear Cell Isolation and Processing (for Human AML Samples)

1. Ficoll Paque™ (Pharmacia, Uppsala, Sweden).
2. 50-ml conical tubes (BD Falcon tubes) (BD Biosciences).
3. Fetal bovine serum (FBS) (PAN-Biotech, Aidenbach, Germany).
4. Iscove's Modified Dulbecco's Media (IMDM) (GIBCO, Invitrogen Corporation, Germany) with 10% FBS is used for washing AML cells.
5. IMDM with 50% FBS and 10% dimethylsulfoxide (DMSO, Sigma, St. Louis, MO) is used as cryopreservative medium.
6. RPMI-1640 medium (GIBCO, Invitrogen Corporation, Germany) supplemented with 20% FBS is used as thawing medium for frozen AML cells as well as flushing of bone marrow cells.
7. Deoxyribonuclease I (DNase I) solution, 1 mg/ml (StemCell Technologies Inc, Cell Systems, Germany), is added at 1 mg/ml to prevent cell clumps when thawing frozen AML MNC.
8. Trypan blue solution 0.4% (Sigma) for detecting nonviable cells.
9. *Hemocytometer.* Neubauer counting chambers (Hausser Scientific, Blue Bell, PA).

2.4. Irradiation of Recipient Mice

1. *Radiation source.* An irradiation facility with a ^{137}Cs γ-irradiation source with chamber air and the provision for whole body irradiation of mice.

2.5. Assessment of Leukemic Engraftment

1. *Heparinized capillaries.* (Microvette CB 300) plastic capillaries for collection of blood, 15 I.E Lithium heparin per ml of blood (Sarstedt, Numbrecht, Germany).
2. *Syringes and needles.* Kendall Monoject 3-ml syringes (Tyco Healthcare, UK) and 0.55 × 25 mm needles (BD Microlance, Drogheda, Ireland) for bone marrow aspiration from living mice.
3. *Flow cytometry.* BD FACS Calibur System (BD Biosciences, Palo Alto, CA) for the analysis of leukemic engraftment by flow cytometry and BD FACSVantage SE System (BD Biosciences, Palo Alto, CA) for the FACS purification of leukemia subpopulations.
4. Avertin (2,2,2-tribromoethanol; Sigma), 10 g, is dissolved in 10 ml of tertiary amyl alcohol (Sigma). 10 ml of this solution is diluted in 390 ml of PBS to give 2.5% Avertin stock solution. Store in the dark at 4°C. For anesthetizing NOD/SCID use 10 μl per gram of body weight (*see* **Note 2**).

2.6. Fluorescence-Activated Cell Sorting (FACS) (for Human AML Samples)

1. *Antihuman monoclonal antibodies.* CD45-PE (BD-Pharmingen, CA) used to detect human cells in murine BM and PB.
2. *FcR-blocking reagent.* Antimouse Fcγ III/II receptor (2.4G2, Stem Cell Technologies, Vancouver, Canada).

3. BD FACS Calibur System (BD Biosciences, Palo Alto, CA) used for acquisition and analysis of fluorochrome-labeled cells.
4. FACS snap-cap tubes (Falcon polystyrene round-bottom tubes, 12 × 75 mm; BD Falcon).
5. Human AB serum (Lonza GmbH, Germany).
6. *FACS buffer*. IMDM 5%FCS with 5 μg/ml propidium iodide (PI) to exclude dead cells.

2.7. Calculation of LPC Frequency

1. *Software*. The L-Calc™ limiting-dilution analysis software and its technical manual can be downloaded from the StemCell Technologies Inc company website. URL: http://www.stemcell.com.

3. Methods

3.1. LSC Estimation in Murine Leukemia Samples

3.1.1. Harvesting Bone Marrow Cells from Leukemic Mice

1. Kill moribund leukemic mice by CO_2 asphyxiation or cervical dislocation. Dip the mouse in a beaker containing 70–75% ethanol for surface sterilization and place on a sterile sheet or disinfected surface.
2. Remove the femur and the tibia with a pair of sterile, sharp scissors and forceps. Clean the bones of muscle mass by removing with the scissors and forceps. Cut open the ends of the bones by making small cuts (*see* **Note 3**).
3. Flush the bones with a 3-ml syringe filled with medium (IMDM or DMEM medium without serum) and attached to a 0.55 × 25 mm needle. Insert the needle directly into the bone cavity from one end and aspirate the marrow by forcing the medium through the bone (*see* **Note 4**). Flush marrow from both directions alternately. The marrow should be flushed rigorously ~3 times per end into a sterile 15- or 50-ml Falcon™ tube; collect marrow of all bones in one tube.
4. Filter the cell suspension by passing through a 40-μm nylon mesh to remove pieces of bone or other particulate matter so as to prevent machine-clogging problems during sorting.
5. Add ammonium chloride RBC lysis solution to the aspirated bone marrow at a ratio of 9:1 v/v for lysis of RBCs. (e.g., 9 ml NH_4Cl solution for 1 ml bone marrow aspirate). Mix well and incubate on ice for 10 min.
6. Centrifuge the collected bone marrow in a table top centrifuge at 300 × *g* for 12 min with low brake. Discard the supernatant and wash cells with cold sterile phosphate-buffered saline (PBS) after RBC lysis. Resuspend the leukemic marrow in 1 ml of cold PBS. Always keep cells on ice.

7. Count the cells in a hemocytometer after incubating the small counting aliquot (~10 μl) with an equal volume of 3% acetic acid solution for 3–5 min (*see* **Note 5**).
8. Kill a syngenic disease-free mouse and extract and count bone marrow cells as described in **Subheading 3.1.1 (steps 1–7)**. This normal bone marrow can be used as radioprotective carrier cells for coinjection with leukemic cells (*see* **Notes 6** and 7).

3.1.2. Sample Preparation

An important thing to note is that this assay can be employed for the:

(a) Assessment of the frequency of LSCs from the leukemic bulk or

(b) Estimation of LSC frequency in separated subpopulations of the leukemic bulk.

While the former technique can be used to determine the overall LSC frequency in the leukemic bulk of different leukemias *(2)*, the latter has been employed to identify which subpopulation enriches for the LSCs, thereby offering clues to the identity of the LSCs in that particular leukemia model *(2, 6)*.

If the determination of LSCs in the leukemic bulk is desired, proceed with **Subheading 3.1.2, step 2** using the leukemic bulk. If, however, the need is to identify which cells in a heterogeneous population enrich for LSC characteristics, include the optional **step 1** in **Subheading 3.1.2**.

1. (**Optional**) Label the cells with fluorescence-conjugated antibodies and sort the different heterogeneous populations from within the leukemic bulk bone marrow cells by FACS (*see* **Note 8**).
2. Perform serial dilutions of leukemic bone marrow cells (unsorted or sorted bone marrow populations) in sterile FACS tubes. Include a wide range of leukemic cell concentrations so as to preferably get leukemia induction at the highest and no leukemic engraftment at the lowest concentrations. This will ensure a proper estimate of the LSC frequency by Poisson distribution statistics. An example is shown in **Fig. 1**. For every cell number, include enough cells for 4–5 mice. For example, for injection of 10,000 cells, include 40,000 cells in the tube for injection into a total of four mice (*see* **Note 9**).
3. Add 5×10^5 nucleated bone marrow cells per mouse for radioprotection of mice (from **Subheading 3.1.1, step 8**) in each injection tube (*see* **Note 10**).
4. In this example with four mice (each receiving 10,000 cells) add 4 times 5×10^5 cells = 2×10^6 cells. Keep all tubes on ice.

a

Analysis Name:		Example Population 1		
Row	Dose	Responses	Tested	Comments
1	100000	4	4	
2	50000	4	4	
3	10000	3	3	
4	5000	4	4	
5	200	3	4	
6	50	0	4	

Analysis Name:		Example Population 2		
Row	Dose	Responses	Tested	Comments
1	100000	4	4	
2	50000	2	4	
3	10000	0	4	
4	5000	0	4	
5	50	0	4	

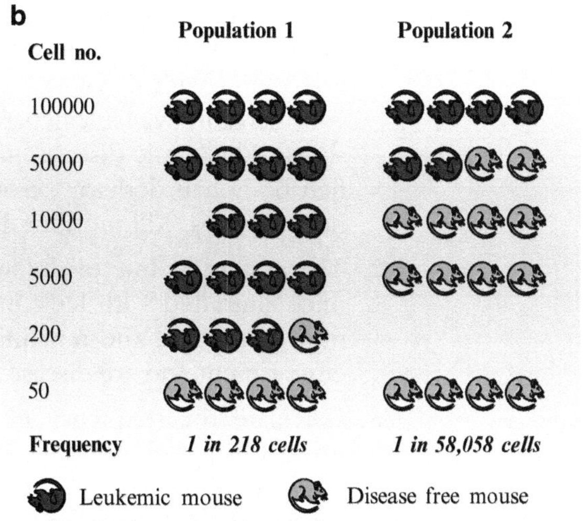

Fig. 1. (**a**) A hypothetical example of an LPC frequency analysis in two different populations using the L-Calc software. This example of an LPC frequency analysis with L-Calc shows an experiment in which titrated numbers of leukemic cells (doses) are injected. The number of leukemic mice and the number of mice tested corresponding to each dose are depicted in the response column and the test column, respectively. It is better but not necessary to include the same doses, but a similar range of doses should ideally be employed for both populations as seen in the example (range: 50–100,000). (**b**) Schematic of LPC frequency analysis in two different populations. In the LPC analysis of Fig. 1a, mice are seen which display leukemic engraftment (dark and inverted) and mice which do not (light and upright). The number of mice engrafted with leukemia is taken as the response and used for the LPC analysis as seen in Fig. 1a. The calculation of LPCs in the two sample populations shown in Fig. 1a reveals the frequency of LPCs to be 1 in 218 leukemic cells in population 1 and 1 in 58,058 cells in population 2. Therefore, in this example, there is a 266-fold enrichment of LPCs in population 1 over population 2.

3.1.3. Mouse Irradiation

1. Irradiate the syngenic or congenic mice (8–12 weeks old) in which the leukemic cells are to be injected at lethal doses of radiation (800–850 cGy of whole body irradiation) (*see* **Note 11**) ~2–3 h prior to injection.

3.1.4. Injection of Mice

1. Transfer the lethally irradiated mouse from the cage and place it gently into a mouse-restraint device exposing the tail.
2. Prewarm the tail of the restrained mouse for vasodilation with a safe heat source such as warm water or heat lamp for 30–60 s (*see* **Note 12**).
3. Mix the cells in PBS by gentle vortexing and bring to room temperature.
4. Load the volume required for injection into the syringe. Prevent air bubble formation or remove air bubbles from the syringe.
5. Locate the vein and insert the needle in the vein near the distal tip of the tail. Inject the fixed volume of PBS containing cells by tail vein injection (*see* **Note 9**).

3.1.5. Calculation of LPC Frequency

1. Observe mice for signs of leukemic engraftment by flow cytometric analysis (*see* **Note 10**) and blood smears from peripheral blood withdrawn at weekly or biweekly intervals using commercially available heparinized capillaries. Check for visual signs of leukemic disease onset, usually characterized by ruffled body hair, lethargy, crouching, disturbed gait, and paleness of limbs. Kill moribund mice.
2. The endpoint time of the assay should be the time when all mice engrafted with leukemic cells are dead or killed (when moribund) and the remaining mice do not show leukemic engraftment and are disease free.
3. Calculate the frequency of the leukemia repopulating unit using the L-Calc limiting dilution analysis software Version 1.1 (StemSoft Inc., Vancouver, Canada). The software is specifically designed for testing the frequency of a cell with a defined responsiveness (in this case, the ability of leukemic engraftment) within a population of cells.
   (a) A new file is automatically opened upon running the L-Calc software. The file has four columns titled dose, responses, tested, and comments. The *dose* is the number of cells injected per mouse, *response* is the number of mice giving the desired response at the end of the assay (i.e., leukemic engraftment), the *tested* column is for the number of mice injected, and the *comments* column for the inclusion of any specific notes for that particular arm.
   (b) Input the number of cells injected per mouse in the *dose* column, the number of mice injected with each individual dose in the *tested* column, and the number of leukemic mice (dead or killed when moribund) in the *response* column (e.g., when three out of four mice in one arm of the experiment are leukemic, the number *tested* is 4 and the *response* is 3). Perform the calculation of the LPC frequency by clicking on the "analyze" button on the toolbar.

(c) **Figure 1a** gives an example for calculation of the LPC by limit-dilution analysis using the L-Calc program. In this example, two populations from within a leukemic bulk are separated and injected at the stated dilutions into cohorts of mice. **Figure. 1b** shows the same experiment schematically.

In the example shown in **Fig. 1**, the comparison of in vivo LPC frequencies between two populations within a leukemic bulk shows that in the "example population 1" the LPC frequency is 1 in 218 cells, which is 266-fold enriched compared to "example population 2" which has an LPC frequency of 1 in 58,058 cells. In this example it can be inferred that the "example population 1" highly enriches for LPCs in this particular type of leukemia.

3.2. LSC Estimation in Human AML Samples

AML peripheral blood (PB) or bone marrow (BM) samples (20–40 ml) are usually collected in 20-ml syringes by trained hospital personnel (*see* **Note 13**). MNC isolated from AML samples are either used directly or enriched for potential stem cells before transplantation. A minimum of 10^7 bulk AML cells, or 2–5 × 10^4 $CD34^+$ $CD38^-$ AML cells are required to successfully induce leukemic engraftment in NOD/SCID mice *(1, 14)*.

3.2.1. Isolation of Low-Density Mononuclear Cells from AML Samples

1. Carefully layer 20 ml of bone marrow sample down the sides f the 50-ml conical tubes on 10 ml of Ficoll Paque (*see* **Note 14**).
2. Centrifuge at 400 × *g* for 30 min in a table top centrifuge at room temperature (RT; 18–21°C). This produces a pellet of red cells at the bottom of the conical tube, a clear layer of Ficoll Paque, a cloudy interface layer of MNC, and an upper plasma layer (*see* **Note 15**).
3. Carefully draw the interface layer with a pipette and collect in another 50-ml tube. Make up the volume to 50 ml by adding IMDM 5% FBS. Centrifuge at 220 × *g* for 10 min at RT. Repeat this washing step.
4. Resuspend cells in 10–20 ml of IMDM 5% FBS and count cells using a Neubauer counting chamber.
5. At this point, the cells may be cryopreserved for future use or processed immediately for antibody staining and cell sorting by FACS or magnetic sorting.

3.2.2. Isolation of Progenitor Cell Population from AML Samples

1. Wash single-cell suspensions of MNC with PBS 2% FBS (centrifuge at 250 × *g* for 10 min) and incubate with 10% human AB serum for 20 min at 4°C (*see* **Note 16**).
2. Wash cells using chilled PBS 2% FBS and aliquot in FACS tubes at a concentration of 10^6 cells per 100 μl for staining with fluorescent-labeled antibodies. Add 20 μl each of CD34 PE and CD38 APC antibodies per 2 × 10^6 cells. Stain a

separate aliquot of 10^6 cells with matched isotype PE and APC control antibodies. Perform the staining for 30 min at 4°C.

3. Wash the labeled cells twice with PBS 2% FBS.
4. Finally, resuspend the cells in FACS buffer at a concentration of $2–5 \times 10^7$ per ml.
5. Define the quadrants on the FACS Vantage cell sorter, using isotype controls. If sorting of $CD34^+$ $CD38^-$ population is desired, define the population by strictly selecting for cells with high CD34 expression and a CD38 expression less than half of the maximum CD38 APC fluorescence of the isotype control.
6. Sort the cells at a maximum rate of 3,000–6,000 cells per second in sterile FACS tubes containing IMDM 50% FCS (collection medium). Before sorting into the FACS tubes with collection medium, invert the closely capped tubes to coat the unimmersed surfaces with the medium. After sorting, immediately place the tubes with sorted cells on ice to preserve their viability.

3.2.3. Transplantation of NOD/SCID mice

1. Irradiate 8–10 week old mice at 350 cGy using a ^{137}Cs source 24 h prior to injection of AML cells (*see* **Note 17** and **Fig. 2**).
2. Inject the freshly prepared or thawed AML MNC ($1–2 \times 10^7$) or $CD34^+$ $CD38^-$ FACS sorted cells ($2–5 \times 10^5$) into the lateral tail vein of mice.
3. When performing a limiting dilution of bulk AML, inject cohorts of 6–8 mice with five different doses of AML cells from a patient sample, ranging from 5×10^4 to 5×10^7. For limiting dilution assay of sorted cells, include a wide range of leukemic cell concentrations so as to preferably get leukemia induction at the highest and no leukemic engraftment at the lowest concentrations. This will ensure a proper estimate of the NOD/SL-IC frequency by Poisson statistics.
4. For intravenous injection, refer to **Subheading 3.1.4** "Injection of Mice." Resuspend cells in IMDM 3% FBS; 200–400 µl is optimal for tail vein injections.
5. Transplanted mice develop signs of disease at variable time points (4–12 weeks). Inspect mice regularly for morbidity. Euthanize moribund mice with CO_2 asphyxiation and dissect out the hind limb and pelvic bones. If necessary BM aspirations may be performed to detect leukemic engraftment at time points prior to the termination of mice.

3.2.4. BM Aspiration by Femur Punctures

BM aspirations may be performed on anesthetized mice to detect leukemic engraftment at 3-week intervals starting from week 3 post-transplantation.

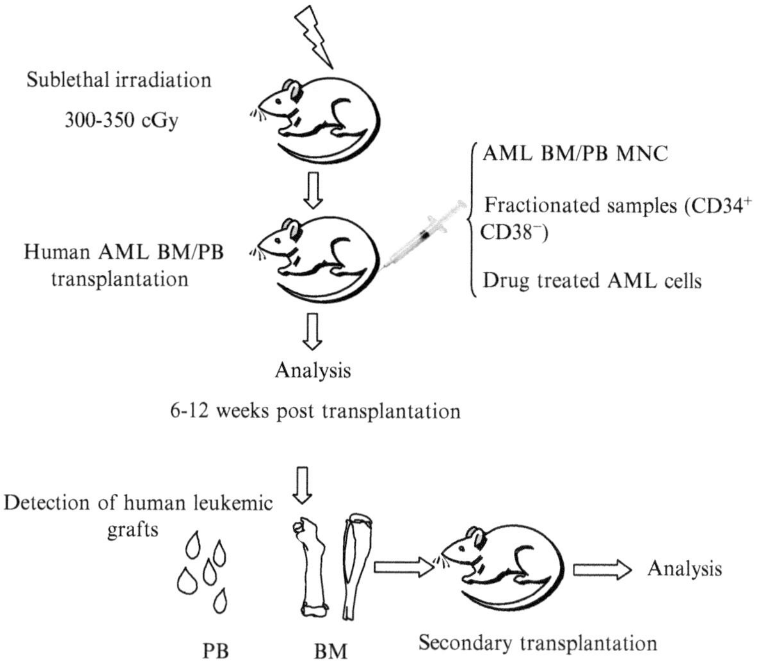

Fig. 2. NOD/SCID mouse transplantation protocol. This is a schematic representation of the transplantation protocol. Mice were preconditioned by sublethal irradiation and injected bulk AML mononuclear cells, purified cells, or drug-treated samples. Analysis of AML engraftment is done starting from week 6 postinjection either by BM aspiration or killing the mice. Secondary transplantations are additionally performed to detect the self-renewal of NOD/SL-IC by injecting BM cells derived from primary mice. AML engraftment is detected by staining mouse BM with antihuman CD45 marker using FACS. As the development of AML in NOD/SCID mice closely resembles the human disease, human AML cells can be detected in peripheral blood, spleen, and other organs of the mice.

1. Anesthetize the mice administering intraperitoneal injections of Avertin (10 μl/g body weight) (*see* **Note 18**).
2. Place the mouse in a supine position on a clean work surface (*see* **Note 19**). Flex the right hind leg at the knee joint, and carefully disinfect the area with 70% ethanol.
3. Hold the femur by using forefinger and thumb and place the needle between the femur condyles, taking care to avoid injury to the knee ligaments (*see* **Note 20**).
4. Insert the needle into the shaft of the femur by gently rotating it under slight pressure. As soon as the needle is correctly inserted, there is a decrease in resistance and inability to move the needles laterally.
5. Collect the BM by withdrawing the plunger, creating a vacuum inside the syringe. Viscous BM will slowly flow into the syringe. Transfer the BM to a 50-ml Falcon tube containing IMDM 3% FBS and 25 U/ml heparin, and proceed for RBC lysis and analysis as described in the following section.

3.2.5. Detection of Leukemic Engraftment

NOD/SCID mice are generally killed at 8 weeks postinjection for the detection of leukemic engraftment *(14)*. However, a different time point up to 12 weeks may be taken. In the case of making secondary transplantations, the primary mice are killed 3–4 weeks postinjection and their BM injected into secondary NOD/SCID that are killed for analysis 8 weeks later.

1. Kill the NOD/SCID mice by CO_2 asphyxiation. Dissect out the hind limbs and pelvic bones.
2. Flush the femurs, tibiae, and pelvic bones with IMDM 15% FCS using syringes and 24-gauge needles.
3. Wash the collected BM cells twice with IMDM 3% FBS and resuspend in 1 ml of the same medium. To this, add 5 ml of chilled ammonium chloride solution (for RBC lyses) and incubate at 4°C for 20 min.
4. Wash the cells twice with IMDM 3% FBS and resuspend in IMDM 2% FBS with 10% human AB serum.
5. Add antimouse Fcγ III/II receptor to the sample (3 μg/ml) and aliquot into FACS tubes for staining with the CD45-PE to detect human AML engraftment.

3.2.6. Secondary Transplantations

Secondary transplantations are performed by injection of BM cells derived from primary transplants that were killed 3–4 week after AML cell injections. The mice are injected and analyzed in the same way as described in previous sections. Secondary transplantations are necessary to prove the in vivo self-renewal of cells injected and hence their true LSC nature.

3.2.7. Calculation of NOD/SL-IC Frequency

When performing a limiting dilution analysis, the frequency of NOD/SL-IC in the starting cell population is calculated from the frequency of mice positive for the presence of AML cells in BM in each cohort of mice. Use the L-Calc software as described in **Subheading 3.1.5** (*see* **Fig. 1a**). Use AML engraftment as the *response*.

4. Notes

1. Although the NOD/SCID-β2m$^{-/-}$ mice engraft better than the NOD/SCID mice, the higher incidence of thymoma development sharply decreases their lifespan.
2. The 2.5% stock solution is stable for at least 1 year when stored at 4°C in dark.
3. It is important to cut just enough of the bone ends so as to expose the bone cavity. Cutting away larger portions will result

in a reduced recovery of bone marrow cells as the ends have especially high concentrations of bone marrow cells.

4. When the bone marrow, especially hypercellular marrow characteristic of leukemia, is flushed out with medium, the medium becomes cloudy. This is a sign that the flushing is being done properly.
5. Incubation with a weak acid is routinely used for counting nucleated white blood cells in blood samples. The treatment with weak acids removes non-nucleated red blood cells. Remember to include the dilution factor accounted for by the addition of the acetic acid solution.
6. Lethally irradiated mice injected with leukemic cells can die of bone marrow failure if there is weak or no leukemic engraftment in the mice. Therefore, radioprotective doses of carrier cells are injected along with leukemic cells. At the end of the assay, therefore, mice which show no leukemia engraftment will not die of bone marrow failure and serve as disease-free readouts.
7. Approximately $1–5 \times 10^7$ nucleated WBCs can be obtained on average from the two hind limb femur and tibias of a single C57BL6 mouse. Since a total of 5×10^5 carrier cells should be injected into each mouse along with the individually determined dilution of leukemic cells, calculate the number of carrier cells needed for the entire limit-dilution assay and kill more disease-free mice if more carrier cells are needed.
8. It has been observed that the LSC population resides in the stem cell or progenitor cell compartments in various murine models of leukemia *(8–10)*, so it would be interesting to stain with markers for subpopulations based on stem cell or progenitor cell characteristics and compare their LSC frequency with the leukemia bulk or the remainder of cells. Alternatively, the LSC frequency in any subpopulation which is expected to enrich for LSCs can be compared to the frequency of LSCs in the remaining cells or the bulk leukemia population.
9. Volumes of 200–400 μl are usually optimal for tail vein injection, so each mouse should receive the desired number of cells in this volume of PBS. It is useful to take 200–300 μl of additional PBS to offset sample loss during injection.
10. It is important to be able to discern leukemic engrafting cells from host cell remnants or engrafting carrier cells by the flow cytometric identification of a specific marker. For example, a fluorescent marker like GFP or strain-specific surface markers like Ly 5.1 and Ly 5.2.

11. We usually do lethal irradiation; the lethal dose for each strain is usually determined individually in each case. For example, the doses we use for the lethal irradiation of C57Bl, C3Hx-PEB, or C3HxC57Bl are standardized at 800–850 cGy.
12. It is recommended to prewarm the tail of mice by dipping it in warm water or by using a heat lamp (e.g., infrared lamp). Take care not to injure the mouse due to scalding or overheating. This dilates the vein and makes it more visible to the eye, aiding in injection. This should be followed by wiping the tail surface with paper tissue soaked in 70% ethanol for surface sterilization.
13. AML BM and PB samples should be obtained from patients after informed consent and with the approval of local ethics committees.
14. Take care not to mix the BM cells with Ficoll Paque when layering.
15. Avoid using brakes at this centrifugation step to prevent mixing of the separated layers.
16. Incubation with human AB serum blocks nonspecific binding of antibodies to human Fc receptors.
17. It is important to provide the mice with acidulated water (pH 2.8–3.2; use HCl acid to decrease pH) containing ciprofloxacin (20 mg/ml of drinking water) after irradiation to prevent the immunodeficient mice from infections.
18. After the procedure, control the body temperature of mice by placing the mice on gel packs prewarmed to 37°C.
19. All handling of NOD/SCID mice are done in sterile conditions.
20. At the beginning of the procedure, before piercing the skin, hold the needle at an angle of 45° to the femur shaft. Once the needle is lodged between the condyles, rotate the syringe upward to a position in line with the femoral axis. This maneuver minimizes the damage to knee ligaments.

References

1. Bonnet, D. and J.E. Dick (1997), *Human acute myeloid leukemia is organized as a hierarchy that originates from a primitive hematopoietic cell.* Nat Med. **3**(7): p. 730–7.
2. Lapidot, T., et al. (1994), *A cell initiating human acute myeloid leukaemia after transplantation into SCID mice.* Nature. **367**(6464): p. 645–8.
3. Al-Hajj, M., et al. (2003), *Prospective identification of tumorigenic breast cancer cells.* Proc Natl Acad Sci U S A. **100**(7): p. 3983–8.
4. Singh, S.K., et al. (2004), *Identification of human brain tumour initiating cells.* Nature. **432**(7015): 396–401.
5. O'Brien, C.A., et al. (2007), *A human colon cancer cell capable of initiating tumour growth in immunodeficient mice.* Nature. **445**(7123): p. 106–10.
6. Deshpande, A.J., et al. (2006), *Acute myeloid leukemia is propagated by a leukemic stem cell with lymphoid characteristics in a mouse model of CALM/AF10-positive leukemia.* Cancer Cell. **10**(5): p. 363–74.

7. Szilvassy, S.J., et al. (1990), *Quantitative assay for totipotent reconstituting hematopoietic stem cells by a competitive repopulation strategy.* Proc Natl Acad Sci U S A. **87**(22): p. 8736–40.
8. Cozzio, A., et al. (2003), *Similar MLL-associated leukemias arising from self-renewing stem cells and short-lived myeloid progenitors.* Genes Dev. **17**(24): p. 3029–35.
9. So, C.W., et al. (2003), *MLL-GAS7 transforms multipotent hematopoietic progenitors and induces mixed lineage leukemias in mice.* Cancer Cell. **3**(2): p. 161–71.
10. Huntly, B.J., et al. (2004), *MOZ-TIF2, but not BCR-ABL, confers properties of leukemic stem cells to committed murine hematopoietic progenitors.* Cancer Cell. **6**(6): p. 587–96.
11. Ailles, L.E., B. Gerhard, and D.E. Hogge (1997), *Detection and characterization of primitive malignant and normal progenitors in patients with acute myelogenous leukemia using long-term coculture with supportive feeder layers and cytokines.* Blood. **90**(7): p. 2555–64.
12. Buick, R.N., J.E. Till, and E.A. McCulloch (1977), *Colony assay for proliferative blast cells circulating in myeloblastic leukaemia.* Lancet. **1**(8016): p. 862–3.
13. Sutherland, H.J., et al. (1991), *Differential regulation of primitive human hematopoietic cells in long-term cultures maintained on genetically engineered murine stromal cells.* Blood. **78**(3): p. 666–72.
14. Ailles, L.E., et al. (1999), *Growth characteristics of acute myelogenous leukemia progenitors that initiate malignant hematopoiesis in non-obese diabetic/severe combined immunodeficient mice.* Blood. **94**(5): p. 1761–72.
15. Pearce, D.J., et al. (2006), *AML engraftment in the NOD/SCID assay reflects the outcome of AML: implications for our understanding of the heterogeneity of AML.* Blood. **107**(3): p. 1166–73.
16. Feuring-Buske, M., et al. (2002), *A diphtheria toxin-interleukin 3 fusion protein is cytotoxic to primitive acute myeloid leukemia progenitors but spares normal progenitors.* Cancer Res. **62**(6): p. 1730–6.
17. Terpstra, W., et al. (1997), *Diphtheria toxin fused to granulocyte-macrophage colony-stimulating factor eliminates acute myeloid leukemia cells with the potential to initiate leukemia in immunodeficient mice, but spares normal hemopoietic stem cells.* Blood. **90**(9): p. 3735–42.
18. Hosen, N., et al. (2007), *CD96 is a leukemic stem cell-specific marker in human acute myeloid leukemia.* Proc Natl Acad Sci U S A. **104**(26): p. 11008–13.

Chapter 4

Methods for Analysis of Brain Tumor Stem Cell and Neural Stem Cell Self-Renewal

Ichiro Nakano and Harley I. Kornblum

Summary

Neural stem cells (NSC) self-renew and are multipotent, producing neurons and glia. Recent studies have shown that brain tumors (BT) contain cells that, like NSC, self-renew and are multipotent, producing the different types of cells found within the brain tumors. These brain tumor stem cells are a kind of cancer stem cell, competent to form tumors that mimic the parent tumor in experimental animals. Studies from our laboratory and others have demonstrated that brain tumor stem cells and NSC share similar mechanisms and pathways for proliferation. For example, we have identified that one of the AMPK/snf1 kinases, maternal embryonic leucine zipper kinase (MELK), is highly expressed in NSC and malignant brain tumors, as well as in brain tumor stem cell-enriched cell cultures. Analysis of transgenic MELK-reporter mice indicated that MELK is expressed in NSC in vivo, and our in vitro studies demonstrated that MELK is required for NSC self-renewal. We have also found that MELK is required for proliferation of putative BT stem cells. Utilizing our studies with MELK as an example, this chapter describes methods to culture NSC and BT stem cells, and to analyze the pathways, which regulate self-renewal of those cells.

Key words: MELK, Brain tumor stem cells, Cancer stem cells, Signaling pathway, Proliferation, Self-renewal, Proliferation, RNA interference, Cell culture

1. Introduction

Neural stem cells (NSC) have the capacity to self-renew and to ultimately produce the major cell types of the central nervous system: neurons, astrocytes, and oligodendrocytes *(1)*. NSC can be found throughout the central nervous system in specialized regions adjacent to the ventricular system, termed periventricular

John S. Yu (ed.), *Cancer Stem Cells,* Methods in Molecular Biology, vol. 568
DOI 10.1007/978-1-59745-280-9_4, © Humana Press, a part of Springer Science + Business Media, LLC 2009

germinal zones, throughout the mammalian lifespan *(1)*. Despite the fact that there is a heterogeneity of NSC phenotypes that exist in different brain and spinal cord areas at different stages of development, each cell classified as an NSC must fulfill the criteria of self-renewal and multipotency *(1)*.

NSC were initially isolated from the murine embryonic and adult murine CNS using what is now known as the neurosphere culture system *(12)*. This method takes advantage of the relative heartiness of stem cells and their resultant selective survival advantage under harsh conditions over other cell types. Neurospheres are floating cellular masses that are grown in the presence of epidermal growth factor (EGF), basic fibroblast growth factor (bFGF), or a combination of the two. Other growth factors that stimulate these receptors also induce the formation of neurospheres *(13, 14)*. When dissociated brain cells are seeded in neurosphere growth medium at high densities, many neurospheres grow that are the product of several (or many) different propagating cells and cell types. However, when appropriate brain areas are dissociated and seeded at very low densities in appropriate tissue culture vessels, each neurosphere is entirely the product of a single progenitor or stem cell *(15, 16)*. When these spheres are placed into conditions that promote differentiation, such as by withdrawing mitogenic growth factors and placing on adherent substrate, one can demonstrate their ability to produce neurons, astrocytes, and oligodendrocytes by immunocytochemistry *(15)*. The production of a sphere from a single cell and the demonstration that these spheres are multipotent (produce all 3 cell types) is one of the current criteria for identification of NSC *(3, 14, 16)*. However, a critical component of this definition is self-renewal. To establish this, using the neurosphere assay, one must demonstrate that dissociation of a sphere into single cells and then that repropagation of these individual cells produces other multipotent spheres *(3, 14, 16)*. Of course, by strict definition, one cannot definitively say that this is true self-renewal, since the properties of the sphere-forming cell may change with each passage.

Self-renewal of stem cells can occur either symmetrically, where a division produces two daughter stem cells, or asymmetrically, where a division produces a daughter stem cell and another progenitor that is restricted in its developmental capacity *(17)*. Using the neurosphere assay, if only asymmetric divisions take place, then one neurosphere will produce only one daughter neurosphere *(18, 19)*. Factors that enhance asymmetric self-renewal will thus produce larger clonal neurospheres rather than more neurospheres *(18, 19)*. Factors that enhance symmetric self-renewal will, on the other hand, produce more neurospheres. In practice, however, most factors that promote self-renewal likely enhance both asymmetric and symmetric self-renewal, as one often observes both larger and more neurospheres in their presence.

The neurosphere culture method has significant limitations. First, it is quite difficult to truly establish "clonality" – that a sphere derives from a single cell. Recent studies have demonstrated that even when cultured at fairly low densities, spheres tend to fuse in culture or envelope stray cells as they propagate *(20)*. Another limitation is that the dissociation of spheres is, in itself, quite stressful to the cells, and could cause a significant underestimate of the "true" stem cell content. Finally, spheres can be quite difficult to transfect or infect when compared to adherent cells. For these reasons and others, several investigators have developed methods to culture NSC as adherent monolayers *(14)*. As we will show later, we have developed methods that combine neurosphere and monolayer methodology to transduce NSC with exogenous genes and small interfering (si) RNAs and to subsequently measure effects on self-renewal.

Using these culture methods, our group and many others have begun to investigate factors and signaling pathways that influence NSC self-renewal *(10, 11, 21, 22)*. Several approaches have been utilized, including the analysis of the effects of exogenously administered compounds and growth factors, manipulation of candidate endogenous genes, and the discovery of novel genes and pathways *(10, 11, 21, 22)*. The major exogenous growth factors that regulate NSC proliferation are EGF and bFGF. We have studied the relative roles of stimulation of FGF and EGF receptors on self-renewal and found that bFGF induces a greater degree of self-renewal than do EGF family members in embryonic cortical NSC (Jackson RL et al., unpublished observations). Several endogenous genes and pathways have been studied. Stimulation of the Wnt-B- catenin pathway promotes self-renewal of murine NSC *(23)*. We have studied the role of the PTEN (phosphatase and tensin homolog deleted on chromosome 10)/PI3K/Akt pathway *(24, 25)*. PTEN is frequently deleted in malignant gliomas *(26–28)*. In a series of studies, we have found that PTEN, which inhibits this pathway, also provides tonic inhibition of NSC self-renewal *(24)*. Other known genes, pathways, and growth factors have been implicated in promoting self-renewal, including Bmi-1 *(3, 18, 19)*, VEGF *(29, 30)*, PDGF *(31)* among others. Finally, we have performed combined genetic subtraction and microarray studies to identify genes and proteins involved in NSC proliferation, in general, as well as self-renewal *(32–34)*. Functional studies have shown that at least three genes identified from these studies, phosphoserine phosphatase (PSP) *(22)*, PBK/TOPK *(35)*, and MELK *(10)*, regulate either NSC or other multipotent progenitor proliferation. We have performed detailed analysis of MELK function, in particular, and found that it is required for embryonic and adult murine and human NSC self-renewal, as indicated by clonal neurosphere formation *(10, 11)*.

Brain tumors are likely to be derived from proliferating cells in the brain and, as such, might be expected to utilize similar methods of proliferation *(2-4, 7)*. Numerous studies have utilized a variety of methods to identify key genes involved in brain tumor proliferation *(36–39)*. It has long been known that stimulation of the EGF and FGF receptors is associated with brain tumor expansion *(9, 14, 24, 40, 41)*. Furthermore, as stated earlier, PTEN is a tumor suppressor gene frequently associated with malignant glioma *(24, 41)*. Each of these pathways is associated with NSC proliferation. Furthermore, analysis of microarrays indicates that several of the genes that we placed at high priority for study in NSC proliferation based on our arrays, such as MELK, PBK/TOPK, ASPM, and FoXM1 *(32, 34, 42)*.

The relationship between brain tumors and NSC has been made even more explicit by the discovery of BT stem cells *(2, 11)*. These cells can be isolated from BT using the same methods as those used to propagate NSC and appear to self-renew as well as to give rise to the different cell types that are found within the parent brain tumor *(8, 43)*. We have performed preliminary analyses that indicate that key NSC genes are expressed by these BT stem cells *(8)*. MELK, for example, is expressed by BT stem cells derived from glioblastoma multiforme and, as is the case for NSC, is required for their self-renewal in vitro *(11)*.

It is our general goal to determine the mechanisms via which BT stem cells self-renew and how these mechanisms overlap with and, hopefully, differ from those of NSC. In order to achieve these goals, we have adapted and developed methods to propagate and study these cells in vitro. This chapter describes this methodology in detail and is meant to allow other investigators to study and compare BT stem cells and NSC in order to discover and evaluate potential new brain tumor therapies.

2. Materials

2.1. Cell Culture

1. Dulbecco's Modified Eagle Medium/Nutrient mixture F-12 (Ham) 1× (DMEM/F12). Stored at 4°C.
2. Neurobasal (Invitrogen). Stored at 4°C.
3. Hanks' Balanced Salt Solution (HBSS) (1×; GIBCO; Cat. #24020).
4. B27 (Gibco BRL) is stored in 0.5-ml aliquots at –20°C. Final concentration is 1× for high-density culture and clonal culture.
5. bFGF (Protech) is stored in 0.1-ml (1 mg/1 ml) aliquots at –20°C. Final concentration is 20 ng/ml for high-density culture and clonal culture.

6. EGF (Protech) is stored in aliquot at –20°C. Final concentration is 50 ng/ml for high-density culture and 20 ng/ml for clonal culture.
7. Penicillin/Streptomycin (Gemini Bioproducts) is stored in 1.0-ml aliquots at –20°C. Final concentration is 100 U/ml for high-density culture and clonal culture.
8. Heparin (Sigma-Aldrich). Final concentration is 5 μg/ml for high-density culture and 2 μg/ml for clonal culture.
9. Fetal human progenitor Cambrex Bio Science. (Walkersville, MD Lot# 4F1565).
10. L-glutamine (Invitrogen) is stored in 0.5-ml aliquots (200 mM) at –20°C. Final concentration is 2 mM for clonal culture.
11. Poly-L-lysine (Sigma-Aldrich) is stored in 0.1-ml aliquots (5 mg/ml) at –20°C.
12. Fetal bovine serum (Gibco BRL) is stored in 0.5-ml aliquots at –20°C.
13. Polyornithin (Sigma Aldrich) is stored in 0.4-ml aliquots (3 mg/ml) at –20°C. Final concentration is 15 μg/ml in DDW.
14. Fibronectin is stored in 0.5-ml aliquots (1.2 μg/ml) at –20°C. Final concentration is 1.2 ng/ml in PBS.

2.2. Transient Transfection

1. Silencer siRNA Construction Kit (Ambion).
2. Lipofectamine 2000 (Invitrogen).
3. OPTIMEM I (1×; GIBCO; Cat. #31985).

2.3. Semiquantitative and Quantitative RT-PCR

1. Improm-II Reverse Transcription System (Promega; Part Number TM236).
2. Taqman Universal PCR MasterMix, No Amp Erase UNG. (ABApplied Bioscience, Part Number 4324018).

2.4. Cell Proliferation Assay

1. BrdU (Roche). Stock concentration; 3 mg/ml (10 mM), Stored at 4°C.
2. Anti-brdU antibody (DAKO) 1:200. Stored at 4°C.
3. Cell Proliferation ELISA (calorimetric) Kit (Roche; Cat. #1444611). Stored at 4°C.

2.5. Cell Death Assay

1. Apoptosis Detection Kit (R&D System; KNX50).
2. Annex V-FITC (R&D Systems). Stored at 4°C.
3. 10× Binding Buffer (R&D Systems). Stored at 4°C.

2.6. Immunocytochemistry

1. Paraformaldehyde (Polyscience, Inc.).
2. Anti-Nestin antibody (1:200; Chemicon).
3. Anti-Msi1 antibody (1:200; Chemicon).

4. Anti-TuJ1 antibody (1:500; Barkeley Antibodies).
5. Anti-GFAP antibody (1:500; DAKO).
6. Anti-O4 antibody (1:40; Chemicon).
7. Phosphor-S6 (1:200; Cell Signaling Technology).
8. Alexa fluorophore-conjugated secondary antibodies (1:2,000; Molecular Probes).
9. 4′6-diamidino-2-phenylindole (DAPI).

2.7. Flow Cytometry

1. Anti-LeX antibody (1:200; Invitrogen).
2. Anti-CD133/1 (AC133)-PE, human 1 ml (Miltenyi Biotec).
3. Anti-CD133/2 (293C3)-PE, human 1 ml (Miltenyi Biotec).
4. Alexa fluorophore-conjugated secondary antibodies (1:2,000; Molecular Probes).

3. Methods

3.1. Cell Culture

3.1.1. Neurosphere Cultures

Cortical telencephalons are removed from E12 CD-1 mice, and cerebral cortices are dissected from older animals (Charles River). Dissection is carried out on a culture dish filled with 1× HBSS (GIBCO), and leptomeningeal membranes are carefully removed from brain tissues. Cells are dissociated with a fire-polished glass pipette several times and are washed with 1× PBS for 4–5 times (*see* **Note 1**). For high-density neurosphere cultures (>50 cells/μl), single cells are resuspended at 50,000 cells per ml in DMEM/F12 medium (Invitrogen) supplemented with B27 (Gibco BRL), 20 ng/ml bFGF (Peprotech), 50 ng/ml EGF, penicillin/streptomycin (Gemini Bioproducts), and heparin (Sigma). For clonal neurosphere cultures (1 cell/μl), cells are resuspended with Neurobasal medium (Invitrogen) supplemented with B27 (Gibco BRL), 20 ng/ml bFGF, 20 ng/ml EGF, penicillin/streptomycin, and heparin. Growth factors are added every 3 days. After 7 days in culture, neurospheres are recovered for further studies.

Multipotent fetal neural progenitor cells are shipped in a frozen vial, and are thawed and placed in neurosphere media (*see* **Note 2**). These human neurospheres maintain stem cell characteristics as shown by immunocytochemistry **(Fig. 1)**. They can self-renew under a clonal condition and differentiate into three major cell types, TuJ1-positive neurons, GFAP-positive astrocytes, and O4-positive oligodendrocytes **(Fig. 1)** (*see* **Note 3**).

To determine whether a particular compound has an effect on the proliferation of NSC in culture, a secondary clonal neurosphere assay is used. The primary spheres are dissociated and plated into 96-well microwell plates in 0.2 ml volume of growth

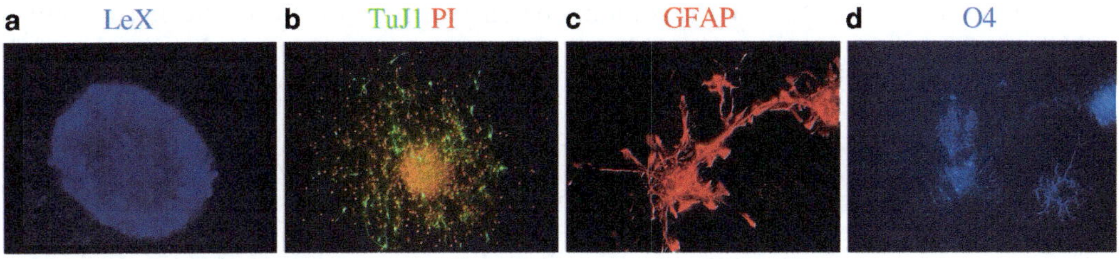

Fig. 1. Neurosphere cultures derived from NSC in mouse embryonic cortices. Under undifferentiation condition, mouse cortical neurospheres are positive for neural progenitor marker, LeX (**a**). Under differentiation condition, clonal neurospheres contain TuJ-1 positive neurons (**b**), GFAP-positive astrocytes (**c**), as well as 04-positive oligodendrocytes (**d**).

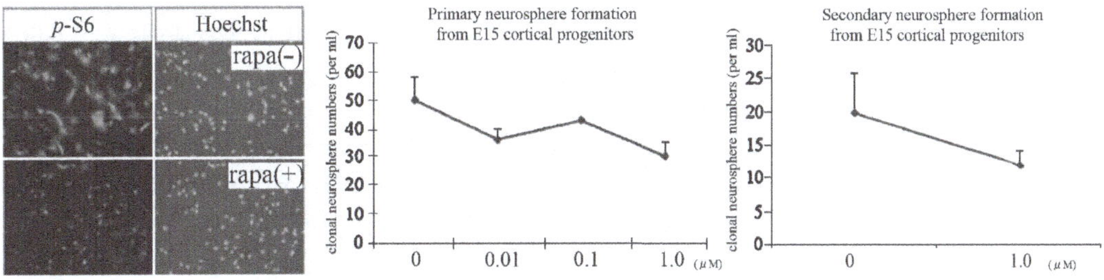

Fig. 2. Treatment of mouse neural progenitors with mTOR inhibitor, rapamycin. Two days' exposure of neural progenitors with rapamycin decreases the expression of phosphor-S6 m the downstream gene of mTOR (*left panel*). Cell nuclei are labeled with Hoechst staining. Both primary (*middle*) and secondary (*right*) neurosphere formations are affected by treatment of rapamycin.

media at 200 cells per microwell, and the resultant sphere numbers are counted after 7 days in culture. For example, the effect of rapamycin treatment on proliferation of NSC can be tested with this method **(Fig. 2)**. The results showed that the numbers of both primary and secondary neurospheres are reduced by rapamycin treatment. The resultant neurospheres from both conditions are tripotent, determined by immunocytochemistry of their differentiated cells (data not shown). These data indicate that rapamycin inhibits proliferation of NSC without affecting lineage commitment of the treated progenitors, which is consistent with the results published by another group independently *(44)*. Inhibition of the mTOR signaling pathway by this rapamycin treatment is confirmed by immunocytochemistry with phospho-S6 antibody (1:300, Cell Signaling) **(Fig. 2)**.

For differentiation, culture medium is replaced with Neurobasal (Invitrogen) supplemented with B27 without bFGF onto poly-L-lysine (PLL)-coated dishes and maintained up to 7 days. Early signs of differentiation can be observed under 24 h, but longer periods in culture allow nascent neurons and glial cells to take on more mature phenotypes. Effect of differentiation can be tested by immunocytochemistry.

3.1.2. GFAP-Positive SVZ Astrocyte-Enriched Cultures

NSC in the adult brain behave differently from NSC in the embryonic brain *(13, 14)*. The current consensus is that, in the adult subventricular zone (SVZ), slowly dividing GFAP-positive NSC give rise to rapidly amplifying GFAP-negative progenitors *(1)*. These progenitors, which express TOPK/PBK, become TuJ-1 positive neuroblasts and travel through the rostral migratory pathway all the way to the olfactory bulb. These GFAP-positive astrocytes can be cultured from the adult SVZ *(45)*. Following exposure of the adult brains, the lateral ventricles are opened. Choroid plexus is carefully removed, and the periventricular area (PV) is scooped out. Dissected tissues are dissociated with fire-polished glass pipettes, and are cultured with DMEM/F12 plus 10% FBS and 1% penicillin/streptomycin in T75 flasks. One day after dissection, half of the culture media is replaced with new media. As cells become confluent (12–14 DIV), they are shaken at 200 × *g* overnight to remove nonadherent cells (neurons and oligodendrocytes), and obtain almost pure GFAP positive cells (>95%), and passaged onto polylysine-coated plates. These GFAP-positive cells from the PV, but not those from the cerebral cortices (CTX), contain NSC (**Fig. 3**) (*see* **Note 4**).

When we change the culture condition from serum-containing media to neurosphere growth medium, GFAP-positive NSC give rise to GFAP-negative progenitors, which can mimic the in vivo

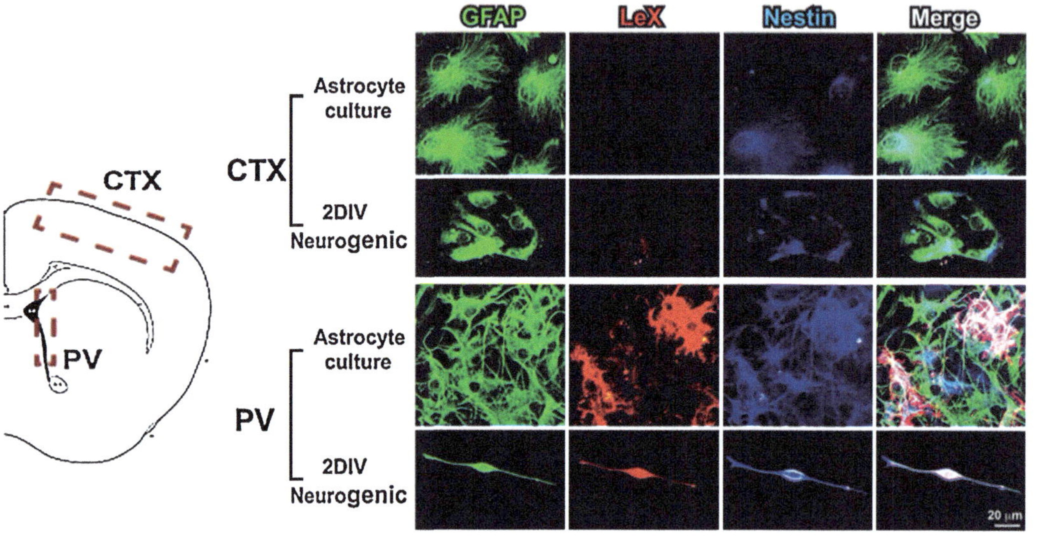

Fig. 3. Comparison of phenotypes of cortical and subventricular zone astrocytes. Primary astrocyte cultures are prepared either from superficial cerebral cortex (CTX) or from periventricular tissue (PV) of adult mice (drawing on the left). Comparison of cell phenotype in primary astrocyte cultures or after transfer to adherent neurogenic conditions for 2DIV. Three-color immunofluorescence shows that under both culture conditions, GFAP-expressing cells derived from adult CTX have a flat polygonal shape and coexpress some nestin, but do not express LeX. In contrast, GFAP-expressing cells derived from adult PV contain an additional small subpopulation of cells that coexpress GFAP, nestin, and LeX. In serum, these triple stained cells are flat polygonal in shape, whereas in neurogenic conditions, triple stained cells have a bipolar shape.

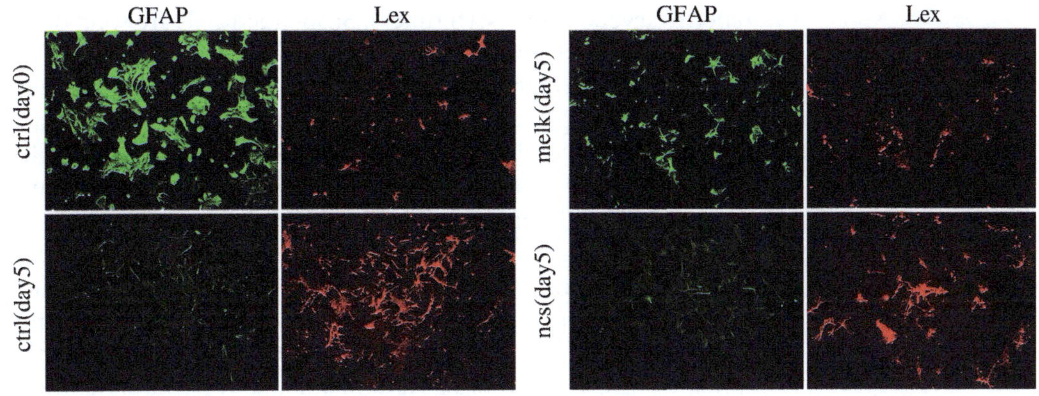

Fig. 4. Effect of MELK siRNA on the transition of GFAP-expressing stem cells into GFAP-negative progenitors in culture. Following transfection of either MELK or Nucleostemin (ncs) siRNA, immunostaining shows GFAP-positive cells (*green in right panels*) and LeX-positive cells (*red in left panels*) before (day 0) and after (day 5) bFGF stimulation.

transition from the GFAP-positive SVZ astrocytes to the rapidly amplifying GFAP-negative progenitors in the adult SVZ. With this culture method, one can determine an effect of a gene or a compound on this transition. For example, we tested function of both MELK and nucleostemin with this culture system *(10)* (**Fig. 4**). MELK, but not nucleostemin, is upregulated during this transition. Inhibition of MELK expression by siRNA resulted in failure to produce GFAP(−) LeX(+) progenitors from GFAP(+) astrocytes (**Fig. 4**).

3.1.3. Brain Tumor Stem Cells

Brain tumor stem cells can be enriched in culture by the neurosphere method *(8)*. Brain tumor spheres can be formed from a single stem cell under a clonal condition; therefore, effects on self-renewal of brain tumor stem cells can be determined by the affected sphere numbers retrospectively.

Tissue Collection and Neurosphere Culture

Brain tumor specimens from patients are obtained immediately after surgical resection in accordance with the protocols approved by the institutional review boards. Generally, 0.5 g of a tumor specimen for each case (0.2 g at the minimum) is required for enrichment of tumor spheres. Tumors are graded by the neuropathologist in accordance with World Health Organization (WHO) established guidelines. Tissues are sliced into four pieces with a blade in N_2 liquid, and three of them are stored in the liquid nitrogen (two for RNA collection and one for protein collection). The last piece is transferred in a 15-ml conical tube with 1 ml of Triplexpress and 1 ml of DMEM/F12. Cells are incubated in 37°C for 30 min with occasional manual shakes. Following incubation, cells in media are shaken up and down by a 10-ml pipette until it is smooth through the pipette. Then the solution is triturated with a large and small fire polished glass pipettes several times (<15 times). Suspended cells are spun down at 800 × *g* for 5 min, resuspended with 2 ml of DMEM/F12, and filtered through a 70-μm strainer (*see* **Note 5**). Following these steps, single cells are put into the neurosphere-forming solution in a 75-cm^2 flask (*see* **Note 6**). Growth factors are added every 3 days. By culturing malignant types of brain tumors, such as high-grade glioma, medulloblastoma, and ependymoma, we can obtain a large number of tumor progenitors as neurospheres (>10^9/50 ml culture) in a week or two. For differentiation induction of tumor progenitors, culture medium is replaced into Neurobasal supplemented with B27 without bFGF onto poly-L-lysine (PLL)-coated dishes and maintained up to 5 days. Many tumors, however, do not yield spheres that differentiate into clearly demarcated neuron and glia-like cells.

Passage of Tumor Progenitors

Brain tumor spheres are formed in a week or two, and some of the cultures start forming adherent and differentiated cells on the bottom of flasks. In order to maintain immature character of brain tumor progenitors, we need to passage them into another culture with fresh medium. The cells are recovered from flasks, spun down at 800×*g* for 5 min, and resuspended with 1 ml of Triplexpress at 37°C for 5 min. Small fire-polished glass pipettes are used for trituration (up to 15 times). They are resuspended in 2 ml of neurosphere-forming media and filtered through a 40-μm strainer. Then the cells are resuspended in media at the desired density (clonal culture: 1 cell/μl and high-density culture: 50–100 cells/μl) (*see* **Note 7**).

3.2. Transient Transfection

One way to determine an effect of a gene on neural and/or brain tumor stem cells is to use transient transfection in a modified neurosphere culture method. Since the effect is transient, we can determine the function of a gene in *proliferation* of stem cells by the number of clonal spheres after treatment. *Self-renewal* needs to be determined by persistent overexpression and/or knockdown of a gene in stem cells, such as occurs with viral infection.

3.2.1. siRNA Synthesis

siRNA constructs are synthesized using the Silencer siRNA Construction Kit following the manufacturer's instructions (Ambion). Targeting sequences can be selected from a tool in the manufacturer's website (http://www.ambion.com/techlib/misc/siRNA_tools.html). The GC content of the target sequences is recommended between 40 and 60%. Similarity to other genes needs to be carefully validated by the Basic Local Alignment Sequence Tool (BLAST) search (http://www.ncbi.nlm.nih.gov/BLAST/). This search should be performed against the library containing the sequences of the whole genome as well as the library of the expressed sequence tag (EST) of the corresponding species.

One of the pitfalls of the siRNA experiments is the potential for "off-target" effects *(46)*. These effects are caused by nonspecific effects on genes other than the target one. In order to avoid the potential noise created by off-target effects, we design 3–4 siRNA constructs targeting at either the coding region or the 3′-untranslated region of a gene of interest (**Fig. 5**, panel A). The phenotype of knockdown should be compared with the level of knockdown induced by each siRNA construct. Also, we validate the dose dependency of siRNA treatment (**Fig. 5**, panel B). Furthermore, an siRNA construct designed to target 3′-untranslated region can be used for transfection together with the overexpression vector of the gene of interest. This *rescue* experiment is aimed at determining whether the phenotype by siRNA treatment is recovered by overexpression of the same gene (**Fig. 5**, panel C). By using the siRNA of 3′-untranslated region, we can avoid the interference between siRNA and exogenously overexpressed gene. When we observe a rescue of the effect of siRNA by the dual transfection, we can presume that the effect by siRNA is due to inhibition of the gene of interest.

3.2.2. Transient Transfection

Cells are transfected using Lipofectamine 2000 (Invitrogen) following the manufacturer's protocol. For transfection of plasmid vectors, the cells are incubated with reagents for 4–6 h with the primary progenitor cells, and for 24 h with most of the cell lines. The efficiency of transfection largely depends on the cell types that

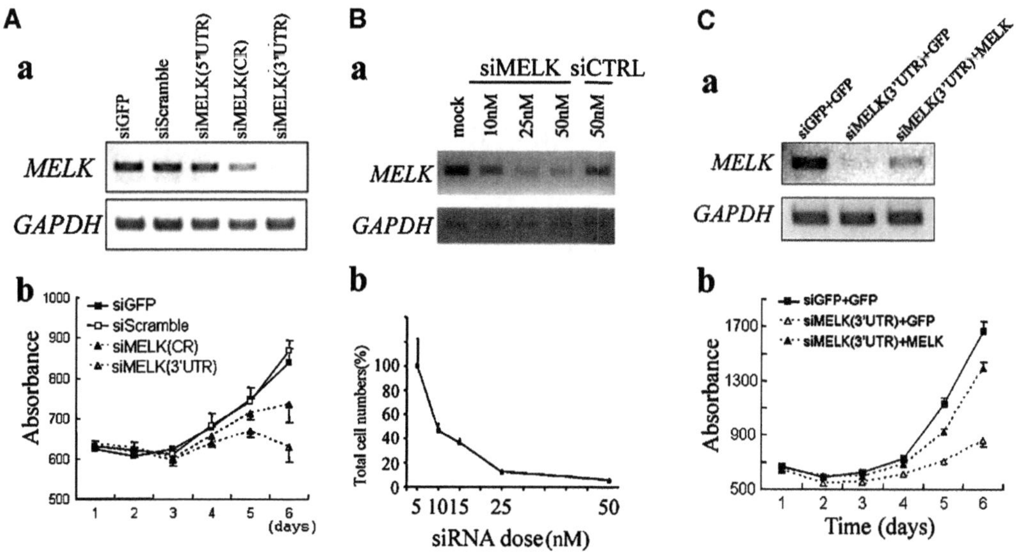

Fig. 5. Transfection of MELK siRNA constructs. (**A**) (**a**) RT-PCR of MELK expression in control siRNA (siGFP and siScramble) or MELK siRNA (CR and 3′UTR)-treated Daoy cell cultures. (**b**) Cell viability is determined by WST assay at the indicated times following MELK or control siRNA treatment. (**B**) RT-PCR for *MELK* expression (**a**) and the corresponding total cell numbers relative to controls (**b**) following treatment with different concentrations of MELK siRNA. (**C**) RT-PCR for *MELK* expression (**a**) and the corresponding total cell numbers (**b**) following transfection with expression vector for EGFP or MELK and siRNA against MELK (3′UTR) or EGFP.

we work with. For transfection of the double-stranded siRNA complex, in most cases, 25 nM of siRNA is high enough concentration to inhibit the expression of the gene of interest without causing the off-target effects *(46)*. Therefore, we generally use 25–50 nM of siRNA for our knockdown experiments.

3.2.3. To Determine the Proliferation Capacity of Transfected Stem Cells

To determine the influence of gene knockdown or overexpression, a modified neurosphere culture system is used **(Fig. 6)**. Neurospheres are propagated for 1 week and then dissociated with trypsin (0.05%) followed by trituration with a fire-polished pipette. The cells are then placed in DMEM/F12 with 2% fetal bovine serum (FBS; Gibco BRL) and plated onto polyornithine/fibronectin coated glass coverslips *(47)*. With 6 h of incubation, the serum-containing medium is replaced with the neurosphere growth medium without heparin. At 24 h post-plating, transfection is performed as described earlier. At this point, virtually all cells adhere to the bottom of the dishes (*see* **Note 8**). To assay the sphere-forming potential of the transfected cells, they are lifted off the plate with trypsin (0.05%), incubated briefly in medium containing 10% FBS to inactivate trypsin, spun down and then placed into Neurobasal media supplemented with B27, bFGF, EGF, and heparin under a clonal density (1 cell/μl) *(48)*. Within 1 week in culture, clonal neurospheres are formed, and the number of clonal spheres corresponds to the proliferation capacity of NSC at the beginning of the secondary neurosphere culture. Alternatively, we can also assess the effect of transfection on differentiation capacity of treated progenitors. Following transfection,

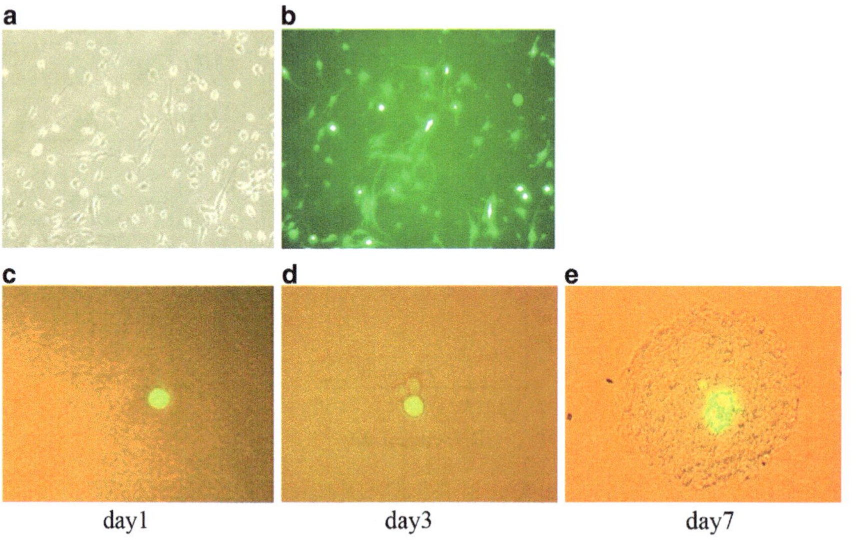

Fig. 6. Transfection of the green fluorescence protein-expressing vector in neural progenitors. A phase-bright picture (**a**) and a fluorescence picture (**b**) of adherent neural progenitors following transfection of GFP-expression vector. (**c–e**) Clonal neurosphere formation of transfected neural progenitors. AU6

adherent progenitors are incubated with differentiation medium (described earlier, or with 2% FBS and 10^6 M retinoic acid) on the coated coverslips for 5 days. Immunocytochemistry using neuronal, astrocytic, and oligodendrocytic markers can determine the effect of transfection on progenitors.

3.2.4. Sphere Diameter Analysis

Symmetrical self-renewal and/or survival effect on NSC can be determined by the number of new clonal neurospheres, whereas effect on asymmetric self-renewal, total neural progenitor proliferation, including NSC and lineage-committed progenitor cells, is reflected by the size of neurospheres, which is proportional to their diameter, an easily measured parameter ***(10, 18)***. In order to measure neurosphere diameter, floating neurospheres are plated onto coverslips and fixed with 4% PFA. Spheres from various conditions are measured using the Microcomputer Imaging Device Program (MCID) or other imaging tool. A minimum cutoff of 40 μm is used in defining a neurosphere, as small clusters of cells can often be present.

3.3. Quantitative RT-PCR

DNAse-treated RNA samples (1 μg) are directly reverse transcribed with ImProm-II RT (Promega). Real-time PCR is performed utilizing a LightCycler rapid thermal cycler system (Roche Diagnostics) according to the manufacturer's instructions. A mastermix of the following reaction components is prepared to the indicated end concentrations: 8.6 μl of water, 4 μl of Betaine (1 M), 2.4 μl of $MgCl_2$ (4 mM), 1 μl of primer mix (0.5 μM), and 2 μl LightCycler (Fast Start DNA Master SYBR Green I: Roche Diagnostics). LightCycler Mastermix (18 μl) is filled in the LightCycler glass capillaries and 2 μl cDNA is added as PCR template. A typical experimental run protocol consisted of an initial denaturation program (95°C for 10 min), amplification and quantification program repeated 45 times (95°C for 15 s, 62°C for 5 s, 72°C for 15 s followed by a single fluorescence measurement). Relative quantification is determined using the LightCycler Relative Quantification Software (Roche Diagnostics), which takes the crossing points (CP) for each target transcript and divides them by the reference GAPDH CP.

3.4. In Vitro Immunostaining

Neural progenitors in various conditions of culture are stained for either βIII-tubulin (TuJ1, 1:500; Barkeley Antibodies), musashi1 (1:200; Chemicon), or nestin (1:200; Chemicon), as markers of neural progenitor (nestin and musashi1), or neuronal (βIII-tubulin) antigenic expression. Others are also stained with mouse antioligodendrocytic O4 IgM (1:40; Chemicon), or rabbit anti-human astrocytic GFAP (1:500; DAKO). In case of immunostaining of floating neurospheres, neurospheres in growth media are replaced onto polylysine coated coverslips with 2% of fetal bovine serum. After 30 min of incubation, spheres are adherent on the bottom of dishes, and they are fixed with 4% paraformaldehyde for 30 min at room temperature.

3.5. Cell Death Assay

One hundred microliters of Annexin V Incubation Reagent (10 µl of 10× Binding Buffer, 10 µl of Propidium lodide, 1 µl of Annexin V-FITC, 79 µl of dH_2O) for each sample of 1×10^5–1×10^6 cells, and 500 µl of 1× Binding Buffer per sample (50 µl of 10× Binding Buffer, 450 µl of dH_2O) are prepared and kept in the dark and on ice. Cells are collected by centrifugation at 500 × *g* for 5 min at room temperature, and washed by resuspending them in 500 µl of 1× PBS (2–8°C), and centrifuged again at 500 × *g* for 5 min. Cells are then gently resuspended in the Annexin V Incubation Reagent at a concentration of 1×10^5–1×10^6 cells/100 µl, and incubated in the dark for 15 min at room temperature. If the number of cells is in the recommended range of 1×10^5–1×10^6, add 400 µl of 1× Binding Buffer to each sample. If the number of cells is not in the recommended range, centrifuge cells at 500 × *g* for 5–10 min and resuspend cells in 100 µl of 1× Binding Buffer. Samples are analyzed by flow cytometry within 1 h for maximal signal.

3.6. Cell Proliferation Assay

Cells are incubated in a 96-well microwell plate in a final volume of 100 µl/well under 37°C for 24 h, and 2 h following addition of 10 µl/well BrdU labeling solution. Labeling medium is removed by tapping and flicking off. 200 µl/well of FixDenat is added to cells and incubate for 30 min at 15–25°C. FixDenat solution is removed thoroughly by tapping and flicking off. Next, 100 µl/well of anti-BrdU-POD working solution is added in each well and incubated for 90 min at 15–25°C. Then, antibody is removed by tapping and flicking off and cells are rinsed well for three times with 200 µl/well of washing solution. Washing solution is removed by tapping; 100 µl/well of Substrate solution is added and incubated at 15–25°C for 5–30 min. Subsequently the color is sufficiently developed for photometric detection. Absorbance is measured by microplate reader (GENios; TECAN). Readout is the optical density at 492 nm (*see* **Note 9**).

Alternatively, cell proliferation can be measured by immunocytochemistry with anti-BrdU antibody. Cells are incubated with BrdU (Roche) for 2 h starting at 48 h post-transfection and are fixed with 4% paraformaldehyde; staining was carried out using sheep anti-BrdU (DAKO 1:200). To obtain the labeling index, the number of BrdU –labeled cells is expressed as a fraction of the total number of cells in a well for at least three wells per condition. Total cell numbers are determined by counterstaining with Hoechst.

3.7. Cell Cycle Analysis

One million cells in 12×15 mm test tube is placed and washed once with cold PBS. After centrifuge, 500 µl of cold PBS is added

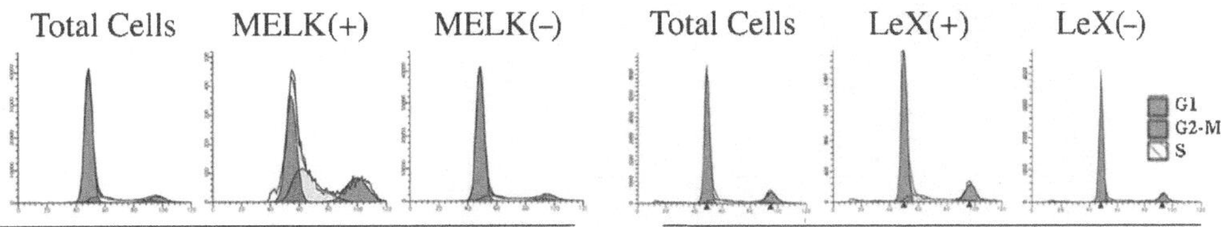

	Total Cells	MELK(+)	MELK(-)	Total Cells	LeX(+)	LeX(-)
G0/G1	79.09%	45.26%	80.41%	74.59%	70.15%	80.63%
S	14.66%	36.21%	13.34%	13.33%	15.00%	9.31%
G2/M	6.25%	18.54%	6.25%	12.08%	14.86%	10.06%

Fig. 7. Cell cycle analysis of neurospheres following separation based on either MELK or LeX expression. Clonal neurospheres are derived from P0 cortices, and dissociated progenitors are separated by cell sorter based on either MELK or LeX expression. The table shows subpopulations of each cell cycle fraction.

and then 500 μl of cold buffered 2% formaldehyde in PBS is mixed. Cells are incubated at 4°C for 1 h. After fixation, cells are washed with 1× PBS once, and 1 ml of 70% ethanol at −20°C is added dropwise to the cell pellet with the tube sitting on a vortex. Cell suspension is incubated overnight at 4°C. Then, cells are centrifuged, and cell pellets are mixed with 1 ml of a solution containing 40 μg/ml of PI and 100 μg/ml of ribonuclease A. With 30 min of incubation in the dark, cell cycle is analyzed by FACScan (Becton-Dickinson). **Figure 7** shows an example of cell cycle analysis with PI labeling. Neural progenitors derived from P0 cortices are divided into MELK(+), (−), and LeX(+), (−) cells. The subdivided cell populations are independently stained with PI. MELK-expressing neural progenitors, but not the rest of cells, are highly cycling neural progenitors.

3.8. Genetically Modified Animal Model

Pmelk. A vector is constructed to express EGFP under the control of the mouse Melk promoter. The promoterless vector (pEGFP: Promega) is digested with *EcoRI* sites to insert a 3.5-kb genomic fragment in the upstream of the start codon of mouse Melk. This genomic fragment contains multiple DNA binding sites, which activate the Melk transcriptional activity. The search for the regulatory element can be performed by PROMOTER-SCAN (http://www-bimas.cit.nih.gov/molbio/proscan/). The specific expression of this transgene is confirmed both by in vitro neurosphere assay and by the transgenic mouse studies (**Fig. 8**). In general, this method is applicable for any “stem cell-genes” in order to characterize stem cells in the normal brain and in the brain tumor.

3.9. Flow Cytometry and Sorting

Flow cytometry and sorting of neural progenitors are performed with a FACS Vantage (Becton-Dickinson) using a purification-mode algorithm. First, gating parameters are set by side and forward scatters to eliminate dead and aggregated cells. An alternative method to exclude dead cells is to use Propium Iodide (PI). Thirty minutes of incubation with PI labels only dead cells.

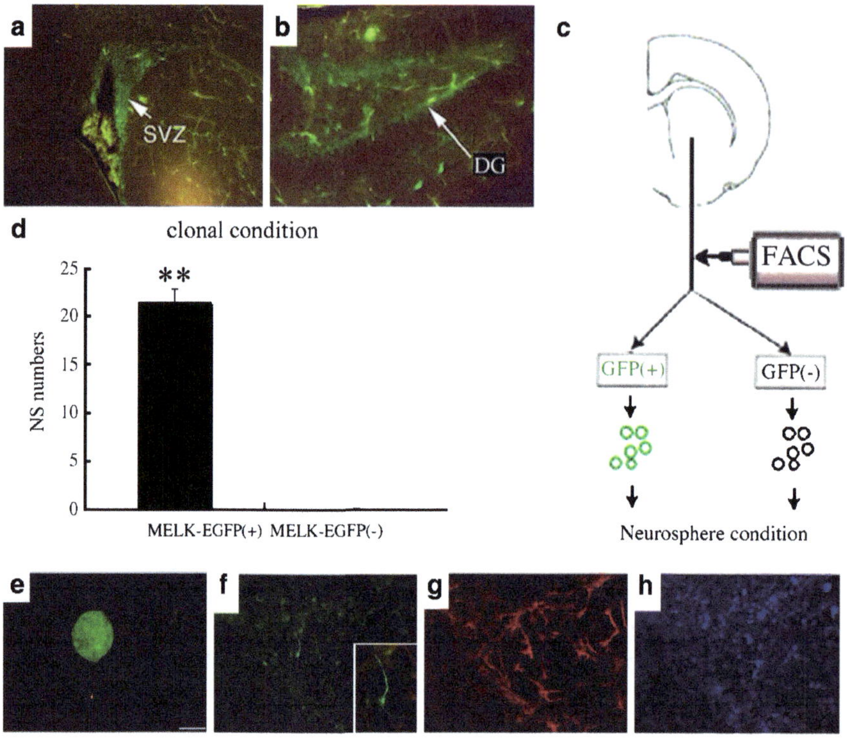

Fig. 8. Analysis of the MELK-expressing cells for their neurosphere-forming capacity. Expression of the MELK reporter expression (*green*) in a P8 transgenic mouse demonstrating specific signals in the SVZ (**a**) and dentate gyrus (DG) of hippocampus (**b**), indicated by arrows. (**c**) Experimental design of direct FACS from P1 cortices. (**d**) Graph shows the number of clonal neurospheres from PMELK-EGFP(+) and PMELK-EGFP(–) cells. ***p* < 0.001, ANOVA followed by post hoc *t* test. Undifferentiated clonal neurospheres (**e**) are differentiated and stained with TuJ1 (**f**), GFAP (**g**), and O4 (**h**). *Inset* in (**f**) shows a magnified positive cell.

Neurospheres derived from GBM are labeled with the CD133 antibody (clone AC133, Miltenyi Biotec) for 30 min and Alexa 488 is used for flow cytometry and sorting. Background signals are determined by incubation of the same set of progenitors without primary antibody. Following cell sorting, we confirm the cell separation by the analytical FACS with FACScan. Generally, 3–40% of cells in GBM specimens express CD133. When tumor cells are incubated with FBS longer than 7–14 days, they lose CD133 content, and %CD133 positive cells drops down to 0–3% (data not shown).

4. Notes

1. In order to obtain high viability, gentle dissociation of mouse cortical cells is required. In order to avoid bacterial contamination, dissociated cells should be washed with 1xPBS at least 3 times prior to the culture.

2. It takes up to 3 months to let the human progenitors grow as neurospheres from the frozen stocks; however, once they start growing, they grow rapidly and can be passaged every 1–2 weeks by mechanical dissociation with fire-polished pipettes and replating in fresh neurosphere medium.
3. The neurosphere-forming capacity of neural progenitors can be enhanced by using the conditioned medium by mouse neurospheres. This conditioned medium is obtained from the secondary neurosphere cultures of mouse cortical progenitors. Conditioned medium is prepared by first centrifuging the cells and then passing the supernatant through a 0.4-μm filter. The filtrate is carefully examined by microscopy for the absence of cells and then frozen at −20°C. Prior to use, a sample from each batch is spun down and cultured to make sure that there are no surviving NSC. When we use conditioned medium for cultures, a third of the culture medium is replaced with the conditioned medium. A great increase of neurosphere formation is observed, especially when we culture progenitors under clonal conditions (<1 cell/μl). This medium is effective for both mouse and human progenitors, as well as for normal and brain tumor progenitors.
4. Adult tissues are more fragile than embryonic ones, and extra caution is required for dissection. It may take longer period of culture in order to obtain confluency of adherent cells. In some cases, growth of these cells depends on the batch of fetal bovine serum.
5. An extensive mechanical dissociation may result in a large number of dead cells in the end without recovering live progenitors. Alternatively, progenitor cells can be isolated by using the difference of the cell weights with the percoal filters.
6. In order to achieve better viability of cells dissected from the tumor specimens, primary cultures can be started with serum-containing medium (DMEM/F12 with 10% fetal bovine serum). However, one needs to be aware of the fact that stimulation of tumor progenitors with serum causes rapid differentiation and loss of "stemness" of the cultured cells.
7. The step of the single cell dissociation is sometimes too harsh for the cells to be maintained in culture. Alternatively, tumor spheres can be "*chopped*" into smaller pieces of spheres, instead of completely dissociated into single cells. In this method, neurospheres are divided into 200 μm in two perpendicular directions and placed into flasks with new medium.
8. When neural or tumor progenitor cells are transfected, the confluency of the adherent cells should be above 70% in order to obtain a high transfection efficiency (>70%). If the density is not high enough (<50%), transfection can be postponed for another day. However, the time window for transfection of adherent progenitors should be 48 h following passage.

9. Effect on proliferation needs to be evaluated by the proportion of BrdU-positive cells among the treated cells, but not by the total signals detected by BrdU labeling in each well. Therefore, the signal intensity needs to be adjusted by the total cell number in each well.

Acknowledgments

This work was supported by NIMH grant MH065756 and the Miriam and Sheldon Adelson Program in Neural Repair Research. IN is supported by the ISCBM-CIRM fellowship and the Khan Family Foundation.

References

1. Alvarez-Buylla, A., and Lim, D. A. (2004) For the long run: maintaining germinal niches in the adult brain. *Neuron* **41**, 683–6.
2. Nakano, I., and Kornblum, H. I. (2006) Brain tumor stem cells. *Pediatr Res* **59**, 54R–8R.
3. Reya, T., Morrison, S. J., Clarke, M. F., and Weissman, I. L. (2001) Stem cells, cancer, and cancer stem cells. *Nature* **414**, 105–11.
4. Dirks, P. B. (2006) Bone morphogenetic proteins inhibit the tumorigenic potential of human brain tumour-initiating cells. *Nature* **444**, 687–8.
5. Bao, S., Wu, Q., McLendon, R. E., Hao, Y., Shi, Q., Hjelmeland, A. B., Dewhirst, M. W., Bigner, D. D., and Rich, J. N. (2006) Glioma stem cells promote radioresistance by preferential activation of the DNA damage response. *Nature* **444**, 756–60.
6. Bao, S., Wu, Q., Sathornsumetee, S., Hao, Y., Li, Z., Hjelmeland, A. B., Shi, Q., McLendon, R. E., Bigner, D. D., and Rich, J. N. (2006) Glioma stem cells promote radioresistance by preferential activation of the DNA damage response. *Cancer Res* **66**, 7843–8.
7. Singh, S. K., Hawkins, C., Clarke, I. D., Squire, J. A., Bayani, J., Hide, T., Henkelman, R. M., Cusimano, M. D., and Dirks, P. B. (2004) Identification of human brain tumour initiating cells. *Nature* **432**, 396–401.
8. Hemmati, H. D., Nakano, I., Lazareff, J. A., Masterman-Smith, M., Geschwind, D. H., Bronner-Fraser, M., and Kornblum, H. I. (2003) Cancerous stem cells can arise from pediatric brain tumors. *Proc Natl Acad Sci U S A* **100**, 15178–83.
9. Phillips, H. S., Kharbanda, S., Chen, R., Forrest, W. F., Soriano, R. H., Wu, T. D., Misra, A., Nigro, J. M., Colman, H., Soroceanu, L., Williams, P. M., Modrusan, Z., Feuerstein, B. G., and Aldape, K. (2006) Molecular subclasses of high-grade glioma predict prognosis, delineate a pattern of disease progression, and resemble stages in neurogenesis. *Cancer Cell* **9**, 157–73.
10. Nakano, I., Paucar, A. A., Bajpai, R., Dougherty, J. D., Zewail, A., Kelly, T. K., Kim, K. J., Ou, J., Groszer, M., Imura, T., Freije, W. A., Nelson, S. F., Sofroniew, M. V., Wu, H., Liu, X., Terskikh, A. V., Geschwind, D. H., and Kornblum, H. I. (2005) Maternal embryonic leucine zipper kinase (MELK) regulates multipotent neural progenitor proliferation. *J Cell Biol* **170**, 413–27.
11. Nakano, I., Masterman-Smith, M., Saigusa, K., Paucar, A., Horvath, S., Watanabe, M., Negro, A., Bajpai, R., Howes, A., Lelievre, V., Washek, J. A., Lazareff, J. A., Freije, W. A., Liau, L. M., Gilbertson, R. J., Cloughesy, T., Geschwind, D. H., Nelson, S. F., Mischel, P. S., Terskikh, A., and Kornblum, H. I. (2007) Maternal embryonic leucine zipper kinase is a key regulator of the proliferation of malignant brain tumors, including brain tumor stem cells. *J Neursci Res* **86**, 48–60.
12. Reynolds, B. A., Tetzlaff, W., and Weiss, S. (1992) A multipotent EGF-responsive striatal embryonic progenitor cell produces neurons and astrocytes. *J Neurosci* **12**, 4565–74.
13. Lie, D. C., Song, H., Colamarino, S. A., Ming, G. L., and Gage, F. H. (2004) Neurogenesis in the adult brain: new strategies for central

nervous system diseases. *Annu Rev Pharmacol Toxicol* **44**, 399–421.

14. Gage, F. H. (2000) Mammalian neural stem cells. *Science* **287**, 1433–8.
15. Kornblum, H., and Geschwind, D. (2001) Molecular markers in CNS stem cell research: hitting a moving target. *Restor Neurol Neurosci* **18**, 89–94.
16. Kornblum, H. I., and Geschwind, D. H. (2001) Molecular markers in CNS stem cell research: hitting a moving target. *Nat Rev Neurosci* **2**, 843–6.
17. Estivill-Torrus, G., Pearson, H., van Heyningen, V., Price, D. J., and Rashbass, P. (2002) Pax6 is required to regulate the cell cycle and the rate of progression from symmetrical to asymmetrical division in mammalian cortical progenitors. *Development* **129**, 455–66.
18. Molofsky, A. V., He, S., Bydon, M., Morrison, S. J., and Pardal, R. (2005) Bmi-1 promotes neural stem cell self-renewal and neural development but not mouse growth and survival by repressing the p16Ink4a and p19Arf senescence pathways. *Genes Dev* **19**, 1432–7.
19. Molofsky, A. V., Pardal, R., Iwashita, T., Park, I. K., Clarke, M. F., and Morrison, S. J. (2003) Bmi-1 dependence distinguishes neural stem cell self-renewal from progenitor proliferation. *Nature* **425**, 962–7.
20. Lee, J. P., Jeyakumar, M., Gonzalez, R., Takahashi, H., Lee, P. J., Baek, R. C., Clark, D., Rose, H., Fu, G., Clarke, J., McKercher, S., Meerloo, J., Muller, F. J., Park, K. I., Butters, T. D., Dwek, R. A., Schwartz, P., Tong, G., Wenger, D., Lipton, S. A., Seyfried, T. N., Platt, F. M., and Snyder, E. Y. (2007) Stem cells act through multiple mechanisms to benefit mice with neurodegenerative metabolic disease. *Nat Med* **13**, 439–47.
21. Ge, W., He, F., Kim, K. J., Blanchi, B., Coskun, V., Nguyen, L., Wu, X., Zhao, J., Heng, J. I., Martinowich, K., Tao, J., Wu, H., Castro, D., Sobeih, M. M., Corfas, G., Gleeson, J. G., Greenberg, M. E., Guillemot, F., and Sun, Y. E. (2006) Coupling of cell migration with neurogenesis by proneural bHLH factors. *Proc Natl Acad Sci U S A* **103**, 1319–24.
22. Nakano, I., Dougherty, J. D., Kim, K. J., Klement, I., Geschwind, D. H., and Kornblum, H. I. (2007) Phosphoserine phosphatase is expressed in the neural stem cell niche and regulates neural stem and progenitor cell proliferation. *Stem Cells* **25**, 1975–84
23. Capela, A., and Temple, S. (2006) LeX is expressed by principle progenitor cells in the embryonic nervous system, is secreted into their environment and binds Wnt-1. *Dev Biol* **291**, 300–13.
24. Groszer, M., Erickson, R., Scripture-Adams, D. D., Dougherty, J. D., Le Belle, J., Zack, J. A., Geschwind, D. H., Liu, X., Kornblum, H. I., and Wu, H. (2006) PTEN negatively regulates neural stem cell self-renewal by modulating G0-G1 cell cycle entry. *Proc Natl Acad Sci U S A* **103**, 111–6.
25. Groszer, M., Erickson, R., Scripture-Adams, D. D., Lesche, R., Trumpp, A., Zack, J. A., Kornblum, H. I., Liu, X., and Wu, H. (2001) Negative regulation of neural stem/progenitor cell proliferation by the Pten tumor suppressor gene in vivo. *Science* **294**, 2186–9.
26. Chow, L. M., and Baker, S. J. (2006) PTEN function in normal and neoplastic growth. *Cancer Lett* **241**, 184–96.
27. Ohgaki, H. (2005) Genetic pathways to glioblastomas. *Neuropathology* **25**, 1–7.
28. Mischel, P. S., and Cloughesy, T. F. (2003) Targeted molecular therapy of GBM. *Brain Pathol* **13**, 52–61.
29. Li, Q., Ford, M. C., Lavik, E. B., and Madri, J. A. (2006) Modeling the neurovascular niche: VEGF- and BDNF-mediated cross-talk between neural stem cells and endothelial cells: an in vitro study. *J Neurosci Res* **84**, 1656–68.
30. Jung, K. H., Chu, K., Lee, S. T., Kim, S. J., Sinn, D. I., Kim, S. U., Kim, M., and Roh, J. K. (2006) Granulocyte colony-stimulating factor stimulates neurogenesis via vascular endothelial growth factor with STAT activation. *Brain Res* **1073–1074**, 190–201.
31. Jackson, E. L., Garcia-Verdugo, J. M., Gil-Perotin, S., Roy, M., Quinones-Hinojosa, A., VandenBerg, S., and Alvarez-Buylla, A. (2006) PDGFR alpha-positive B cells are neural stem cells in the adult SVZ that form glioma-like growths in response to increased PDGF signaling. *Neuron* **51**, 187–99.
32. Easterday, M. C., Dougherty, J. D., Jackson, R. L., Ou, J., Nakano, I., Paucar, A. A., Roobini, B., Dianati, M., Irvin, D. K., Weissman, I. L., Terskikh, A. V., Geschwind, D. H., and Kornblum, H. I. (2003) Neural progenitor genes. Germinal zone expression and analysis of genetic overlap in stem cell populations. *Dev Biol* **264**, 309–22.
33. Terskikh, A. V., Easterday, M. C., Li, L., Hood, L., Kornblum, H. I., Geschwind, D. H., and Weissman, I. L. (2001) From hematopoiesis to neuropoiesis: evidence of overlapping genetic programs. *Proc Natl Acad Sci U S A* **98**, 7934–9.
34. Geschwind, D. H., Ou, J., Easterday, M. C., Dougherty, J. D., Jackson, R. L., Chen, Z., Antoine, H., Terskikh, A., Weissman, I. L., Nelson, S. F., and Kornblum, H. I. (2001)

A genetic analysis of neural progenitor differentiation. *Neuron* **29**, 325–39.

35. Dougherty, J. D., Garcia, A. D., Nakano, I., Livingstone, M., Norris, B., Polakiewicz, R., Wexler, E. M., Sofroniew, M. V., Kornblum, H. I., and Geschwind, D. H. (2005) PBK/TOPK, a proliferating neural progenitor-specific mitogen-activated protein kinase kinase. *J Neurosci* **25**, 10773–85.
36. Wong, M. L., Kaye, A. H., and Hovens, C. M. (2007) Targeting malignant glioma survival signalling to improve clinical outcomes. *J Clin Neurosci* **14**, 301–8.
37. Salhia, B., Tran, N. L., Symons, M., Winkles, J. A., Rutka, J. T., and Berens, M. E. (2006) Molecular pathways triggering glioma cell invasion. *Expert Rev Mol Diagn* **6**, 613–26.
38. Tomera, J. F. (2000) Glioma: Novel considerations and treatment modalities. *Drugs Today (Barc)* **36**, 355–67.
39. Ware, M. L., Berger, M. S., and Binder, D. K. (2003) Molecular biology of glioma tumorigenesis. *Histol Histopathol* **18**, 207–16.
40. Fox, I. J., Paucar, A. A., Nakano, I., Mottahedeh, J., Dougherty, J. D., and Kornblum, H. I. (2004) Developmental expression of glial fibrillary acidic protein mRNA in mouse forebrain germinal zones – implications for stem cell biology. *Brain Res Dev Brain Res* **153**, 121–5.
41. Mellinghoff, I. K., Wang, M. Y., Vivanco, I., Haas-Kogan, D. A., Zhu, S., Dia, E. Q., Lu, K. V., Yoshimoto, K., Huang, J. H., Chute, D. J., Riggs, B. L., Horvath, S., Liau, L. M., Cavenee, W. K., Rao, P. N., Beroukhim, R., Peck, T. C., Lee, J. C., Sellers, W. R., Stokoe, D., Prados, M., Cloughesy, T. F., Sawyers, C. L., and Mischel, P. S. (2005) Molecular determinants of the response of glioblastomas to EGFR kinase inhibitors. *N Engl J Med* **353**, 2012–24.
42. Horvath, S., Zhang, B., Carlson, M., Lu, K. V., Zhu, S., Felciano, R. M., Laurance, M. F., Zhao, W., Qi, S., Chen, Z., Lee, Y., Scheck, A. C., Liau, L. M., Wu, H., Geschwind, D. H., Febbo, P. G., Kornblum, H. I., Cloughesy, T. F., Nelson, S. F., and Mischel, P. S. (2006) Analysis of oncogenic signaling networks in glioblastoma identifies ASPM as a molecular target. *Proc Natl Acad Sci U S A* **103**, 17402–7.
43. Singh, S. K., Clarke, I. D., Terasaki, M., Bonn, V. E., Hawkins, C., Squire, J., and Dirks, P. B. (2003) Identification of a cancer stem cell in human brain tumors. *Cancer Res* **63**, 5821–8.
44. Torroglosa, A., Murillo-Carretero, M., Romero-Grimaldi, C., Matarredona, E. R., Campos-Caro, A., and Estrada, C. (2007) Nitric oxide decreases subventricular zone stem cell proliferation by inhibition of epidermal growth factor receptor and phosphoinositide-3-kinase/Akt pathway. *Stem Cells* **25**, 88–97.
45. Imura, T., Nakano, I., Kornblum, H. I., and Sofroniew, M. V. (2006) Phenotypic and functional heterogeneity of GFAP-expressing cells in vitro: differential expression of LeX/CD15 by GFAP-expressing multipotent neural stem cells and non-neurogenic astrocytes. *Glia* **53**, 277–93.
46. Cheng, J. C., Horwitz, E. M., Karsten, S. L., Shoemaker, L., Kornblum, H. I., Malik, P., and Sakamoto, K. M. (2007) Report on the workshop "New Technologies in Stem Cell Research," Society for Pediatric Research, San Francisco, California, April 29, 2006. *Stem Cells* **25**, 1070–88.
47. Sun, Y., Nadal-Vicens, M., Misono, S., Lin, M. Z., Zubiaga, A., Hua, X., Fan, G., and Greenberg, M. E. (2001) Neurogenin promotes neurogenesis and inhibits glial differentiation by independent mechanisms. *Cell* **104**, 365–76.
48. Wachs, F. P., Couillard-Despres, S., Engelhardt, M., Wilhelm, D., Ploetz, S., Vroemen, M., Kaesbauer, J., Uyanik, G., Klucken, J., Karl, C., Tebbing, J., Svendsen, C., Weidner, N., Kuhn, H. G., Winkler, J., and Aigner, L. (2003) High efficacy of clonal growth and expansion of adult neural stem cells. *Lab Invest* **83**, 949–62.

Chapter 5

Prospective Identification of Cancer Stem Cells with the Surface Antigen CD133

Kyeung Min Joo and Do-Hyun Nam

Summary

Cancer cells do not share equal tumor-initiating potential. Only cancer stem cells (CSCs) can initiate cancer, which is important clinically because they should be eradicated to treat cancer patients. The purpose of experimental methods for identification of CSC is to isolate CSCs among various kinds of cancer cells in cancer masses. To identify CSCs, cancer masses derived from patients should be dissociated into single cells. Dissociated cells are classified into several groups according to expression status of one or several surface proteins using magnetic cell sorting (MACS) or fluorescence-activated cell sorting (FACS) methods. Sorted cells are cultured in a specialized culture medium for stem cells or inoculated into the primary cancer site of immunodeficient mice. In this chapter detailed experimental methods will be described and glioblastoma will be used as an example of solid cancers.

Key words: Cancer stem cell, Marker, Glioblastoma, Dissociation methods, Primary culture, MACS, FACS, Animal model

1. Introduction

Cancer masses harbor various types of cells. Some of them are cancer cells while others are normal cells derived from nearby noncancerous tissues (i.e., endothelial cells, tumor associated lymphatic cells, fibroblasts, etc.). Of these, cancer cells themselves further show diverse morphology and protein expression. Thus it is suspected that a cancer mass is composed of several kinds of cancer cells that have different characteristics. Glioblastoma cells are also heterogenous; their size is diverse, some show astrocytic differentiation, and others express the proteins localized in mature neurons. These disparate cancer cells are suspect to have different tumor-initiating potential. If all cancer cells share high tumor-

John S. Yu (ed.), *Cancer Stem Cells*, Methods in Molecular Biology, vol. 568
DOI 10.1007/978-1-59745-280-9_5, © Humana Press, a part of Springer Science+Business Media, LLC 2009

initiating potential, only several cancer cells would make cancer mass in mice. However, millions of cells need to reproduce the original cancer in immunodeficient mice *(1, 2)*. Clinically cells harboring tumor-initiating potential are important because they should be eradicated to treat cancer patients *(3–6)*. Therefore the purpose of experimental methods for identification of cancer stem cell is to isolate cancer cells that have tumor-initiating potential among various kinds of cancer cells in cancer masses.

To identify cancer stem cells (CSCs), several steps of experiments are required (1) cancer masses derived from patients should be dissociated into single cells mechanically and chemically; (2) dissociated cells are classified into several groups according to expressional status of one or several surface proteins using magnetic cell sorting (MACS) or fluorescence-activated cell sorting (FACS) method; and (3) various numbers of sorted cells are cultured in a specialized stem cell culture medium for stem cells or inoculated into the primary cancer site of immunodeficient mice (**Fig. 1**). If one group of cells can make cancer masses in mice that have similar histological and molecular characteristics with

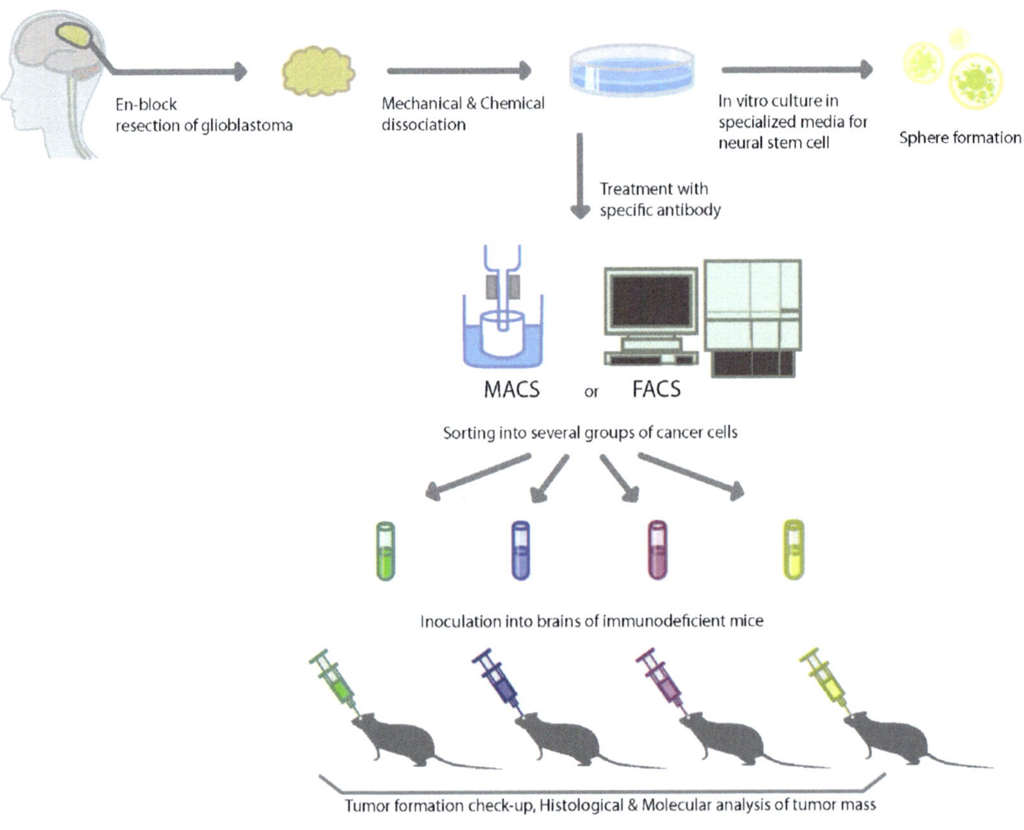

Fig. 1. Experimental methods for the identification of cancer stem cell.

the original cancer, cancer-initiating cells can be expected to be present in that group, and it can further be expected that these cells would possess surface proteins that can be used as markers of CSCs. In this chapter, the detailed experimental methods for isolating CSCs and identifying cancer stem cell markers (from cancer sample collection in operating rooms to *in vivo* sorted cancer cell inoculation study) will be described with several photographs. Since recent reports identified CD133 as an unequivocal marker for glioblastoma CSCs *(1, 2)*, this chapter will focus on isolation of CSCs from glioblastoma to serve as a representative procedure for all solid cancers.

2. Materials

2.1. Preparation of Cancer Samples

1. 4°C sterile normal saline solution (0.9% Sodium Chloride solution).

2.2. Dissociation of Primary Cancer into Single Cells

2.2.1. Equipment

1. Scissors.
2. Forceps.
3. Pasteur pipettes.
4. 100-mm^2 culture dishes.
(*See* **Note 1** for Nontissue culture-grade Petri dish).
5. 75-cm^2 culture flasks.
6. Cell strainers (pore size = 40 μm).
7. Conical tubes (15 or 50 ml).
8. Effendorf tubes (E-tube, 1.5 ml).

All equipment should be sterilized prior to start dissociation.

2.2.2. Enzyme Mixture Solution

	Stock concentration (mg/ml)	Working concentration (mg/ml)	Volume (ml)
Collagenase	200	0.4	0.1
Dispase	200	0.5	0.125
DNase I	100	0.2	0.1
Dulbecco's phosphate-buffered saline (DPBS)			49.675

Collagenase, dispase, DNase I stock. Store at −20°C. Stocks can be made according to manufacturer's instructions.

Enzyme mixture solution: Make fresh as required, store at 4°C.

2.2.3. Ficoll Gradient Centrifugation

1. Ficoll-Paque PLUS (1.077 density).
2. 4°C sterile DPBS.

Ficoll-Paque PLUS: Store at room temperature.

2.3. Specialized Culture Media for Cancer Stem Cell

2.3.1. "NBE" Condition (7)

1. Neurobasal medium supplemented with
   (a) N2 supplement (0.5×).
   (b) B27 supplement without vitamin A (0.5×).
   (c) L-Glutamine.
   (d) Penicillin/Streptomycin.
   (e) Basic fibroblast growth factor (bFGF, 50 ng/ml).
   (f) Epidermal growth factor (EGF, 50 ng/ml).
2. N2 supplement, B27 supplement without vitamin A: Store at −20°C. *bFGF, EGF.* Store at −70°C (20,000× stock solution).

Neurobasal medium. Store at 4°C.

3. "NBE" condition: Make fresh as required, store at 4°C.

2.4. Sorting of Dissociated Cancer Cells Using Their Surface Proteins

2.4.1. Magnetic Cell Sorting

Equipment

1. Separating columns (MS or LS Columns, Miltenyi Biotec Inc.).
2. Magnet which can produce strong magnetic field (Miltenyi Biotec Inc.).
3. MultiStand (Miltenyi Biotec Inc.).

MACS Buffer and Antibodies

1. MACS buffer: PBS supplemented with 0.5% BSA and 2 mM EDTA (pH 7.2).
2. PE-conjugated anti-human CD133 antibody.
3. Microbeads-conjugated anti-PE antibody.

MACS buffer: Store at 4°C after filtration.

Antibodies. Stable at 4°C for up to 1 month. Stable at −20°C for up to 1 year (do not freeze and thaw repeatedly).

2.4.2. Fluorescence-Activated Cell Sorting

Equipment

1. A FACS machine (i.e., FACSAria, BD Biosciences).

FACS Buffer and Antibodies

1. *FACS buffer.* PBS supplemented with 0.5% BSA and 2 mM EDTA (pH 7.2).
2. PE-conjugated anti-human CD133 antibody.

FACS buffer. Store at 4°C after filtering.

Antibodies. Stable at 4°C for up to 1 month. Stable at −20°C for up to 1 year (do not freeze and thaw repeatedly).

2.5. Transplantation of Sorted Cells into Immunodeficient Mice

2.5.1. Equipment

1. Hamilton syringes.
2. Screw guide.
3. Dummy cannula.

4. Hand Drill.
5. Stereotactic device.
6. Surgical instruments.

All equipment should be sterilized.

2.5.2. Animals and Others

1. Immunodeficient mice.
2. Bone wax.
3. *Anesthetics.* ketamine hydrochloride and xylazine.

Immunodeficient mice: Maintained in the SPF condition.

Anesthetics. Store at room temperature.

3. Methods

3.1. Preparation of Cancer Samples

Resection and preparation of primary tumor is the most important procedure because the status of removed cancer mass determines the yield of the experiment. Healthier cancer cells can be dissociated from well-manipulated cancer mass and more CSCs can be sorted using healthy cells. Therefore, the surgeon should keep in mind that CSCs are derived from well-manipulated cancer samples.

Primary glioblastomas usually have gliotic margins between the tumor and the nearby normal brain (tumor-associated normal brain). Dissecting along the gliotic margin enables en-block resection of glioblastomas; total tumor mass can be removed as a whole. En-block resection provides sufficient tumor tissue for pathological diagnosis and further laboratory experiments. However, massive tumor necrosis is often associated with glioblastoma which can have a negative effect on the following experimental procedures since there would be fewer viable cells in the necrotic portion.

En-block resection is difficult in cases of deep-seated glioblastomas or tumor masses with poor dissection margin. To remove these tumors, some neurosurgeons use a cavitron ultrasonic aspirator (CUSA). However, CUSA makes it impossible to collect glioblastoma cells. Therefore in such cases, a tissue collector with a 10-French suction tip (bone collector, **Fig. 2**) is recommended. It enables the collection of glioblastoma cells without loss, and it dissociates tumor mass into single cells mechanically, allowing the following mechanical dissociation step to be skipped. Take caution, however, because cells can be contaminated by a large amount of blood cells when a tissue collector with a 10-French suction tip is applied.

To maintain fresh tumor sample, samples should be placed in 4°C sterile normal saline solution (0.9% Sodium Chloride solution) immediately after the removal, and experimental procedures

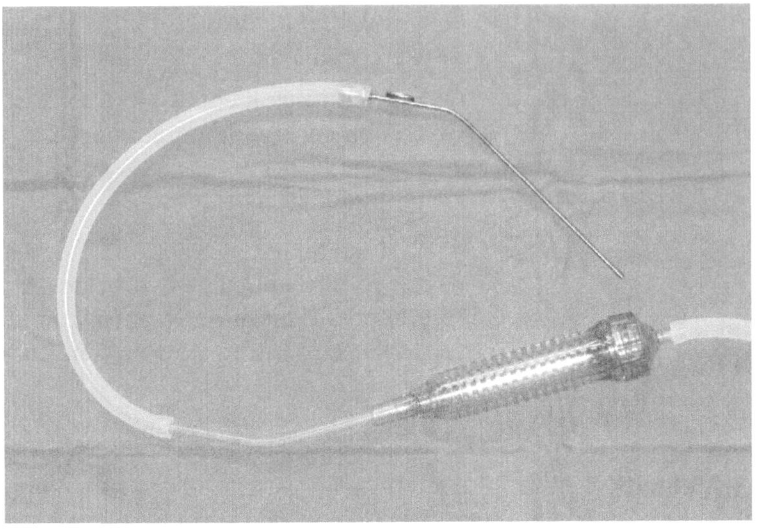

Fig. 2. A photograph of a tissue collector with a 10-French suction tip.

(mechanical and chemical dissociation) should be started as soon as possible because tumor cells can be damaged even in that condition.

3.2. Dissociation of Primary Cancer into Single Cells

Primary cancer mass should be dissociated into single cells to perform other laboratory experiments. Both mechanical and chemical methods are used in this process; cancer tissue architectures are grinded up mechanically and extracellular matrix proteins are digested using several kinds of enzymes such as collagenase and dispase. The use of DNase I reduces the amount of DNA fibers which induce aggregation and clumping of cells. Note, however, that mechanical and chemical insults to primary cancer mass can negatively influence cancer cell survival. Proteins expressed on the plasma membrane can also be damaged, which would make it difficult to characterize the cancer cells. Therefore, an effective dissociation of primary cancer masses with minimal damage to the cancer cells and their surface proteins must be achieved in this process. In particular, one has to select the dissociation enzyme carefully since it can digest potential cancer stem cell markers. For instance, CD133, a candidate cancer stem cell marker of glioblastoma *(1, 2)*, has many amino acid residues that can be digested by trypsin, indicating that trypsin must not be used in dissociation of glioblastoma samples. Therefore more specific enzymes, such as collagenase, dispase, and DNase I, are preferred, even though they are less effective in digestion of extracellular proteins.

Here is an example of dissociation procedures of primary glioblastoma samples.

All procedures should be performed in a clean bench and in ice to preserve the viability of the cancer cells.

1. Transfer an extracted glioblastoma sample into 100-mm^2 culture dish and then wash the sample several times with ice-cold DPBS. While washing the sample, remove blood cloths and gross necroses.
2. Divide the sample into 1-cm^3 blocks using a scissor and transfer them into 1.5-ml E-tubes. Samples should be in ice-cold DPBS during the procedure to prevent tissues from being dried.
3. Dissociate the blocks mechanically by chopping them many times with scissors as shown in **Fig. 3a**.
4. After primary mincing, collect the samples in a 15-ml conical tube or a 1.5-ml E-tube and triturate them further by sucking them in a Pasteur pipette multiple times as shown in **Fig. 3b** (*see* **Note 2**. for air bubbles).
5. Wash the sample several times with ice-cold DPBS.
6. Digest the sample using the enzyme mixture solution (*see* **Note 3**. for alternative enzymes). The ratio of enzyme mixture solution and mechanically dissociated sample volume is usually 1:1. The enzymatic digestion is persisted for 30–60 min at 37°C. After the first 30 min of incubation, mechanical mixing or mild trituration can be applied.
7. Wash the chemically digested sample several times with ice-cold DPBS.
8. Filter the mechanically and chemically dissociated sample through a cell strainer (40 μm pore size) and collect single cancer cells as shown in **Fig. 4**.
9. Collected cancer cells can be contaminated by red blood cells (RBCs). To eliminate RBCs, carefully layer the cancer cell suspension on 15 ml Ficoll-Paque PLUS (1.077 density) in a 50-ml conical tube.
10. Centrifuge it at 400 × *g* (2,000 rpm) for 20 min at room temperature without brake.

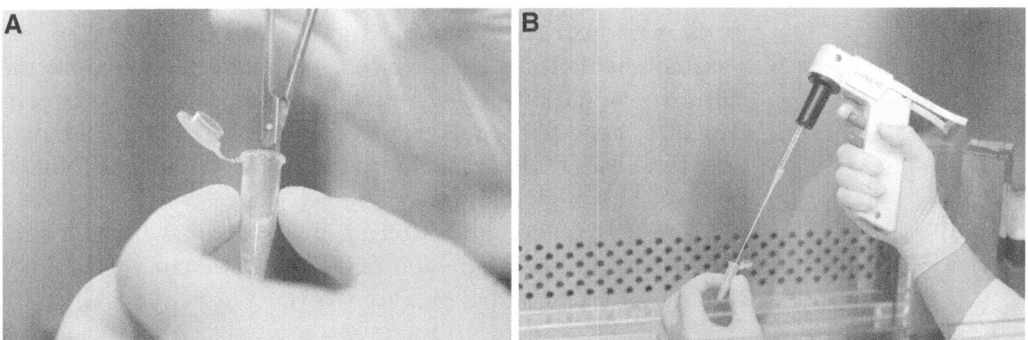

Fig. 3. Mechanical dissociation of cancer samples. **A**. mechanical dissociation using a scissor. **B**. mechanical dissociation using a pasteur pipette.

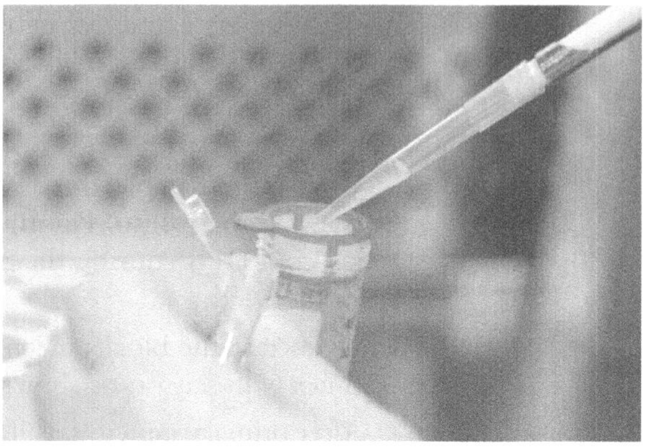

Fig. 4. Filtration of sample through a cell strainer to collect single cancer cells.

11. Take the cancer cell layer (middle) between DPBS (upper) and Ficoll-Paque PLUS (lower) using a Pasteur pipette.
12. Wash the cancer cells with ice-cold DPBS for several times.
13. If cell debris still remains, allow the cancer cell suspension to stand for over 15 min at 4°C and remove 70% of upper portion of suspension containing unwanted cell debris.
14. Count cancer cell number. Counting dead cells can be avoided by using Trypan blue staining.

See **Note 4** for individualization of procedures.

3.3. Specialized Culture Media for Cancer Stem Cells

Stem cells usually lose the stem cell characteristics and differentiate when they are maintained in media containing fetal bovine serum (FBS). Therefore, serum-free media are used in the culture of stem cells, including neural stem cells *(8, 9)*. According to the cancer stem cell hypothesis, CSCs treated with FBS also differentiate losing their cancer-initiating potential. It is reported that genetic alternations are accumulated and protein expression characteristics are changed by FBS *(7)*. To prevent CSCs from losing their cancer-initiating potential and to preserve the genetic and biological characteristics of the primary cancer, a specialized media without FBS is employed in the *in vitro* culture of dissociated glioblastoma cells, which is originally developed for the culture of neural stem cells. Although this medium does not contain FBS, basic fibroblast growth factor (bFGF) and epidermal growth factor (EGF) are included because they induce proliferation of CSCs of glioblastoma.

The "NBE" condition (serum-free Neurobasal media supplemented with basic FGF and EGF) is a well-established culture condition for CSCs of glioblastoma *(7)*.

3.3.1. Procedures

1. 1×10^6 dissociated glioblastoma cells in the "NBE" condition (1×10^5 cells/ml) are cultured in a 75-cm^2 flask.

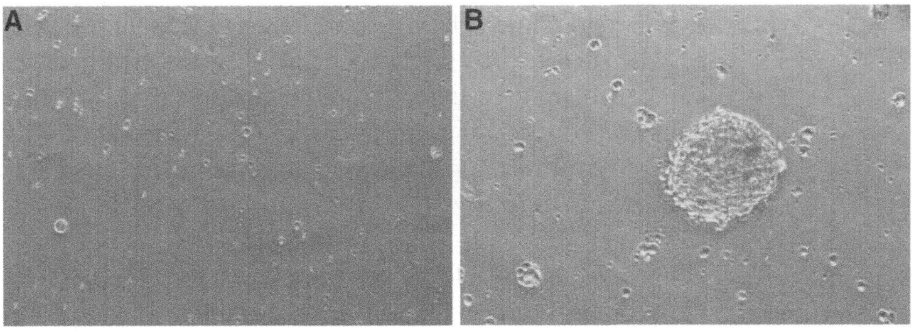

Fig. 5. Growth of dissociated glioblastoma single cells in the "NBE" condition. **A**. initial dissociated glioblastoma cells. **B**. spheres made by dissociated glioblastoma cells *in vitro*.

2. Cultured glioblastoma cells can be frozen in 10% Dimethyl sulfoxide (DMSO) culture medium without growth factor. 1× 10^6 cells are usually frozen in a freezing vial.

Dissociated glioblastoma cells (**Fig. 5a**) proliferate in the "NBE" condition making spheres (**Fig. 5b**). Spheres are usually formed within 1 week.

3.4. Sorting of Dissociated Cancer Cells Using Their Surface Proteins

A cancer mass is composed of various kinds of cancer cells but only one or several kinds of cells among them are reported to possess tumor-initiating potential *(1, 2, 10–15)*. Therefore, cancer cells should be classified and sorted into several groups to identify cells that can initiate cancer *in vivo*. Protein expression patterns on the plasma membrane can be used in this process. Membrane proteins have merits as classification markers; they can be easily detected by specific antibodies (antibodies cannot penetrate the plasma membrane freely. Therefore, the plasma membrane should be permeabilized using detergents such as Triton X and Saponin to detect intracellular proteins. Cells cannot survive this permeabilizing procedure). Good experimental methods, such as MACS and FACS methods, have been developed to sort cells according to the presence of surface proteins

There are many reports suggesting that glioblastoma is a neural stem cell disease, and some data indicate that glioblastoma originates from a transformed neural stem cell *(16, 17)*. Therefore, CSCs of glioblastoma are postulated to possess similar molecular and biological characteristics with neural stem cells. Based on this concept, several surface proteins specifically expressed on the neural stem cells were tested as cancer stem cell markers. CD133, a recently reported cancer stem cell marker of glioblastoma, is one of them *(1, 2)*.

3.4.1. MACS

MACS method uses magnetic force to sort cells *(18)*. Antibodies are conjugated with extremely small, superparamagnetic particles (microbeads). When a high-gradient magnetic field is applied, cells attached to microbeads-conjugated antibodies are retained in the field while others pass through it. Therefore, cells expressing the surface protein that specifically interacts with the antibody can be isolated. In this section, procedures that sort glioblastoma cells according to the expression of CD133 using an MS Column (Miltenyi Biotec Inc.) and its MultiStand (Miltenyi Biotec Inc.) will be introduced.

Procedures

All procedures should be performed in a clean bench for following experiments and on ice to preserve viability of cancer cells.

1. Resuspend up to 1×10^8 glioblastoma cells in ice-cold MACS buffer.
2. Wash the cells several times with ice-cold MACS buffer.
3. Treat the cells with PE-conjugated anti-human CD133 antibody (10 μl PE-conjugated anti-human CD133 antibody/ 5×10^5 cells in 90 μl ice-cold MACS buffer) for 30 min at 4°C. (cf. Antibody concentration is required to be adjusted.)
4. Wash the cells several times with ice-cold MACS buffer.
5. Treat the cells with Microbeads-conjugated anti-PE antibody (20 μl Microbeads-conjugated anti-PE antibody/5×10^5 cells in 80 cl ice-cold MACS buffer) for 30 min at 4°C. (cf. Antibody concentration is required to be adjusted.)
6. Wash the cells several times with ice-cold MACS buffer.
7. Resuspend the antibody-attached glioblastoma cells in 500 μl of ice-cold MACS buffer.
8. Preparation of an MS Column as in **Fig. 6**.
9. Apply cell suspension onto the prepared MS Column **(Fig. 6)**.

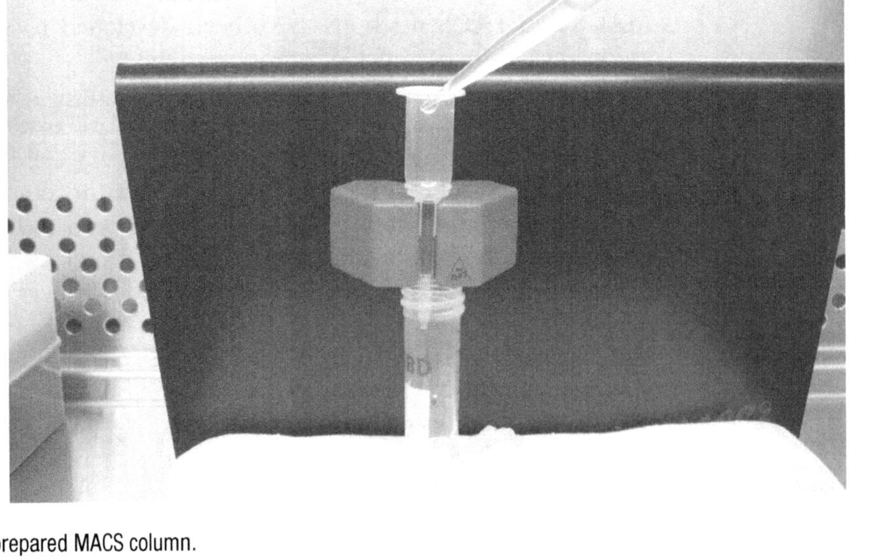

Fig. 6. A prepared MACS column.

10. Collect unlabeled cells that pass through. Wash the MS Column with 500 μl ice-cold MACS buffer three times. Add buffer each time when the column reservoir is empty. Collect total effluent. This is the unlabeled cell fraction.
11. Remove the MS Column from the separator and place it on a new collection tube.
12. Apply 1 ml ice-cold MACS buffer onto the MS Column. Immediately flush out fraction with the magnetically labeled cells by firmly applying the plunger supplied with the column.
13. Check the purity of each cell fraction using a flow cytometry.

Only two fractions, CD133-positive and negative, can be separated by MACS method. Therefore, MACS method cannot be employed if cancer cells should be sorted according to the expressional status of multiple surface proteins. In that case, FACS method can be used.

3.4.2. FACS

FACS method sorts cells using fluorescence *(19)*. Therefore, cells are required to be treated with fluorescence-emitting chemical-conjugated antibodies. In contrast to the microbead, many kinds of fluorescence emitting chemicals such as FITC, PE, and APC have been developed. These chemicals emit fluorescence with different wavelengths. If surface proteins are conjugated with antibodies with different fluorescent materials, cells can be isolated into multiple groups according to the expressional status of the multiple surface proteins. However, to detect multiple fluorescence of each cell and sort cells according to the signal, a special machine is required (i.e., FACSAria, BD Biosciences, **Fig. 7**).

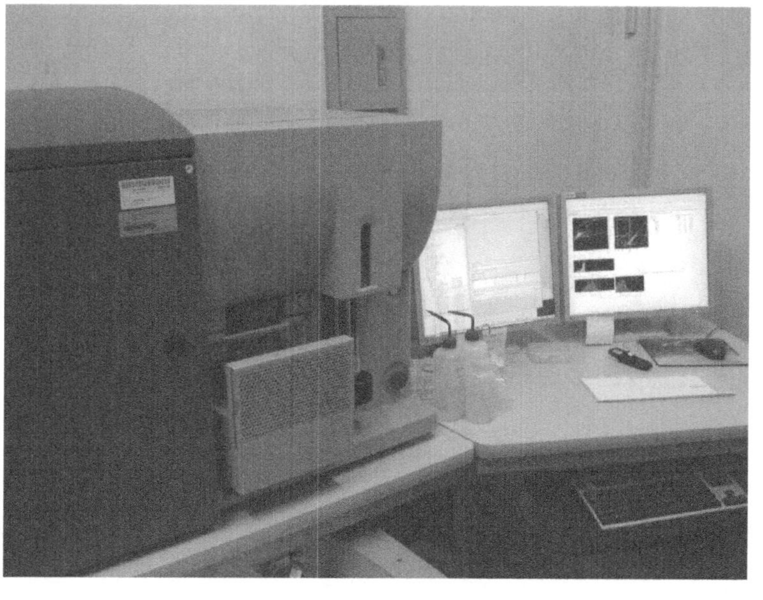

Fig. 7. A photograph of a FACSAria machine.

Staining processes of cells with antibodies are exactly the same with MACS methods (*see* **Note 5** for dead cell staining method for FACS). The only difference is that different kinds of antibodies are used. The working mechanism of FACS machine and its operating methods are complicated and beyond the scope of this chapter. Therefore, researchers should contact with technical experts of their institute prior to use.

3.5. Transplantation of Sorted Cells into Immunodeficient Mice

Although CSCs have many unique characteristics compared with the other cancer cells, the most important characteristics is *in vivo* tumor-initiating potential. Therefore, the *in vivo* tumor-initiating potential of sorted cancer cells should be checked. Because human cancer cells provoke immune rejection of mice, they should be transplanted into immunodeficient mice such as nude mice and severe combined immunodeficient (SCID) mice. The interaction between cancer cells and surrounding noncancer cells is important in the development and progress of cancers *(20)*. To supply more similar environments to cancer cells, it is preferred to inject cancer cells into their orthotopic site; i.e., glioblastoma cells should be administrated into the brains of immunodeficient mice.

3.5.1. Procedures

1. Anesthetize animals with intraperitoneal injection of ketamine hydrochloride (100 mg/kg) and xylazine (10 mg/kg).
2. Shave heads of animals.
3. Place animals in a rodent stereotactic device **(Fig. 8)**. Place the front teeth on the bite bar and clamp the nose.

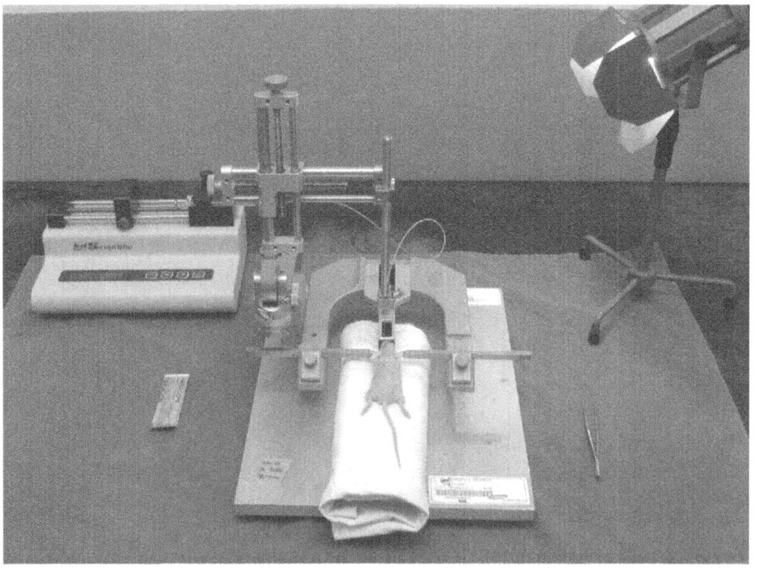

Fig. 8. A photograph of a rodent stereotactic device.

4. Disinfect the scalp with a solution of 70% ethanol and povidone iodine.
5. Make a midline incision on the dorsal aspect of the head and expose the bregma that serves as the zero point (0 AP, 0 lat.) of the coordinates.
6. Implant a screw into a drill hole made 2 mm left and 1 mm anterior to the bregma.
7. Inject the glioblastoma cells into the white matter at a depth of 2 mm using a 10 μl Hamilton syringe connected to the manipulating arm of the stereotactic device. Glioblastoma cells in 10 μl of HBSS are delivered over 10 min by a microinfusion pump. (cf. The number of cells required is to be adjusted depending on the experimental conditions.)
8. Withdraw the needle and flush the skull with sterile normal saline solution and fill the hole with a small amount of bone wax.
9. Clip the scalp and disinfect it with a solution of 70% ethanol and povidone iodine.

After the implantation, cancer incidence should be checked periodically. If there is a cancer, it needs to be examined histologically and molecularly to confirm that it shares the characteristics of its original cancer.

3.6. Conclusion

When sorted cells make cancer masses in mice that have similar histological and molecular characteristics with the original cancer, it can be concluded that those cells harbor CSCs and the surface proteins that are used in the sorting of those cells are markers for CSCs. CSCs are important because they should be eradicated to treat cancer patients. When CSCs are isolated as described here, it would allow more precise characterization of cancer cells that would lead to the development of cancer stem cell-targeting therapeutic methods.

4. Notes

1. Nontissue culture-grade Petri dishes can be used in the dissociation of primary cancer. They prevent cancer cells from attaching to the dishes.
2. It is essential to use pipettes carefully during the mechanical dissociation of primary cancer. Bubble formation could make negative influences on the survival of cancer cells.
3. Dispase, Collagenase, and DNase I are commonly used in the chemical dissociation of primary cancer. Because they

are specific proteinases, their enzymatic potency is relatively weak. Weak potency is helpful maintaining viability of cancer cells; however, they cannot work properly in cases that there is much extracellular matrix or primary cancer harbors large necrosis. More potent nonspecific proteinases, such as trypsin and papain, can be useful solutions. In this section, procedures using papain will be introduced *(21)*. Materials; Papain (13.2 U in 1 ml DMEM Preheating of enzymes is important) and DNase (10,000 U in 1 ml PBS). Procedures; 1. Digest the mechanically dissociated sample using Papain for 30 min at 37°C. The ratio of papain solution and mechanically dissociated sample volume is usually 1:1; 2. After the incubation, mechanical mixing or mild trituration is applied; 3. Add DNase (Final concentration = 200 U/ml) and incubate the sample for 15 min at 37°C.

4. Tumor masses of glioblastomas have diverse characteristics. Therefore, the dissociation procedures need to be modified according to the characteristics of each glioblastoma. When a mechanically dissociated sample has few RBCs, the Ficoll gradient centrifugation can be omitted. The duration for enzyme digestion can be modified and mechanical mixing or mild trituration can be applied during the chemical digestion.

5. The ultimate purpose of the dissociation of primary cancer into single cells is to isolate pure viable cancer cells from surgical cancer samples. Viable cells can be isolated by FACS *(22)*. FACS using 7-Amino-actinomycin D (7-AAD) has merits (1) Staining of dead cells using 7-AAD takes only 20 min; and (2) Multiple surface markers can be analyzed simultaneously (e.g., surface marker differences between viable cells and dead cells can be compared). If dissociated cancer cells are stained using anti-CD133 antibody and 7-AAD, one can sort out pure viable CD133-positive or CD133-negative cells. Materials; 7-Amino-actinomycin D (7-AAD, 7-AAD stock solution: Dissolve 1 mg 7-AAD using 50 μl absolute methanol and add 950 μl 1× PBS. Store at 4°C. Light should be avoided). Procedures; 1. Stain the dissociated cells as outlined in **Subheading 3.4.2**. 2. After the last washing step, resuspend the cells in 1 ml of FACS buffer. 3. Add 1–2 μl of the 7-AAD stock solution to each tube and Keep them at 4°C for 20 min in a dark room. 4. Analyze stained cells using a FACS machine.

References

1. Singh, S. K., Hawkins, C., Clarke, I. D., Squire, J. A., Bayani, J., Hide, T., Henkelman, R. M., Cusimano, M. D., and Dirks, P. B. (2004) Identification of human brain tumour initiating cells. *Nature*. 432, 396–401.
2. Singh, S. K., Clarke, I. D., Terasaki, M., Bonn, V. E., Hawkins, C., Squire, J., and Dirks, P. B. (2003) Identification of a cancer stem cell in human brain tumors. *Cancer Res*. 63, 5821–5828.

3. Piccirillo, S. G., Reynolds, B. A., Zanetti, N., Lamorte, G., Binda, E., Broggi, G., Brem, H., Olivi, A., Dimeco, F., and Vescovi, A. L. (2006) Bone morphogenetic proteins inhibit the tumorigenic potential of human brain tumour initiating cells. *Nature.* 444, 761–765.
4. Stupp, R. and Hegi, M. E. (2007) Targeting brain-tumor stem cells. *Nat. Biotechnol.* 25, 193–194.
5. Bao, S., Wu, Q., McLendon, R. E., Hao, Y., Shi, Q., Hjelmeland, A. B., Dewhirst, M. W., Bigner, D. D., and Rich, J. N. (2006) Glioma stem cells promote radioresistance by preferential activation of the DNA damage response. *Nature.* 444, 756–760.
6. Dean, M., Fojo, T., and Bates, S. (2005) Tumour stem cells and drug resistance. *Nat. Rev. Cancer.* 5, 275–284.
7. Lee, J., Kotliarova, S., Kotliarov, Y., Li, A., Su, Q., Donin, N. M., Pastorino, S., Purow, B. W., Christopher, N., Zhang, W., Park, J. K., and Fine, H. A. (2006) Tumor stem cells derived from glioblastomas cultured in bFGF and EGF more closely mirror the phenotype and genotype of primary tumors than do serum-cultured cell lines. *Cancer Cell.* 9, 391–403.
8. Rietze, R. L., Valcanis, H., Brooker, G. F., Thomas, T., Voss, A. K., and Bartlett, P. F. (2001) Purification of a pluripotent neural stem cell from the adult mouse brain. *Nature.* 412, 736–739.
9. Nunes, M. C., Roy, N. S., Keyoung, H. M., Goodman, R. R., McKhann, G., 2nd, Jiang, L., Kang, J., Nedergaard, M., and Goldman, S. A. (2003) Identification and isolation of multipotential neural progenitor cells from the subcortical white matter of the adult human brain. *Nat. Med.* 9, 439–447.
10. Bonnet, D. and Dick, J. E. (1997) Human acute myeloid leukemia is organized as a hierarchy that originates from a primitive hematopoietic cell. *Nat. Med.* 3, 730–737.
11. Matsui, W., Huff, C. A., Wang, Q., Malehorn, M. T., Barber, J., Tanhehco, Y., Smith, B. D., Civin, C. I., and Jones, R. J. (2004) Characterization of clonogenic multiple myeloma cells. *Blood.* 103, 2332–2336.
12. Al-Hajj, M., Wicha, M. S., Benito-Hernandez, A., Morrison, S. J., and Clarke, M. F. (2003) Prospective identification of tumorigenic breast cancer cells. *Proc. Natl. Acad. Sci. U. S. A.* 100, 3983–3988.
13. O'Brien, C. A., Pollett, A., Gallinger, S., and Dick, J. E. (2007) A human colon cancer cell capable of initiating tumour growth in immunodeficient mice. *Nature.* 445, 106–110.
14. Ricci-Vitiani, L., Lombardi, D. G., Pilozzi, E., Biffoni, M., Todaro, M., Peschle, C., and De Maria, R. (2007) Identification and expansion of human colon-cancer-initiating cells. *Nature.* 445, 111–115.
15. Schatton, T., Murphy, G. F., Frank, N. Y., Yamaura, K., Waaga-Gasser, A. M., Gasser, M., Zhan, Q., Jordan, S., Duncan, L. M., Weishaupt, C., Fuhlbrigge, R. C., Kupper, T. S., Sayegh, M. H., and Frank, M. H. (2008) Identification of cells initiating human melanomas. *Nature.* 451, 345–349.
16. Vescovi, A. L., Galli, R., and Reynolds, B. A. (2006) Brain tumour stem cells. *Nat. Rev. Cancer.* 6, 425–436.
17. Sanai, N., Alvarez-Buylla, A., and Berger, M. S. (2005) Neural stem cells and the origin of gliomas. *N. Engl. J. Med.* 353, 811–822.
18. Zborowski, M. and Chalmers, J. J. (eds.) (2007) Magnetic Cell Separation. Elsevier, Amsterdam, The Netherlands.
19. Macey, M. G. (ed.) (2007) Flow Cytometry and Cell Sorting. Humana, Totowa, NJ.
20. Fidler, I. J. (2003) The pathogenesis of cancer metastasis: the 'seed and soil' hypothesis revisited. *Nat. Rev. Cancer.* 3, 453–458.
21. Panchision, D. M., Chen, H. L., Pistollato, F., Papini, D., Ni, H. T., and Hawley, T. S. (2007) Optimized flow cytometric analysis of central nervous system tissue reveals novel functional relationships among cells expressing CD133, CD15, and CD24. *Stem Cells.* 25, 1560–1570.
22. Schmid, I., Uittenbogaart, C. H., Krall, W. J., Braun, J., and Giorgi, J. V. (1992) Dead cell discrimination with 7-amio-actinomycin D in combination with dual color immunofluorescence in single laser flow cytometry. *Cytometry.* 13, 204–208.

Chapter 6

Neurosphere Culture and Human Organotypic Model to Evaluate Brain Tumor Stem Cells

Hugo Guerrero-Cázares, Kaisorn L. Chaichana, and Alfredo Quiñones-Hinojosa

Summary

The brain tumor stem cell (BTSC) hypothesis is based on the premise that there is a subpopulation of cells within tumors with tumorigenic and pluripotent properties. BTSC are believed to be responsible for both the initiation of brain tumors and their resistance to current therapeutic modalities. This new paradigm stresses the need for adequate techniques to culture and characterize this special population of cells. Furthermore, the use of different cell migration assays offers the possibility to evaluate the processes involved in glioma metastasis. In this chapter, we summarize a method to culture, analyze the cellular characteristics, and study the invasion of BTSCs using a neurosphere assay, cryostat sectioning, and human organotypic brain cortex migration assay, respectively.

Key words: Brain tumor stem cells, Neurospheres, Human organotypic cultures, GBM, Cell migration

1. Introduction

The invasion of tumor cells within normal tissue is thought to be a multifactorial process, requiring the expression of specific proteins, activation of various enzymes, and formation of different types of cell interactions *(1)*. The diffuse infiltration of glioblastoma multiforme (GBM) cells into the healthy brain parenchyma makes complete surgical resection nearly impossible. In fact, there is a recurrence incidence of 99% following gross total resection of these tumors *(1–3)*. Nevertheless, not all the tumor cells have the ability to form a new tumor *(4–6)*. There is increasing evidence that suggests this tumor-initiating ability resides only in a specific subpopulation of cells with characteristics similar to

John S. Yu (ed.), *Cancer Stem Cells,* Methods in Molecular Biology, vol. 568
DOI 10.1007/978-1-59745-280-9_6,

normal neural stem cells (NSC) *(5, 7, 8)*. These cells have been aptly named brain tumor stem cells (BTSC) because they, like NSC, possess self-renewal and multipotential properties, with the added ability to initiate tumor growth *(8)*. Therefore, the development of a BTSC migration model that accurately recapitulates what occurs in the human brain is essential for the study of tumor invasion.

The first findings that showed evidence of glioma-derived BTSC were obtained by Steindler and colleagues *(7)*. With the use of single-cell cultures in a methyl-cellulose (MC) matrix and the addition of epidermal growth factor (EGF) and fibroblast growth factor (FGF), they showed that glioma-derived cells were able to form clones in the MC matrix *(7)*. These clonal cells were also able to express markers specific for glial or neuronal cells *(7)*. Subsequently, several groups have also shown that these cells, like NSCs, have self-renewal and multipotential capabilities *(4–6, 9–11)*. In addition, they had the capability of forming tumors at low cell concentrations (100–1,000 cells). More importantly, they formed tumors that recapitulated the histological characteristics of the parent tumor when implanted into an animal model *(8, 11, 12)*. Interestingly, cells within other tumors, including medulloblastomas *(4, 8)* and ependymomas *(13)*, also possess these same BTSC characteristics. These findings have led many to believe that brain tumors are initiated and maintained by a small population of BTSC that possess self-renewal, multipotentiality, and tumor-initiating capacity *(14)*.

Advances in research have created the need for experimental techniques to study both NSC and BTSC. Neurosphere assays are currently the standard for identifying these unique stem cell populations *(15–17)*. These assays utilize a selective serum-free culture system that allows NSC and BTSC to proliferate and generate multipotent floating cell clusters called neurospheres *(15–17)*. The neurosphere assay protocols, however, are not uniform and vary significantly between studies. Therefore, the use of specific culture and passaging protocols, as well as different characterization methods, is necessary to correctly identify, maintain, and characterize a true BTSC population *(15, 17)*.

The characterization of BTSC neurospheres using immunocytochemistry (ICC) is difficult due to their floating condition, size, and fragility. As a result, different techniques have been implemented for their staining. This includes the use of a cytospin device (Thermo scientific, USA) to centrifuge the neurospheres against a glass slide *(9)* or manually adhering neurospheres to a plate *(18)* for future staining, as well as flotation staining protocols *(15)*. These techniques have significant disadvantages because they deform the neurosphere architecture and prevent clear staining and visualization of the neurospheres. The use of cryostat sectioning of neurospheres, however, gives the best reported resolution without affecting the neurosphere architecture *(19)*. This method also offers the added benefit of obtaining multiple sections from the same neurosphere. We will describe the techniques we use to section BTSC neurospheres with a cryostat, which will allow for effective characterization of these cells using immunocytochemistry.

In addition to the study of BTSC neurospheres, investigating tumor migration and invasion is essential. Understanding how brain tumor-derived cells invade normal tissue is necessary to develop effective strategies for preventing tumor recurrence, which can largely be attributed to their invasive abilities. The most commonly used approaches to study brain tumor cell migration and/or invasion in vitro include the wound healing assay *(20)*, microliter-scale migration assay *(21)*, spot assay *(22)*, and transwell migration assay *(23, 24)*. These methods, however, do not accurately represent the human brain matrix, the natural environment in which the cells migrate. The brain slice invasion assay allows the study of tumor cell invasion using actual brain matrix *(25, 26)*. We will therefore summarize methods used to study BTSC migration using brain slice or organotypic cultures from human intraoperative specimens.

In this chapter, we will describe the techniques we use to identify and maintain GBM-derived BTSCs, as well as some of the methods for characterizing neurospheres and studying BTSC migration.

2. Materials

2.1. Neurosphere Culture from Brain Tumors

1. Laminar flow culture hood.
2. Dissecting microscope.
3. HBSS plus Ca and Mg (Gibco/BRL, Bethesda, MD).
4. HBSS without Ca and Mg (Gibco/BRL, Bethesda, MD).
5. *Neurosphere culture media.* D-MEM/F12 (1:1)(Invitrogen, Carlsbad, Ca) plus 1× B27 supplement (Gibco/BRL, Bethesda, MD), 1× Antibiotic–antimycotic (Invitrogen, Carlsbad, Ca) and 20 ng/ml of Epidermal Growth Factor (EGF) (Prepotech Inc. Rocky Hill, NY) and 20 ng/ml basic Fibroblastic Growth Factor (bFGF) (Prepotech Inc. Rocky Hill, NY). The neurosphere media can be prepared without the growth factors and stored at 4°C. The growth factors are stored in aliquots at −20°C. Complete neurosphere media is made in small volumes (50 ml) and stored for no more than 2 weeks at 4°C.
6. Trypsin–EDTA (Gibco/BRL, Bethesda, MD).

7. Microsurgical instruments (forceps, scissors, and scalpel).
8. Wide- and narrow-tipped fire-polished Pasteur pipettes. Pasteur pipette tips are narrowed by exposing them briefly to the flame of a Bunsen burner while spinning the pipette; check the tip frequently to get the desired caliber.
9. Hemocytometer or cell counter machine.

2.2. Neurosphere Cutting

1. 4% paraformaldehyde (4% PFA) in 0.1 M Phosphate buffer. 4% PFA is prepared fresh every time and stored at 4°C for no longer than 1 week.
2. 30% sucrose (Sigma-Aldrich St. Louis, MO) in 0.1 M PBS.
3. OCT compound (Sakura Finetek USA, Torrance, CA, USA).
4. Peel-A-Way® Embedding Mold (Polysciences Inc. Warrington, PA) for cryosectioning.
5. Cryostat.

2.3. Human Organotypic Model

1. *Organotypic culture media.* Minimal Essential Medium (MEM) (Sigma, St. Louis, MO, USA) containing 25% heat-inactivated horse serum (Gibco/BRL, Bethesda, MD), 25% HBSS (Gibco/BRL, Bethesda, MD) with 25.8 mg/ml of glucose, and 12 mg/ml of 4-(2-hydroxyethyl)-1-piperazineethanesulfonic acid (HEPES) buffer (Sigma, St. Louis, MO, USA) and 1% 0.2 M glutamine (Gibco/BRL, Bethesda, MD) pH 7.2
2. McIlwain Tissue Chopper (Mickle laboratory engineering Co. Ltd. UK)
3. 12-mm culture plate inserts (Millipore, Billerica, MA)
4. Angled dissecting microscope or surgical microscope
5. Stereotactic frame
6. Glass micropipette needle (Drummond Scientific Co. Broomall, PA). micropipettes are pulled with a needle pipette puller (David Kopf Instruments Tujunga, CA) and beveled with a 48,000 Micropipette Beveler (World Precision Instruments Sarasota Fl). The final pipette must have a sharp tip of 30–50 μm diameter by 400–500 μm length.

3. Methods

3.1. Neurosphere Culture from Brain Tumors

1. Tumor sample must be transported in saline solution, PBS, or cell culture media, and kept on ice until cultured. Before starting the culture, the following materials are prepared as follows: warm DMEM/F12 plus 10% heat-inactivated FBS and 1% of 100× antibiotic–antimycotic, neurosphere culture media, and

trypsin–EDTA 0.25% to 37°C in a water bath; warm HBSS medium with calcium chloride and magnesium chloride and HBSS medium without calcium chloride and magnesium chloride to room temperature; place the dissecting microscope (previously cleaned with EtOH), the sterile surgical instruments in a beaker with 96% EtOH (*see* **Note 1**), and a 10-mm Petri dish with 7 ml HBSS + Ca + Mg into a culture hood.

2. The tumor sample is then placed in the sterile Petri dish with HBSS + Ca + Mg, and necrotic tissue and blood vessels from the tumor are removed under the dissecting microscope (*see* **Note 2**). The clean sample is then divided into three pieces for (a) protein extraction, using an appropriate cryovial and snap freezing the sample in liquid nitrogen; (b) RNA extraction, placing the tumor in a 1.5-ml RNAse-free tube and adding 1 mL of RNA for storage at 4°C; (c) cell culture.
3. The clean sample that is going to be used for cell culture is dissociated enzymatically by adding 2 ml of Trypsin–EDTA and mechanically by cutting the tumor into small pieces using microdissecting scissors.
4. The pieces of tissue suspended in 2 ml of Trypsin–EDTA are then placed into a 15-mL conical tube to be homogenized with a sterile wide-tipped fire-polished Pasteur pipette. Pipette up and down gently until the solution becomes blurry (do not let the tissue remain in contact with trypsin for more than 10 min to prevent low cell viability). Trypsin is then inhibited by adding 3 mL of DMEM/F12 + 10% FBS media. The tissue is homogenized further with a sterile narrow-tipped fire-polished Pasteur pipette, and any remaining nonhomogenized pieces of tissue are removed by passing the cell suspension through a 40-µm cell strainer.
5. Cells are counted with a hemocytometer or a cell counter machine and cell viability is determined with the use of trypan blue.
6. Cells are centrifuged for 5 min at 180 RCF (Relative Centrifugal Force) at 4°C and serum-containing media is decanted. Pre-warmed neurosphere media is added to the cells pellet to get a final concentration of 4×10^4 cells per ml.
7. The cell suspension is added to nonadherent cell culture flasks (5 ml per flask) and placed in an incubator at 37°C and 5% CO_2. Culture media is changed twice a week (*see* **Note 3**) and neurospheres are passaged every 1 or 2 weeks, depending on the growth rate of each sample (*see* **Note 4**).

3.2. Neurosphere Embedding for Cryostat Sectioning

1. Neurospheres from the culture flasks are taken with a 20-µl pipette set to 2 µl and placed in a PCR tube. They are fixed by adding 50 µl of cold 4% paraformaldehyde for 30 min at room temperature.

2. Neurospheres are then centrifuged at 150 RCF for 5 min at 4°C and supernatant is removed carefully. 30% sucrose is added and the neurospheres are left in this solution for 30 min. At this moment an empty tissue embedding mold is taken and OCT compound is added to have a flat surface so as to place the neurospheres.
3. Neurospheres are centrifuged again at 150 RCF for 5 min at 4°C and supernatant is removed carefully.
4. The neurospheres are then resuspend in 50 μl of OCT compound and placed on the top of the frozen OCT in the mold, giving them enough time to freeze (*see* **Notes 6 and 7**).
5. 5-μm slices are obtained with a cryostat for further immunostaining (*see* **Note 8**) (**Fig. 1b–d**).

3.3. Human Organotypic Model (27)

1. A 24-well culture plate is prepared by adding 500 μl of organotypic culture media and placing one 12-mm Millicell insert into each well, taking care to not create bubbles underneath the membrane.

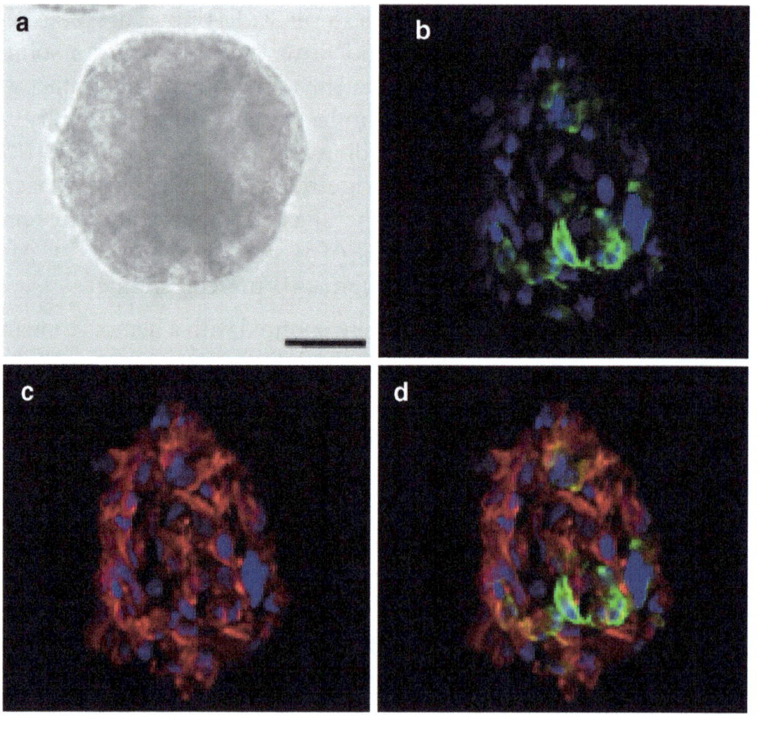

Fig. 1. GBM-derived neurosphere. (**a**) In bright field prior to the cryosectioning protocol described; (**b–d**), immunostained against GFAP (*green*) and Nestin (*red*). Dapi was used as a nuclear marker. The finding of more differentiated cells in the core of the neurosphere has been reported previously ***(9, 15)***, showing the cell heterogeneity within the neurospheres.

2. After preparing the plate, a tumor sample (collected and transported as described on section A-1) is placed on a Sylgard plate and cut into rectangular pieces (5–20 mm/1–2 mm). The pieces are then sectioned into 350-μm thick slices using the tissue chopper and placed into a Petri Dish containing high-glucose HBSS (see **Notes 9** and **10**)
3. Each slice is put into the 12-mm Millicell inserts with the help of a paint brush (*see* **Note 11**). The plate with the organotypic cultures is then placed into an incubator at 37°C and 5% CO_2. The organotypic culture media is changed every 2 days taking care to not disturb the explant (*see* **Note 12**).
4. After 2 days of culture, a single-cell suspension of 1 × 10^5 GFP labeled cells per μl is prepared to be injected into the organotypic explants.
5. The culture plate with the tissue explants is placed in the hood on a rodent stereotactic frame, under the angled dissecting microscope to obtain the best visualization of the tissue explant inside the millicell insert (**Fig. 2**).
6. 1 μl of the cell suspension is taken into the glass micropipette needle and then injected into the tissue slice with the help of the stereotactic frame, with particular attention to not penetrate the tissue with the needle.

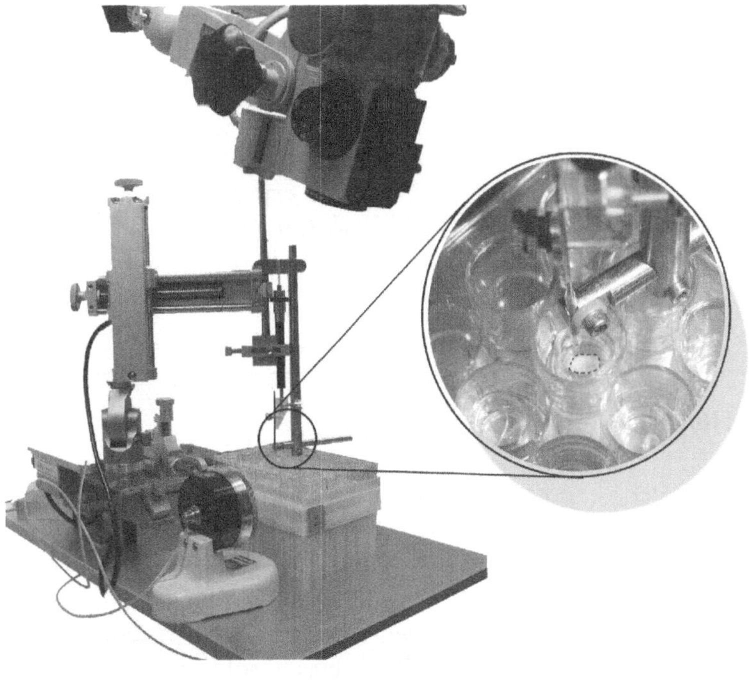

Fig. 2. Setup for the injection of human GBM-derived cells into the human organotypic cultures. The 24-well plate is placed on the stereotactic frame and the surgical microscope is used to visualize both the glass needle and the tissue. Cell injections are performed with a microinjector attached to the glass needle.

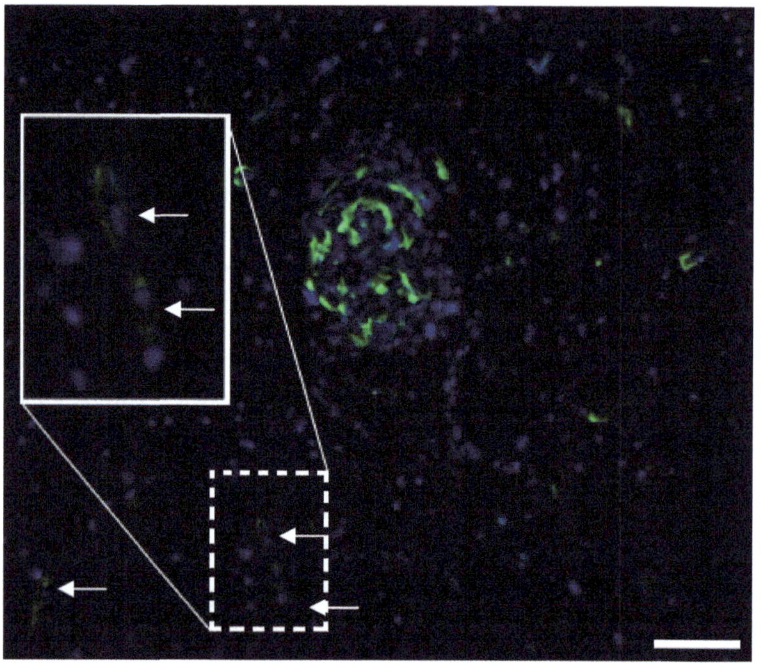

Fig. 3. Migration of human GBM-derived cells in human brain tissue. GFP-labeled GBM cells were injected as described in the methods section, then immunostained after 6 days. DAPI was used as a nuclear marker. Some GFP-labeled GBM cells (*arrows*) were found to have migrated away from the injection site as shown in the *inset*. Scale bar = 100 μm.

7. The cells are injected at a rate of 1 μl per minute while looking under the microscope (*see* **Note 13**), and the plate is placed back into the incubator for 2 weeks, changing the culture media every 2 days.
8. Once the experiment is concluded, the tissue explants are fixed with 4% paraformaldehyde and prepared for cryosectioning and immunostaining (**Fig. 3**).

4. Notes

1. In order to protect the tips of the surgical instruments, it is recommended some cotton be placed in the bottom of a beaker and filled with 96% EtOH.
2. Necrotic tissue can be identified by its dark color. Blood vessels need to be removed to reduce the presence of contaminant cells such as fibroblasts. Nevertheless, tumor sample are often highly vascularized, which makes it difficult to remove the vessels. In this case, try to avoid culturing the vascularized area if the sample is large enough.

3. To change the media, avoid the use of centrifuge since this can cause the formation of cell clumps and form structures similar to neurospheres. Preferentially leave the flasks in a vertical position to let the neurospheres precipitate, take out half of the cell culture volume, and replace it with fresh neurosphere media. Some cells can attach to the bottom of the flask and not form neurospheres. When this happens, take out the total volume of the flask and place it in a new one to avoid contact with differentiated cells
4. The passage of neurospheres with protocols that involve the use of enzymes is widely accepted. Some groups, however, have observed a faster neurosphere growth rate when passaged by the use of mechanical trituration instead of enzymatic digestion. To passage the neurospheres and form a single-cell suspension without the use of enzymes, centri-fuge the neurosphere cell culture for 5 min at 180 × *g*, discard the supernatant, and resuspend the pellet in 200 μl of neurosphere media. Triturate the pellet by pipetting using a p200-μl pipette tip, where several passes are needed to break the neurospheres. In our experience, this takes on average 200 times.
5. Freeze the mold with OCT compound by placing it on dry ice with ethanol. Once frozen, mark the surface of the frozen OCT compound with a permanent marker. This will help to identify the place where the neurospheres are when cutting in the cryostat.
6. Right before adding the OCT-neurosphere solution, take out the mold with frozen OCT from the dry ice. This will allow the OCT-neurosphere suspension to come out from the tip without freezing before getting to the mold.
7. To resuspend the neurospheres in OCT, add 50 μl of the embedding compound and set the micropipette to 45 μl, and slowly pipette up and down without creating bubbles.
8. Start cutting until the marks are visible. After the marks are visible, collect the slices and prepare for immunostaining.
9. When using the tissue chopper, the tissue piece may have a tendency to move as the tissue is being cut. Make sure the cutting surface is dry.
10. After cutting the tissue, some of the pieces will still be adhered to one another. Use a microsurgical scalpel to cut the adherent portions of the tissue.
11. When transferring the tissue to the Millicel inserts, try to transfer with a minimal amount of media as excess media will prevent the tissue from adhering to the membrane.

12. During the media changes, make sure to avoid placing media on top of the Millicel membrane as this may cause the tissue to detach.
13. Giving the reduce volume that can be injected into the tissue slice, special attention needs to be put on the moment when the cell suspension fills the injection place. At that point take out the needle and aspirate any suspension that could have came out to prevent the deposit of cells on the top of the tissue.

Acknowledgments

The authors would like to thank Ms. Alyssa Choi for her contributions to the neurosphere-staining technique and Mr. Frank Attenello and Ms. Grettel Zamora-Berridi for their help with the organotypic culture injections. This work was supported by NIH K08NS055851, Children's Cancer Foundation, and the American Society of Clinical Oncology.

References

1. Demuth, T., J. L. Rennert, D. B. Hoelzinger, L. B. Reavie, M. Nakada, C. Beaudry, S. Nakada, E. M. Anderson, A. N. Henrichs, W. S. McDonough, D. Holz, A. Joy, R. Lin, K. H. Pan, C. J. Lih, S. N. Cohen, and M. E. Berens. 2008. Glioma cells on the run - the migratory transcriptome of 10 human glioma cell lines. *BMC Genomics* 9: 54.
2. Filippini, G., C. Falcone, A. Boiardi, G. Broggi, M. G. Bruzzone, D. Caldiroli, R. Farina, M. Farinotti, L. Fariselli, G. Finocchiaro, S. Giombini, B. Pollo, M. Savoiardo, C. L. Solero, and M. G. Valsecchi. 2008. Prognostic factors for survival in 676 consecutive patients with newly diagnosed primary glioblastoma. *Neuro Oncol.*10:79–87
3. McFerrin, M. B., and H. Sontheimer. 2006. A role for ion channels in glioma cell invasion. *Neuron Glia Biol* 2: 39–49.
4. Hemmati, H. D., I. Nakano, J. A. Lazareff, M. Masterman-Smith, D. H. Geschwind, M. Bronner-Fraser, and H. I. Kornblum. 2003. Cancerous stem cells can arise from pediatric brain tumors. *Proc Natl Acad Sci U S A* 100: 15178–83.
5. Singh, S. K., I. D. Clarke, T. Hide, and P. B. Dirks. 2004. Cancer stem cells in nervous system tumors. *Oncogene* 23: 7267–73.
6. Singh, S. K., C. Hawkins, I. D. Clarke, J. A. Squire, J. Bayani, T. Hide, R. M. Henkelman, M. D. Cusimano, and P. B. Dirks. 2004. Identification of human brain tumour initiating cells. *Nature* 432: 396–401.
7. Ignatova, T. N., V. G. Kukekov, E. D. Laywell, O. N. Suslov, F. D. Vrionis, and D. A. Steindler. 2002. Human cortical glial tumors contain neural stem-like cells expressing astroglial and neuronal markers in vitro. *Glia* 39: 193–206.
8. Singh, S. K., I. D. Clarke, M. Terasaki, V. E. Bonn, C. Hawkins, J. Squire, and P. B. Dirks. 2003. Identification of a cancer stem cell in human brain tumors. *Cancer Res* 63: 5821–8.
9. Bar, E. E., A. Chaudhry, A. Lin, X. Fan, K. Schreck, W. Matsui, S. Piccirillo, A. L. Vescovi, F. DiMeco, A. Olivi, and C. G. Eberhart. 2007. Cyclopamine-mediated hedgehog pathway inhibition depletes stem-like cancer cells in glioblastoma. *Stem Cells* 25: 2524–33.
10. Galli, R., E. Binda, U. Orfanelli, B. Cipelletti, A. Gritti, S. De Vitis, R. Fiocco, C. Foroni, F. Dimeco, and A. Vescovi. 2004. Isolation and characterization of tumorigenic, stem-like neural precursors from human glioblastoma. *Cancer Res* 64: 7011–21.
11. Lee, J., S. Kotliarova, Y. Kotliarov, A. Li, Q. Su, N. M. Donin, S. Pastorino, B. W. Purow, N. Christopher, W. Zhang, J. K. Park, and H. A. Fine. 2006. Tumor stem cells derived from glioblastomas cultured in bFGF and EGF more closely mirror the phenotype and genotype of primary tumors than do serum-cultured cell lines. *Cancer Cell* 9: 391–403.

12. Lee, L. M., E. A. Seftor, G. Bonde, R. A. Cornell, and M. J. Hendrix. 2005. The fate of human malignant melanoma cells transplanted into zebrafish embryos: assessment of migration and cell division in the absence of tumor formation. *Dev Dyn* 233: 1560–70.
13. Taylor, M. D., H. Poppleton, C. Fuller, X. Su, Y. Liu, P. Jensen, S. Magdaleno, J. Dalton, C. Calabrese, J. Board, T. Macdonald, J. Rutka, A. Guha, A. Gajjar, T. Curran, and R. J. Gilbertson. 2005. Radial glia cells are candidate stem cells of ependymoma. *Cancer Cell* 8: 323–35.
14. Quinones-Hinojosa, A., and K. Chaichana. 2007. The human subventricular zone: a source of new cells and a potential source of brain tumors. *Exp Neurol* 205: 313–24.
15. Singec, I., R. Knoth, R. P. Meyer, J. Maciaczyk, B. Volk, G. Nikkhah, M. Frotscher, and E. Y. Snyder. 2006. Defining the actual sensitivity and specificity of the neurosphere assay in stem cell biology. *Nat Methods* 3: 801–6.
16. Chaichana, K., G. Zamora-Berridi, J. Camara-Quintana, and A. Quinones-Hinojosa. 2006. Neurosphere Assays: Growth Factors and Hormone Differences in Tumor and Non-tumor Studies. *Stem Cells.*
17. Gritti, A., R. Galli, and A. L. Vescovi. 2008. Clonal analyses and cryopreservation of neural stem cell cultures. *Methods Mol Biol* 438: 173–84.
18. Reynolds, B. A., and S. Weiss. 1992. Generation of neurons and astrocytes from isolated cells of the adult mammalian central nervous system. *Science* 255: 1707–10.
19. Parmar, M., A. Sjoberg, A. Bjorklund, and Z. Kokaia. 2003. Phenotypic and molecular identity of cells in the adult subventricular zone. in vivo and after expansion in vitro. *Mol Cell Neurosci* 24: 741–52.
20. Nobes, C. D., and A. Hall. 1999. Rho GTPases control polarity, protrusion, and adhesion during cell movement. *J Cell Biol* 144: 1235–44.
21. McDonough, W., N. Tran, A. Giese, S. A. Norman, and M. E. Berens. 1998. Altered gene expression in human astrocytoma cells selected for migration: I. Thromboxane synthase. *J Neuropathol Exp Neurol* 57: 449–55.
22. Wichterle, H., M. Alvarez-Dolado, L. Erskine, and A. Alvarez-Buylla. 2003. Permissive corridor and diffusible gradients direct medial ganglionic eminence cell migration to the neocortex. *Proc Natl Acad Sci U S A* 100: 727–32.
23. Albini, A., Y. Iwamoto, H. K. Kleinman, G. R. Martin, S. A. Aaronson, J. M. Kozlowski, and R. N. McEwan. 1987. A rapid in vitro assay for quantitating the invasive potential of tumor cells. *Cancer Res* 47: 3239–45.
24. Iwasaki, K., L. R. Rogers, G. H. Barnett, M. L. Estes, and B. P. Barna. 1993. Effect of recombinant tumor necrosis factor-alpha on three-dimensional growth, morphology, and invasiveness of human glioblastoma cells in vitro. *J Neurosurg* 78: 952–8.
25. Ohnishi, T., H. Matsumura, S. Izumoto, S. Hiraga, and T. Hayakawa. 1998. A novel model of glioma cell invasion using organotypic brain slice culture. *Cancer Res* 58: 2935–40.
26. Valster, A., N. L. Tran, M. Nakada, M. E. Berens, A. Y. Chan, and M. Symons. 2005. Cell migration and invasion assays. *Methods* 37: 208–15.
27. Chaichana, K. L., V. Capilla-Gonzalez, O. Gonzalez-Perez, G. Pradilla, J. Han, A. Olivi, H. Brem, J. M. Garcia-Verdugo, and A. Quinones-Hinojosa. 2007. Preservation of glial cytoarchitecture from ex vivo human tumor and non-tumor cerebral cortical explants: A human model to study neurological diseases. *J Neurosci Methods* 164: 261–70.

Chapter 7

Methodologies in Assaying Prostate Cancer Stem Cells

Hangwen Li, Ming Jiang, Sofia Honorio, Lubna Patrawala, Collene R. Jeter, Tammy Calhoun-Davis, Simon W. Hayward, and Dean G. Tang

Summary

The cancer stem cell (CSC) theory posits that only a small population of tumor cells within the tumor has the ability to reinitiate tumor development and is responsible for tumor homeostasis and progression. Tumor initiation is a defining property of putative CSCs, which have been reported in both blood malignancies and solid tumors. In order to test whether any given human tumor cell population has CSC properties, the relatively enriched single cells have to be put into a foreign microenvironment in a recipient animal to test their tumorigenic potential. Furthermore, various in vitro assays need be performed to demonstrate that the presumed CSCs have certain biological properties normally associated with the stem cells (SCs). Herein, we present a comprehensive review of the experimental methodologies that our lab has been using in assaying putative prostate cancer (PCa) SCs in culture, xenograft tumors, and primary tumor samples.

Key words: Cancer stem cells (CSCs), Prostate cancer, Clonal and clonogenic assays, Side population, Self-renewal, Transplantation sites, Sphere-formation assays

1. Introduction

CSCs were first identified in leukemia *(1)*. Leukemic stem cells (LSCs), although constituting a minority (~0.1%) of the total cell population, are the only cells that can transfer the disease to NOD/SCID mice. In the past 5 years, putative CSCs, or tumor-initiating cells, have been reported for many human solid tumors (**Table 1**), including brain tumors *(2)*, melanoma *(3)*, and cancers of the breast *(4, 5)*, colon *(6–9)*, pancreas *(10, 11)*, liver *(12)*, lung *(13)*, and head and neck *(14)*. Several important principles have emerged from these studies.

John S. Yu (ed.), *Cancer Stem Cells,* Methods in Molecular Biology, vol. 568
DOI 10.1007/978-1-59745-280-9_7, © Humana Press, a part of Springer Science+Business Media, LLC 2009

Table 1
CSC studies in human solid tumors (2003–2008)

Tumor type	Samples	Marker	Mice	Transplantation	Results	Ref.
Breast cancer	9 (1 primary; 8 met.)	$CD44^+CD24^{-/lc}ESA^+$FACS	NOD/SCID mice pretreated with VP16	Mammary fat pad	>50-fold enrichment in tumorigenicity	4
Breast cancer	4 xenotransplants (from 2 primary; 2 met.)	$ALDH^+$ FACS	NOD/SCID mice	Humanized mammary fat pad	500 $ALDH^+$ cells generate T; 20 $ALDH^+CD44^+CD24^-Lin^-$ cells generate T	5
Brain tumors	7 primary tumors	$CD133^+$ (MACS)	6–8 week NOD/SCID	Intracranial injection	$CD133^+$ more tumorigenic	2
Prostate cancer	7 (4 primary, 1 benign, 2 LN mets)	$CD44^+$ $\alpha2\beta1^{hi}CD133^+$ (MACS) purified from long-term cultured cells		no tumor experiments	Marker + cells more clonogenic	28
Colon cancer	17 (6 primary, 10 liver & 1 retroperitoneal met.)	$CD133^+$ (double MACS)	8 week NOD/SCID irradiated	Renal capsule	1 CSC/57,000 T. cells 1 CSC/262 CD133+ cells	6
Colon cancer	19 primary (5 Dukes A)	$CD133^+$ (FACS or MACS)	SCID	Subcutaneous	3,000 CD133+ cells generate T	7
Colon cancer	21 primary CRC	CD133+	5–6 week nude mice	Subcutaneous	2,500 CD133+ cells generate T 25 CD133+-derived spheres generate T	8
Colon cancer	2 primary, 6 xenografts	$EpCAMCD166^+CD44^+$ (FACS)	6–8 week NOD/SCID	Subcutaneous	150 $EpCAMCD166^+CD44^+$ cells generate T	9
Pancreatic cancer	10 (2 primary; 2 met.)	$CD44^+CD24^+ESA^+$	NOD/SCID	Subcutaneous + pancreas	>100-fold enrichment	10

Pancreatic cancer	11 (6 met.); sorting for 7 L3.6pl metastatic line	$CD133^+$ (MACS) $CD133^+CXCR4^+$ (FACS)	8–12 week nude mice	pancreas	500 $CD133^+$ cells generate T The $CD133^+CXCR4^+$ pop. mediates met.	11
Head & Neck	25 primary (3 recurrences) 9 for sorting (4 primary+5 xenografts)	$CD44^+Lin^-$ (FACS)	NOD/SCID & $Rag2^{-/-}$	Subcutaneous	5,000 $CD44^+Lin^-$ cells generate T Only 13/25 HNSCC samples gave tumors	14
Melanoma	7 (1 primary; 4 LN & 2 visceral met.)	$ABCB5^+$ (MACS)	NOD/SCID	Subcutaneous	1 MMIC/1 million bulk T cells 1^{ary} xeno: 1 MMIC/160,000 $ABCB5^+$ cells 2^{ary} xeno: 1 MMIC/120,000 $ABCB5^+$ cells	3
Lung cancer	19 (18 primary; 1 met.)	$CD133^+$ (FACS)	4 week SCID or nude	Subcutaneous	10^4 CD133+ cells generate T	13
Liver cancer	28 primary (only 13 used)	CD45-CD90+ (MACS)	SCID	Intrahepatic	CD45–CD90+ more tumorigenic	12

Met metastasis; *T* tumor; *FACS* fluorescence activated cell sorting; *MACS* magnetic-beads assisted cell sorting; *CRC* colorectal cancer; *LN* lymph node; *xeno* xenograft; *MMIC* malignant melanoma initiating cell

First, most CSCs have been identified using cell surface markers for the corresponding normal tissue stem/progenitor cells, suggesting that normal and cancer SCs share some phenotypic markers.

Second, interestingly, although no markers may be truly SC specific, CD44 and CD133 have been used to identify many types of CSCs. For example, CD44 has been used to enrich for breast, colon, pancreatic, liver, and head and neck CSCs whereas CD133 for CSCs in lung and colon cancers and glioma (**Table 1**). Some other markers may be tumor specific, e.g., breast CSCs have a (CD44+)CD24- phenotype *(4)* whereas pancreatic CSCs possess the (CD44+)CD24+ phenotype *(10)*.

Third, in a particular tumor, CD44 and CD133 may identify distinct and/or overlapping populations of tumor stem/progenitor cells. For instance, both CD133 *(6–8)* and CD44 *(9)* have been utilized as the positive selection marker for colon CSCs. The same two markers have also been employed to independently select for pancreatic CSCs *(10, 11)*. In both cases, the interrelationship (inclusive, exclusive, or hierarchical) between the CD133 and CD44 selected CSCs remains unclear. These observations *(6–11)* emphasize the important point that the CSC population is likely heterogeneous, as elucidated in LSCs *(15)*, and also raise the possibility that combining CD44 and CD133 may enrich for more primitive CSCs.

Fourth, CSCs are only operationally or functionally defined. Perhaps one of the most important criteria is that putative CSCs possess an enhanced ability to initiate serially transplantable tumors that phenotypically recapitulate patient tumor histology *(16, 17)*. In all of the mentioned CSC studies (**Table 1**), "naked" tumor cells were injected into the immunodeficient mice, implying that putative CSCs possess an intrinsic ability to establish a "niche" in a foreign microenvironment.

Fifth, nevertheless, reconstitution of CSC activity and tumor development of human tumor cells in mice represents an extremely challenging task *(17, 18)* involving numerous variables associated with both tumors (availability, heterogeneity, stage/grade, size, quality, digestion/purification/implantation methods, etc.; **Table 2**) and recipient mice (strains, degree of immune deficiency, preconditioning, injection/implantation sites, etc.; **Table 3**). Consequently, different tumors have a wide variety of "empirical" details that cannot be interpreted readily and reconciled scientifically. For instance, although some tumorigenic subsets were implanted "orthotopically," many others were injected at ectopic sites, in particular, subcutaneously (s.c.) or under the kidney capsule (KC) (**Table 1**).

Table 2
Variables affecting the outcome of human prostate cancer tissue/cell grafting and establishment in mice

Recipient mice	Strains (*see* **Tables 3**) Preconditioning (pre-injection of etoposide; irradiation; injection of anti-NK antibodies, "humanizing" implantation sites, etc.)
Tumors (samples)	Type (primary tumors; hormone-refractory tumors (HRPC); poly treatment-refractory tumors (TRPC); metastases) Tumor grade (Gleason 6–10) and stage Tumor size (0.3–3-g) Tumor involvement (3–80%) Quality (fibrous, necrosis, hemorrhagic, etc.) Heterogeneity (multifoci; mixed with benign tissues)
Processing	Transport media/time Digestion media/time Final cell viability
Purification	Marker(s): CD44, CD133, CD44CD133, ALDH, ALDHCD44, etc. Methods: FACS, MACS, double-MACS, etc; time used Final purity
Injection mix	"naked" cells in Matrigel (25–50%) or collagen Rat urogenital sinus mesenchyme (rUGM) Rat seminal vesicle mesenchyme (rSVM) Irradiated fibroblasts; carcinoma-associated fibroblasts (CAFs) Human mesenchymal stem cells (hMSCs)
Injection/implant. site	Subcutaneous (s.c) Prostate (DP, VP, AP) Kidney capsule (KC); KC upon recombination with rUGM (TR/KC) Others (lymph node, i.p., bone, tail vein, intracardiac, etc.)

Sixth, as predicted, CSCs seem to be more resistant to antitumor therapeutics, including chemotherapy and radiation *(11, 19–21)*. Of clinical significance, the abundance of CSCs significantly increases in breast cancer patients who have received prior therapies *(19)*.

Table 3
Mouse strains commonly used in CSC studies[a]

Strain	Immune deficiency in	Comments
Athymic nu/nu (*Foxn 1 nu*)	Mature T cells (lacks thymus) Ab formation that requires CD4+ helper T cells CD4+ and/or CD8+ T cell-mediated immune responses Delayed-type hypersensitivity responses (which require CD4+ T cells) Killing of virus-infected or malignant cells (requiring CD8+ T cells) Graft rejection (requires both CD4+ and CD8+ T cells)	*foxn 1* gene mutation
SCID	Lacks both T and B cells	*Prkdc*^{scid} (chr. 16) gene mutation
	Lymph nodes without follicles Rudimentary thymic medulla without a cortex A small spleen lacking white pulp	no V(D)J recombination Bgrd: C57BL/6J, BALB/cByJ C3H/SmnJ or NOD/LtSz
NOD/SCID (NOD/LtSz-Prkdc<scid>/J)	*scid* mutation backcrossed to NOD/Lt for ten generations Functional deficit in NK cells Absence of circulating complement Defects in the differentiation and function of APCs NOD/LtSz-*scid* mice remain insulitis and diabetes-free High incidence of thymic lymphomas Mean lifespan of NOD/LtSz-*scid* (NOD/LTSZ-*Prkdc*^{scid}) mice is 8.5 months	
NOD.SCID/Ncr	Received by NCI in 2004 from NIH; the scid mutation has been transferred onto a diabetes-susceptible Non-Obese Diabetic background	
Humanized NOD/SCID	NOD/SCID mice transplanted with hHSCs or whose transplantation site (e.g., mammary fat pad) has been "humanized" by implanting human stroma	
NOD/SCID/MPSVII	Also lack β-glucuronidase (GUSB) activity, an enzyme expressed only in human; engraftment of human cells in this mouse model can be efficiently tracked in situ using a colorimetric substrate GUSB	

NOD/LtSz-*Prkdc*^{scid}/J *B2m*$^{tm\ 1\ Unc}$	Generated by backcrossing the Class I deficient *B2m* targeted mutation on to the NOD/LtSz-*Prkdc*^{scid} strain (NOD/SCID/B2m^{null} mice)
Rag2$^{-/-}$	Contains a disruption of the recombination activating gene 2 (Rag2) Homozygous mice exhibit total inability to initiate V(D)J rearrangement and fail to generate mature T or B lymphocytes. Otherwise, the Rag2 mouse has apparently normal hematopoiesis
Rag2$^{-/-}$/γ c$^{-/-}$	Combining Rag2 deficiency with common cytokine receptor γ chain mutations completely alymphoid (T-, B-, NK-); no spontaneous tumor formation exhibit normal hematopoietic parameters

[a]Most inbred strains express normal levels of natural killer (NK) cells, hemolytic complement and myeloid function precluding long-term repopulation of homozygous *scid* mice with human cells. In addition, *scid* mutant mice on some backgrounds produce immunoglobulin and functional T cells at low levels

Adult rodent and human prostates are thought to contain SCs albeit their precise phenotypic and biological properties have not been defined *(17, 22–25)*. Pioneering studies from Drs. Collins/Maitland provided the evidence that putative human prostate epithelial SCs bear the CD44 + $\alpha 2\beta 1^{hi}$–CD133+ phenotype *(26, 27)*. In 2005, Collins et al reported that prostate tumor cells with the same surface phenotype represent potential prostate CSCs, although tumor experiments were not reported in this study *(28)*. Using several xenograft prostate tumors (Du145, LAPC4, and LAPC9), we have shown that the CD44+ cell population is enriched in prostate CSCs and that PCa cells are organized as a tumorigenic hierarchy *(17, 29–32*; **Fig. 1**). *First*, putative CSCs that can initiate serially passageable spheres and serially transplantable tumors are marked by CD44 expression and constitute the minority. Most of the CD44+ cells in the spheres or tumors are not proliferating. Importantly, essentially all metastatic activity resides in the CD44+ cell population *(30)*. *Second*, the side population (SP) also contains tumorigenic cells and 97% of the SP cells are CD44+. *Third*, in contrast to the SP and CD44+ cells, the α2β1+ and ABCG2+ PCa cells identify fast proliferating tumor progenitors. *Fourth*, essentially all ABCG2+ and >80% α2β1+ cells are encompassed in the CD44+ population. Therefore, the CD44 + α2β1+ cell population is highly enriched in tumor-initiating cells, whereas the CD44–α2β1 cells virtually

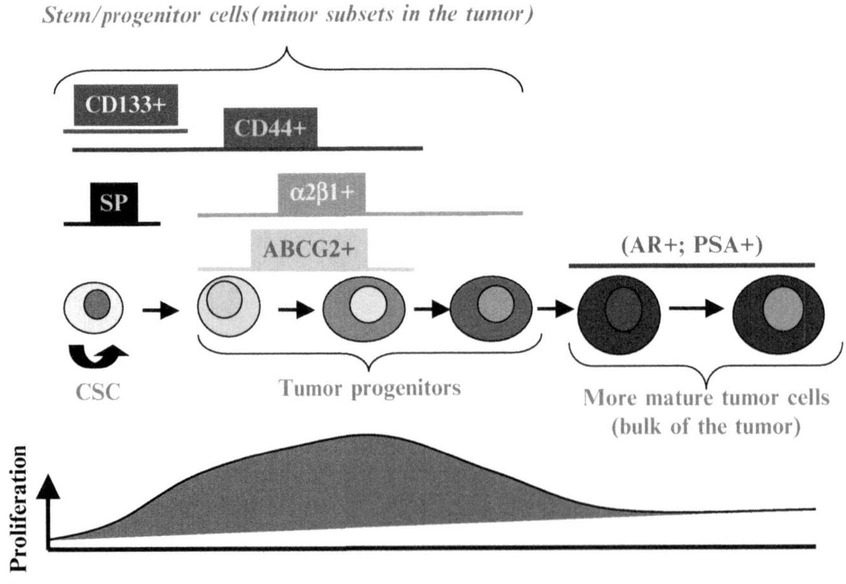

Fig. 1. A model of hierarchical organization of tumorigenic PCa cells. See text for details.

lack tumorigenicity. *Fifth*, most CD44 + PCa cells are AR- and can give rise to AR+ cells in the spheres and tumors, thus indicating their ability to self-renew and undergo asymmetric division. Recent work by others has confirmed the presence of stem-like PCa cells in cell lines *(33, 34)*.

2. Basic Materials and Reagents

2.1. Cell Culture Reagents

1. Iscove's Modified Dulbecco's Medium (IMDM) (Gibco, Bethesda, MD) supplemented with 15% heat-inactivated fetal bovine serum (FBS, Gemini, West Sacramento, CA).
2. RPMI 1640 medium (Gibco, Bethesda, MD) supplemented with 7% FBS.
3. Dulbecco's Modified Eagle's Medium (DMEM, Gibco, Bethesda, MD) supplemented with 7% FBS.
4. 0.05% Trypsin–EDTA (Gibco).
5. Dulbecco's Phosphate-Buffered Saline (PBS) 1× sterile, without calcium and magnesium (Mediatech, Herndon, VA).
6. Erythrosin B (ATCC, Manassas, VA).
7. *Modified Prostate Epithelial Basal Medium (PrEBM)*. 500 ml of PrEBM (CC-3165; Clonetics, Walkersville, MD) supplemented with insulin (Sigma I-6634; 4 mg insulin in 8 ml distilled water. Add 40 μl 1 N HCl. Filter sterilize), hydrocortisone (Sigma H-0396; 250 μg), epidermal growth factor (EGF; 5 μg; Peprotech, Rocky Hill, NJ), bovine pituitary extract (BPE; 25 μg/ml; Invitrogen, Carlsbad, CA), Penicillin/Streptomycin (Gibco; 1/100 dilution), and Fungizone (Gibco; 1/2,000 dilution, freshly added).
8. *Collection medium*. DMEM with 5% Knockout Serum Replacement (KO serum; Gibco), 1% Penicillin/Streptomycin, and 1% Fungizone.
9. *Collagenase solution*. DMEM, 5% KO serum, Collagenase Type I (Sigma, C-0310) at 200 units/ml, 1× DNAse (Sigma, DN-25; 100× stock is 10 mg/ml and is kept at –20°C). Filter sterilize prior to use.
10. *Trypsin solution*. 2.5% trypsin (17–160E; BioWhittaker, Walkersville, MD), dilute 10× in PBS, add 1× DNAse.
11. 0.25 mg/ml trypsin/EDTA (Clonetics).
12. Trypsin-neutralizing solution (TNS, Clonetics) (*see* **Note 1**).
13. Hemocytometer.
14. Dimethyl sulfoxide (DMSO, Sigma).

15. Vitrogen 100 (Cohesion, Tyngsboro, MA); dilute 1:50 in sterile ddH_2O and coat coverslips or flasks for 1 h at 37°C or overnight at room temperature; wash twice with sterile ddH_2O and air dry in the hood.

2.2. Tumor-Processing Reagents

1. Razor blade (Surgical Blades, No. 22).
2. Accumax/EDTA (AM105; Innovative Cell Technologies, San Diego, CA).
3. Histopaque-1077 (Sigma, St Louis, MO).
4. Percoll (GE Healthcare Bio-Sciences, Piscataway, NJ).
5. 40-μm nylon mesh or 40-μm nylon cell strainer (BD Falcon, Bedford, MD).
6. Lineage-depletion kit for mouse stromal cells (Miltenyi Biotec, Auburn, CA) (*see* **Note 2**).
7. Biotin anti-$H2K^d$ antibody (BD Pharmingen, Franklin Lakes, NJ).
8. *Lineage-depletion mix for human primary tumors*. Lineage cocktail 1 (lin 1)-FITC combined with anti-CD140b-PE and anti-CD31-FITC antibodies (BD Pharmingen, Franklin Lakes, NJ).
9. Antibiotin or antifluorochrome (anti-PE or anti-FITC) microbeads, as appropriate (Miltenyi).
10. MACS separation apparatus and MACS MS separation columns (Miltenyi Biotec).

2.3. Fluorescence-Activated Cell Sorting (FACS)

1. *Staining buffer*. Sterile PBS supplemented with 0.5% bovine serum albumin (BSA, Sigma, A-4161) and insulin (Sigma, I-6634; final concentration 5 μg/ml) (*see* **Note 3**); may be replaced by modified PrEBM media.
2. Antihuman CD44 antibodies from BD Pharmingen.
3. Antihuman CD133 antibodies from Miltenyi Biotec.
4. For other antibodies and isotype controls, see ref. 29–32.
5. FcR Blocking Reagent, human (Miltenyi Biotec).
6. 7-AAD (Molecular Probes, Carlsbad, CA).

2.4. Side Population (SP) Analysis

1. Hoechst dye 33342 (bis-benzimide) from Sigma, diluted at 1 mg/ml in H_2O. Filter sterilize and freeze in small aliquots.
2. Verapamil (50 μmol/l) or reserpine from Sigma.
3. Waterbath at precisely 37°C.

2.5. BrdU Pulse and BrdU Staining

1. 5-bromo-2′-deoxyuridine (BrdU) from Sigma.
2. *Monoclonal anti-BrdU antibody*. Bu20a hybridoma supernatant (1:5–1:10 dilution).

3. *PDL* (poly-d-*lysine hydrobromide; Sigma P-6407*). 5 mg/ml stock in sterile H_2O; final concentration at 50 μg/ml.
4. Goat whole serum (Vector Laboratories, Burfingame, CA). Stored at −20°C.

2.6. Clonal, Clonogenic, and Sphere-Formation Assays

1. Ultralow attachment plates (Corning, Lowell, MA).
2. Mouse Swiss 3 T3 fibroblasts (ATCC).
3. Mitomycin C (Sigma).
4. Matrigel, phenol-red free (BD, Franklin Lakes, NJ).
5. BBD Agarose (BD).
6. Dispase solution from Invitrogen.
7. Collagenase Type I (200 units/ml) from Sigma.
8. 27-G needles and syringes.
9. 0.005% Crystal Violet.

2.7. Xenograft Tumors

1. Ketamine/Xylazine (Ketaset; Henry Schein Veterinary, Melville, NY).
2. Isoflurane (ABBOTT Animal Health, North Chicago, IL).
3. Banamine (Flunixamine) from Henry Schein Veterinary.
4. Ophthalmic ointment (Vetropolycin) (Pharmaderm Animal Health, Melville, NY).
5. Testosterone Propionate powder (T-1875, Sigma).
6. Sterile saline (Hospira, Lake forest, IL).
7. Dissecting microscope.
8. Hamilton syringe needle (Hamilton 80500, Reno, NV).
9. Ultra-Fine II Insulin syringe, ½ cc, 31G needle, from BD.
10. 4/6–0 gut sutures.
11. 1:500 acetic acid solution (200 μl glacial acetic acid per 1 g tendon in 100 ml water).
12. *Buthanesia (Schering-Plough, Kenilworth, NJ)*. 0.3 ml per rat in a syringe; dilute 1:2 with 1× PBS.
13. Microdissection tools (scissors and fine forceps soaked in ethanol).
14. 27-G ½in. needles.
15. 1-cc syringes.
16. Pasteur pipettes.
17. Beveled glass slides (soak in ethanol prior to use).
18. 4-chambered Petri dishes.
19. Collagenase V (187 units/ml) from Sigma.
20. 1 M NaOH.
21. 100% acetone from Sigma.

3. Methods

CSCs are stem-like cells in a tumor and are only functionally and operationally defined. We have proposed the following criteria to functionally define the putative CSCs *(17)*. *First*, the presumptive CSC must be *prospectively* purified. When purifying candidate populations of CSCs, lineage selection should be performed to remove "irrelevant" cells such as stromal and blood cells that may contain other SCs, including mesenchymal SCs and HSCs. *Second*, in vivo tumorigenicity experiments must be done to show that such cell populations, freshly purified and without extensive expansion in vitro, are enriched in tumor-reinitiating cells. When feasible, serial tumor xenotransplantation should be carried out to determine whether the tumors derived from the putative CSCs could be transplanted for multiple generations. *Third*, the reconstituted as well as serially xenotransplanted tumors should histologically resemble the original patient tumor. *Fourth*, importantly, the presumptive CSC population, or a subpopulation within, has to be studied to show that they possess certain intrinsic biological properties (extensive proliferative capacity, self-renewal, differentiation, etc.) normally associated with the SCs. Only when these conditions are fulfilled can one confidently claim that the candidate population of tumor cells under investigation is *enriched* in potential CSCs or tumor-initiating cells. It is important to bear in mind that such tumorigenic populations are still likely to be heterogeneous with true CSCs representing perhaps a very small fraction.

Practically, CSC studies involve (a) sample preparation, (b) candidate cell purification, and (c) in vitro and (d) in vivo analysis and characterizations of CSC properties. In the following we present some basic experimental methodologies for each of these aspects applied to prostate CSC studies. Some protocols can also be found in our publications *(17, 29–32)* and on our website (http://sciencepark.mdanderson.org/tanglab/protocols).

3.1. Sample (Single Cell) Preparation

3.1.1. Cultured Cell Processing

1. Cell quality is essential for CSC-related assays. Cultured cells should be in log phase and cell density should be no more than 80%. A yellow culture medium indicates that cells have used up nutrients and most likely have been starved. Only healthy-looking cells should be used for experiments.
2. Warm up 0.05% Trypsin–EDTA and other media to 37°C.
3. Remove culture media from culture flasks (dishes).
4. Add 5 ml (for a T-75 flask) of Trypsin/EDTA and incubate for 3–5 min at 37°C in a CO_2 incubator. Check for cell detachment.

5. When desired detachment is achieved, add equal volume of serum-containing medium (for cancer cells) or TNS (for primary/nontumor cells).
6. Centrifuge cells at 130 × *g* for 5 min at room temperature (RT).
7. Aspirate the medium.
8. Resuspend in medium or PBS to make single-cell suspension and count cell number (*see* **Note 4**).

3.1.2. Xenograft Prostate Tumor Processing

1. Harvest xenograft tumor in a sterile environment and weigh it in a pretared sterile 35-mm Petri dish.
2. Add a small volume (1–2 ml) of cold culture medium to the Petri dish and chop it into small pieces using a razor blade. Further mincing can be done using dissecting scissors to obtain a thick soup of tumor cells/pieces.
3. Transfer the soup to a 50-ml tube using a cut 1-ml pipette tip. Add ~20× the volume of cold medium, invert to mix, and spin at 130 × *g* for 5 min. Tissue pieces and dissociated cells should pellet down while most of debris should remain in the supernatant.
4. Aspirate off the supernatant and add 30–50 ml of cold PBS. Invert to mix and spin again at 130 × *g* for 5 min. Discard supernatant. Repeat if necessary.
5. Resuspend the cells/chunks in Accumax/EDTA solution at a concentration of 10 ml/0.5 g of original tissue. Incubate at RT for 30 min on a shaker (*see* **Notes 5** and **6**).
6. At the end of incubation, briefly vortex the tissue solution and then set the tube straight for about 2 min to allow the largest chunks to settle to the bottom. Transfer the supernatant containing the dissociated cells to a fresh tube and spin at 130 × *g* for 5 min (*see* **Note 7**).
7. While the cells are spinning, prepare a sterile 40-μm nylon mesh or cell strainer by placing it on a 50-ml centrifuge tube and gently prewetting it with medium. At the end of centrifugation, discard the supernatant and resuspend cells in a small volume of medium (1–3 ml). Slowly, pass the cells through the mesh without applying too much pressure. If the mesh becomes clogged, replace with a new, prewetted one and continue until all the cells have gone through. Wash the mesh 3× with medium. Only single cells or doublets should pass through the mesh.
8. Count the total cell number. Live cells can be distinguished from dead cells on a hemocytometer using a viability dye such as erythrosin B or trypan blue at a 1:1 ratio of cells to dye. If the cells are very concentrated, a 1:10–1:100 dilution can be done to achieve a more accurate count.

9. Live cells are separated from dead cells, debris, and red blood cells through use of a Histopaque-1077 gradient. Briefly, cell suspension is adjusted to 1–1.5 million cells (dead + live) per ml. In a 15-ml tube, add 3 ml of Histopaque. Carefully load the same volume of cell suspension on the Histopaque so that two layers are clearly observed. The loading should be done with a 1-ml pipette tip as a larger pipette tends to disrupt the interface between the two layers. Multiple tubes can be used depending on the total cell number. Alternately, a 50-ml tube can be used with 15 ml of Histopaque followed by 15 ml of cell suspension at a concentration of no more than 1×10^6 cells/ml (dead + live) (*see* **Note 8**).
10. Spin the cells on Histopaque at 400 × *g* for 30 min at RT.
11. At the end of centrifugation, an opaque layer of cells should be observed at the interface of the two layers. These are the live cells. The dead cells and debris should pass through the Histopaque layer and collect at the bottom of the tube. Using a 1-ml pipette tip, carefully remove the live cell band and transfer to a new tube. Combine the live cells from all tubes and spin at 130 × *g* for 5 min.
12. Resuspend in an appropriate volume (0.1–5 ml) of medium and count the cell number.
13. Spin again at 130 × *g* for 5 min.
14. To deplete the xenograft tumor cell population of mouse (host) lineage-positive cells, a lineage-depletion kit from Miltenyi is used. Mouse cells can also be removed using the anti-$H2K^d$ antibody that recognizes the mouse major histocompatibility class I (MHC-I) antigens. In either case, the procedure is done at 4°C in the dark (in the fridge). All solutions should be precooled.
15. During centrifugation, make staining buffer.
16. Suspend cell pellet in 40 μl of staining buffer (*see* **Note 9**).
17. Add 10 μl of biotin-antibody cocktail or biotin anti-$H2K^d$.
18. Mix well and incubate for 10 min at 4°C in the dark.
19. Add 30 μl of buffer.
20. Add 20 μl of antibiotin microbeads.
21. Mix well and incubate for additional 15 min at 4°C in the dark.
22. Wash cells by adding 10–20× labeling volume in cold PBS and centrifuge at 130 × *g* for 5 min. Pipette off supernatant completely.
23. Resuspend up to 10^7 cells in 500 μl of buffer (*see* **Note 10**).
24. For the separation, assemble the MACS apparatus by attaching the magnet on the stand and placing the column in the

magnetic field. Prepare the column by rinsing with 500 μl of buffer. Make sure that there is a 15-ml collection tube below the column.

25. Apply the cell suspension to the column. Allow the cells to pass by gravity and collect the effluent as fraction with unlabeled cells, representing the enriched tumor cell fraction depleted of mouse cells. The purity of the cells can be assessed by immunolabeling using human-specific antibodies, e.g., ESA (or epithelial specific antigen; also called Ber-EP4), human nuclei, and human-specific mitochondria or lamin.
26. Wash the column with 500 μl of buffer (3×).
27. Spin down the effluent at 130 × *g* for 5 min and discard supernatant.
28. Purified cells should be resuspended in appropriate medium and another count should be made to obtain the final cell number.
29. To freeze down cells, we use the regular culture medium supplemented with 10% DMSO and 10–40% FBS (use more serum for nontumor cells).

3.1.3. Human Primary Prostate Tumor Processing

1. Tumor tissues must be collected in a sterile container with collection medium as soon as possible after removal from the patient/donor.
2. Rinse tumor piece(s) once with DMEM + 5% KO serum to remove traces of blood.
3. Before digesting the tumor, small pieces are sliced out for the following purposes: (a) Formalin-fixed and paraffin-embedded for histology and IHC; (b) Snap-frozen in OCT for cryosectioning; (c) Snap-frozen for RNA extraction; (d) Snap-frozen for protein extraction; and (e) Implanted subcutaneously (s.c.) or in anterior prostate (AP) or under the kidney capsule (KC) in male NOD/SCID mice supplemented with a testosterone (T) pellet, aiming to establish primary (1°) xenograft tumors.
4. For each patient tumor sample obtained, a piece of "normal" or benign tissue distant from the tumor (i.e., uninvolved region) is also secured. For comparative purposes, the benign tissue is also subjected to (a)–(d) described earlier. The benign tissue can be processed in a similar way to the processing of tumor tissues described later to obtain benign prostate epithelial cells for various studies.
5. Proceed with **steps 2–4** in **Subheading 3.1.2**.
6. Resuspend the minced tissue pieces in Collagenase I solution using 10 ml/0.5 g of tissue. Agitate gently at 37°C overnight (14–16 h) on an orbital shaker (*see* **Note 11**).

7. At the end of the incubation, pipette up and down gently with a 5-ml pipette to break up any large clumps. The suspension should contain mostly organoids and appear cloudy with tiny tissue pieces. Spin at 670 × *g* for 5 min.
8. Discard supernatant and add 0.25% trypsin solution to the pellet (10 ml/1 g tissue). Shake gently on an orbital shaker at 37°C for 15–30 min.
9. Neutralize trypsin activity by adding 2 ml of FBS. Spin at 670 × *g* for 5 min.
10. Resuspend pellet in 3 ml of PrEBM. Serially triturate the suspension first with a 1-ml pipette, followed by 18-G and 21-G needles. The cells should now be mostly in single-cell suspension.
11. Pass the single-cell suspension on a prewetted 40-μm nylon mesh. Do not apply too much pressure. If the mesh becomes clogged, replace with a new, prewetted one and continue until all the cells have gone through. Wash the mesh 3× with medium.
12. Count the total cell number.
13. The live cells are separated from dead cells and debris by the use of a discontinuous Percoll gradient. Briefly, the cell suspension volume is calibrated at ~1–1.5 million total cells (dead and live) per ml. In a 50-ml tube, pipette 10 ml of heavy Percoll and gently apply 10 ml of light Percoll. Gently load 15 ml of cell suspension upon the Percoll gradient. Multiple tubes should be used depending on the total cell number (*see* **Note 12**).
14. Spin the cells at 250 × *g* for 20 min at 4°C with no brake.
15. Debris stays in the light Percoll. Cells should collect at the interface of the Light Percoll (red) and Heavy Percoll, but with human primary samples the interface is not very clear. After centrifugation, remove the cell debris layer (top 10 ml). Collect the cell layer (in the following 15 ml) and transfer to a new tube. Combine the live cells from all tubes, top off with PBS, and spin at 380 × *g* for 5 min.
16. Resuspend in an appropriate volume (0.1–5 ml) of medium and count the cell number.
17. To deplete the human tumor cell population of hematopoietic and stromal cells, an immunomagnetic lineage-depletion step is performed. We use a depletion mix that includes antibodies to several markers associated with human leukocytes (lineage cocktail 1 contains a combination of antibodies (CD3, CD14, CD16, CD19, CD20, and CD56) that stain lymphocytes, monocytes, eosinophils, and neutrophils) and stromal cells (CD140b antibody reacts with platelet-derived

growth factor (PDGF) receptor β, that is expressed on fibroblasts and smooth muscle cells; CD31 antibody binds endothelial cells). This procedure is done at 4°C in the dark. All solutions should be precooled. Basic protocol is provided in the manufacturer's product sheet.

18. Make staining buffer as described for xenograft tumor processing.
19. Suspend cell pellet in 100 μl of buffer (*see* **Note 13**).
20. Add 10 μl of Lin1-FITC cocktail + 10 μl of anti-CD140b-PE + 10 μl of anti-CD31-FITC.
21. Tap the tube to mix and incubate for 15 min at 4°C in dark. Tap every 5 min to keep the cells in suspension.
22. Wash cells by adding 10–20× labeling volume in cold PBS and centrifuge at 380 × *g* for 5 min. Pipette off supernatant completely.
23. Add 100 μl of buffer to cell pellet.
24. Add 20 μl of anti-FITC microbeads + 20 μl of anti-PE microbeads.
25. Tap to mix and incubate for 15 min at 4°C in dark. Tap the tube every 5 min to keep the cells in suspension.
26. Wash cells by adding 10–20× labeling volume in cold PBS and centrifuge at 380 × *g* for 5 min. Pipette off supernatant completely.
27. Resuspend up to 10^7 cells in 500 μl of buffer. Scale up for higher numbers.
28. For the separation, assemble the MACS apparatus by attaching the magnet to the stand and placing the column in the magnetic field. Prepare the column by rinsing with 500 μl of buffer. Make sure that there is a 15-ml collection tube below the column.
29. Apply cell suspension to the column. Allow the cells to pass by gravity and collect the effluent as fraction with unlabeled cells, representing the enriched lineage-negative epithelial cell fraction.
30. Wash the column with 500 μl buffer 3×.
31. Spin down the effluent at 380 × *g* for 5 min and discard supernatant.
32. At this juncture, the cell suspension is ready for in vitro analyses (e.g., clonogenicity), transplantation in vivo, and/or further cell fractionation and enrichment by MACS or FACS. While many markers can be negatively selected against using the MACS technique, direct positive selection can only be done for one marker in each experiment. For dual selection, FACS is recommended.

3.2. Candidate CSC Purification from Single-Cell Suspensions

3.2.1. Marker-Based Purification by FACS

1. Centrifuge the single-cell suspension at 380 × *g* for 5 min to bring the cells in 90 µl of cold staining buffer (*see* **Note 14**). Remove 10 µl from cell suspension and add to 90 µl of buffer in another tube. This is to set aside some cells for the isotype control staining.
2. Add 20 µl of FcR blocking reagent to the remaining 80 µl of cell suspension and incubate at 4°C in the dark for 10 min.
3. Add 10 µl of the antibody of interest (such as CD44-FITC) to the cell suspension, mix by tapping, and incubate at 4°C in the dark for 15 min. The tube should be gently tapped every 5 min to prevent the cells from settling down. For the control tube, use 1 µl of appropriate isotype control (e.g., since anti-CD44-FITC is a mouse IgG2b , the control would be FITC-mouse IgG2b isotype control immunoglobulin).
4. If double staining is needed, repeat **step 3** by adding 10 µl of the second antibody tagged with a different fluorophore. Single-staining tube controls should also be prepared.
5. Add 5 ml of cold PBS and spin at 380 × *g* for 5 min. Resuspend in cold serum-free medium (+1% antibiotics) at a concentration of 2 × 10^6 cells/ml. To hand over to the FACS facility, the cell suspension should be transferred to 5-ml polystyrene tubes at no more than 2.5 ml/tube. Collection tubes should also be prepared using 5-ml polypropylene tubes containing 1 ml of FBS (+1% antibiotics). This provides a cushion for the cells when they come through the flow machine at high speeds. Tubes should be labeled " + " or "–"depending on the cell number expected.
6. Add 7-AAD to the sort tube10 min before analysis (*see* **Note 15**). The stock concentration of 7-AAD is 100 µg/ml (100×, final = 1 µg/ml).
7. All reagents and samples must be submitted for FACS analysis on ice and shielded from light. FACS is carried out anywhere from 1,000 to 2,000 cells/s, depending on the size and aggregation properties of the cell type and the fluorophores utilized.
8. Once FACS analysis has been carried out, the "+" and "–" cells should be spun at 380 × *g*, resuspended in 0.5–3 ml of regular culture medium, and counted (*see* **Note 16**).
9. At the end of every sort, it is generally a good idea to do a postsort analysis as a quality control *(29–31)*. This is best done by rerunning an aliquot of the sorted cells through the flow cytometer and determining the purity. However, if the number of sorted cells is too low, to conserve the sample, it is better to simply look at a drop of the suspension under the fluorescence microscope and determine the percent purity.

Some dyes (e.g., PE), however, are not easily visible under the microscope and a postsort analysis can only be done using flow (*see* **Note 17**).

10. In our experience, FACS generally results in cell populations of high purities. For example, we routinely obtain CD44+ PCa cells at >95% purity and CD44– cells at ~100% purity *(30)*.

3.2.2. Marker-Based Purification by MACS

1. To the pellet obtained postlineage selection, add 100 μl staining buffer + 10 μl of anti-CD44-FITC, mix by tapping, and incubate at 4°C in the dark for 15 min. The tube should be gently tapped every 5 min to prevent the cells from settling down (*see* **Note 18**).
2. Add 10× volume of cold PBS and spin at 380 × *g* for 5 min. Discard supernatant.
3. Add 100 μl of staining buffer + 20 μl of anti-FITC microbeads, mix by tapping, and incubate at 4°C in the dark for 15 min.
4. Add 10× volume of cold PBS and spin at 380 × *g* for 5 min. Discard supernatant.
5. Resuspend pellet in 500 μl up to 10^7 cells and pass through a pre-equilibrated MACS column as described in **Subheading** 3.1.3. In this case, we will be using the positive as well as the negative fractions. The negative cells will be in the collection tube. To collect the positive fraction, remove the column from the magnetic field, transfer to a new tube, add 1 ml of buffer to the column, and push with the plunger. Add another 1 ml of buffer to the column; replace the plunger and push.
6. In our experience, MACS-based sorting generally results in lower purities (compared to FACS). Although >90% pure marker-negative cell population can be routinely achieved, the marker-positive population often has a purity of 40–60%, especially with patient tumor-derived cells. In this case, several rounds of MACS purification can be attempted to further enrich for the marker-positive population (*see* **Note 19**).

3.2.3. CSC Enrichment by the SP Technique (See Note 20)

1. Ensure that a waterbath is precisely set at 37°C. The medium needs to be prewarmed and should contain no more than 5% FBS.
2. Cells are resuspended at 10^6 cells per ml in prewarmed medium. Mix well.
3. Add Hoechst 33342 to a final concentration of 5 μg/ml (a 200× dilution of the stock) (*see* **Note 21**).

4. For the control, remove 1 ml of the cells/dye mixture and place in another tube. Add verapamil at 10 μg/ml (a 100× dilution of the stock). Verapamil is a calcium channel blocker that inhibits the ABC transporters and, therefore, the SP should be abolished or inhibited by verapamil. Reserpine can also be used for this purpose.
5. Mix the cells well, and place in the 37°C waterbath for exactly 120 min. Make sure that the staining tubes are well submerged in the bath water to ensure that the temperature of the cells is maintained at 37°C. Tubes should be mixed every 15 min during incubation (*see* **Note 22**).
6. After 120 min, add excess cold medium to stop the ABC transporter activity and spin the cells for 5 min. Resuspend cells in cold medium without serum at 10^6 cells/ml.
7. All subsequent procedures should be carried out at 4°C.
8. Add 7-AAD to the suspended cells and mix for about 10 min before FACS analysis.
9. Use the FACS to obtain SP and non-SP (also called MP, Main Population) cells.
10. It is important to note that different flow cytometers often give different SP profiles **(Fig. 2)**. For example, when samples are run on a Beckman–Coulter Altra [UV laser: Coherent Innova Enterprise II Ion (Mdl 621), water-cooled laser, 50 mW (351–364 nm Em); optics: Jet in Stream; Hoechst blue/splitter/Hoechst red: 515SP/540DSP/608LP], the SP displays as a distinct population on the side **(Fig. 2**; *29***)**. However, when the same cells are analyzed on a BD Biosciences FACSAria SORP [UV laser: Lightwave Xcyte, solid-state laser, 20 mW (355 nm Em); optics: Jet-in-Air; Hoechst blue/splitter/Hoechst red: 450/50 (425 nm–475 nm)/ 635LP/670LP], the SP displays as a tail of the MP **(Fig. 2)**. The SP profile on MoFlo cytometer resembles that the FACSAria *(35)*.
11. Whether the SP displays as a "side" population or a "tail," it is important to determine its specificity by setting up an experimental control, in which verapamil (or other inhibitors of ABC transporters) should abrogate or significantly reduce the SP **(Fig. 2)**. Since the SP phenotype in many SC populations is mediated, at least partly, by ABCG2 (also called BCRP; *29*), the specificity of the SP can thus be confirmed by incubating cells with the inhibitors of ABCG2 function, such as novobiocin (a nonspecific inhibitor of the ATPase activity of the ABC transporters) and fumitrimorgin (FTC; a specific substrate inhibitor of ABCG2; generally used at 10 μg/ml added 30 min before the dye labeling).

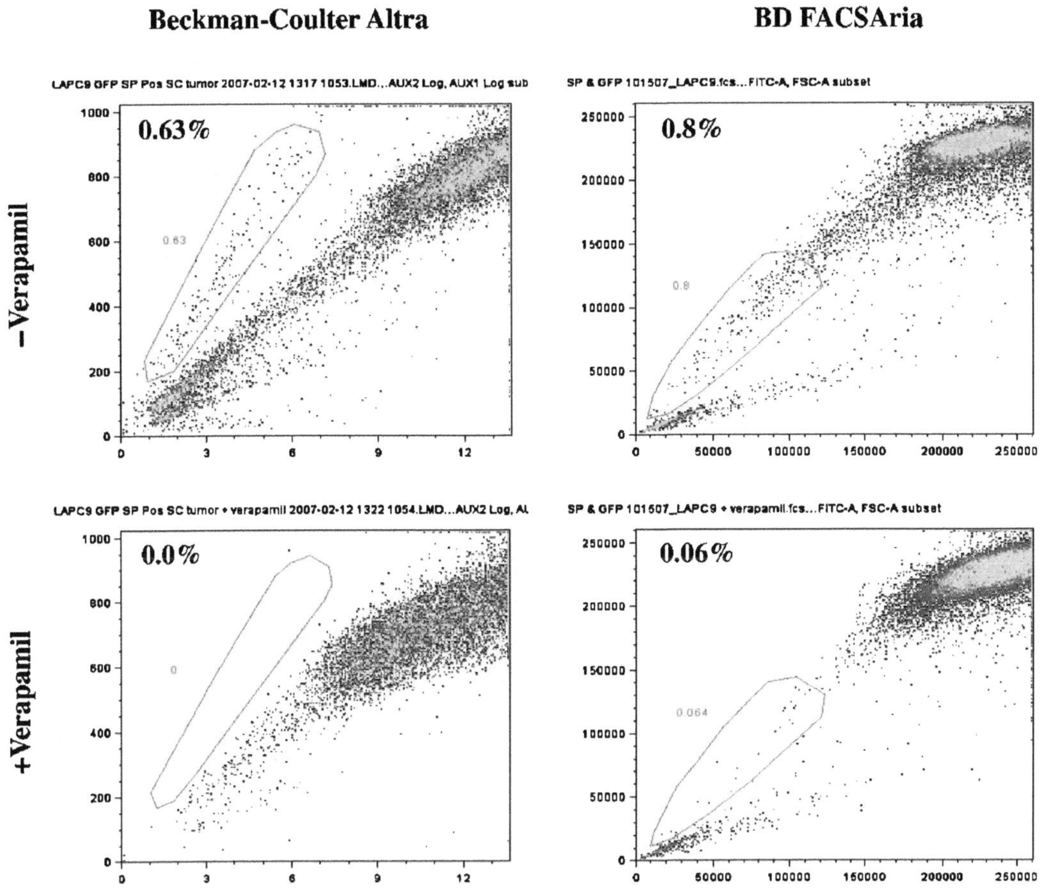

Fig. 2. SP analysis of LAPC9 tumor cells *(29)* on two different flow cytometers. See text for details.

12. Recently, Invitrogen-Molecular Probes has introduced the Vybrant DyeCycle Violet (DCV) stain, which does not require a high-wattage Argon laser for UV excitation (used for Hoechst dye), thus eliminating the necessity of exposing the (MP) cells to UV light during cell sorting. This system offers a distinct advantage of avoiding one of the major concerns associated with the Hoechst 33342-based SP analysis, in which chronic accumulation of the Hoechst dye in the MP cells may result in cytotoxicity to these cells thus obviating the validity of the "negative" control.

3.3. In Vitro Assays for Prostate CSC Properties

3.3.1. Assessment of Proliferative Capacity (Rate) by BrdU Incorporation Assay

1. The simplest experiment to assess the proliferative capacity of candidate cell populations is to plate equal numbers of marker-positive and marker-negative cells and count cell numbers at defined time points after plating. The data can be plotted as cell numbers (or cumulative population doublings) as a function of culture (passage) time, which is typically called a growth curve (see ref. *36* for examples).

2. To directly assess the cell proliferation rate (capacity), BrdU incorporation assays *(29, 36)* can be performed, which is briefly described as follows.
3. Grow cells on either glass coverslips or slide flasks coated with PDL (poly-d-lysine hydrobromide; Sigma P-6407; mw 70,000–150,000; 5 mg/ml stock in sterile H_2O. Final concentration at 50 mg/ml). For primary cells, 3–5,000 cells are grown on each 13 mm^2 glass and 5–10,000 cells on each slide flask.
4. To the culture medium of the log-phase cells add BrdU (5-bromo-2′-deoxyuridine; Sigma B-5002; mw = 307.1; stock solution = 10 mM in DMEM; final concentration = 2.5–10 μM) for 2–48 h, depending on the experimental purposes. For a single-time pulse, BrdU is added for the final 4 h of the culture period.
5. At the end of pulse, fix cells in 2–4% PFA at RT for 5′.
6. 70% ethanol + 2% acetic alcohol for 5′ at RT.
7. 70% ethanol in PBS (w/o Ca^{++}/Mg^{++}) at –20°C for 20′.
8. 1% Triton in PBS at RT for 20′.
9. 6 M HCI in 1% Triton at RT for 15′ (denaturation).
10. 0.1 M $Na_2B_4O_7$ in 1% Triton at RT for 10′.
11. Blocking with goat whole serum (15–30%) depending on the secondary antibody.
12. *Primary anti-BrdU.* Bu20a hybridoma supernatant (1:5–1:10 dilution) in blocking solution. For 1 h (RT) to overnight (4°C), followed by washing.
13. Secondary antibody (1:1,000) for 1 h at RT, followed by washing.
14. *Nuclear counterstaining.* Hoechst dye 33342 bisBenzimide (Sigma B32261) at 4 mg/ml stock solution in H_2O used at 200–1,000 dilution (15′ at RT). Note that another Hoechst dye 33258 (and DAPI) can be used to label live cell nuclei, as they are membrane permeable (0.1–1 μg/ml).
15. Mount slides in a mounting medium containing an antifade reagent and seal with nail varnish.
16. Count 500–1,000 nuclei under a fluorescence microscope and determine the labeling index as ($BrdU^+$ cells/total nuclei)/100.
17. *Determine the cumulative labeling index.* The labeling index plotted against the BrdU pulse time on Cricket graph, using y = ax ±b, where y is the labeling index, a the slope, × the cumulative BrdU labeling time, and b the intercept on the y axis. The linear region of the curve (2–24 h) is used to deduce T_c.

18. The estimate of cell-cycle time (T_c) = 1/a and the rough S-phase time (T_s) = bT_c.
19. This protocol can be adapted to simultaneously label another cell surface protein or intracellular antigen by using a regular immunolabeling procedure, although the harsh conditions used to denature DNA may destroy the antigens to be labeled.
20. Invitrogen has recently developed the Click-iT EdU (5-ethynyl-2′-deoxyuridine) system, which relies on a copper-catalyzed covalent reaction between an azide and an alkyne. The azide's small size allows efficient detection of incorporated EdU under mild conditions, thus eliminating the harsh DNA denaturation and allowing simultaneous analysis of several molecules.

3.3.2. Label-Retaining Cells (LRCs) in Cells, Spheres, and Tumors

1. The BrdU pulse-labeling strategy can be adapted to identify slow-cycling cells. The principle is as follows. After a long period of BrdU "pulse," i.e., from 1 day to weeks, most or all cells with proliferative capacity will be labeled (i.e., will incorporate the thymidine analog BrdU into the DNA). Then the samples will be washed free of BrdU. Following an extended period of culture or maintenance in the absence of BrdU (i.e., the "chase"), samples will be terminated and used to stain for BrdU, as described earlier. Progenitors with a faster cycling kinetic (i.e., shorter cell-cycle time) will gradually dilute out BrdU and eventually become BrdU-negative whereas the putative SCs, which are generally quiescent, will stay BrdU-positive and be identified as the LRCs.
2. The LRCs in the bulge of mouse epidermis and the proximal tubules of mouse prostate have been shown to possess SC properties. The LRC strategy has been utilized in both cultures and tumor cell spheres *(30)* to characterize putative CSCs. We briefly describe methods in tracking the LRCs in the xenograft tumors.
3. Make a 20-mM BrdU solution by dissolving 100 mg BrdU into 16.2 ml dH_2O (*see* **Note 23**).
4. When a 5-mm tumor is palpated (approximately 1–1.5 months after cells are injected), mice are injected intraperitoneally (i.p) using tuberculin syringes and the following dosage calculations: Dose (volume of 20 mM BrdU solution) = weight × (0.06 mg BrdU/g)/(6.14 mg/ml) = weight × 0.0098 ml/g, where weight = mouse body weight in grams.
5. Inject mice four times over 48 h. The chase period is extended until the tumor reaches the harvest size of approximately 1 cm in diameter (chase varies from 20 to 68 days).

6. Harvest tumors.
7. Tumor pieces are fixed in formalin and embedded in either paraffin or OCT for immunohistochemistry of BrdU (and other antigens if necessary).

3.3.3. Clonal Aassays

Preparing the Swiss 3T3 Cells for Use as a Feeder Layer

1. Plate mouse Swiss 3T3 cells on a 6-well plate in DMEM 7% FBS.
2. Once plate is confluent, treat with 10 μg/ml mitomycin C (stock 0.5 mg/ml, dissolved in H_2O) in complete medium for 2 h at 37°C. Alternatively, bulk-treated 3T3 cells can be frozen and aliquots retrieved for use.
3. After treatment, wash plate with prewarmed PBS at least 3 times, aspirating after each wash.
4. Add desired prewarmed culture medium. Plate cells of interest.

Plate Tumor Cells for Clonal Assays

1. Plate 100–1,000 cells/well in a 6-well dish (actual numbers are determined from pilot experiments) in 2 ml media per well.
2. Primary PCa cells are plated on mitomycin C-treated Swiss 3T3 feeder cells.
3. Clones with >32–50 cells (depending on the type of cells) are scored within 1–2 weeks.
4. Count the number of clones. The percentage of cells that initiate a clone is presented as cloning efficiency.
5. Similar clonal analysis can be performed by simply plating any cultured or xenograft-derived tumor cells at clonal density in uncoated or collagen-coated dishes or wells. In this case, only tightly packed holoclones *(32)* should be scored.

3.3.4. Clonogenic and Sphere-Formation Assays

Clonogenic Assay in Soft Agar

1. *Base agar*. Melt 1% Agarose (w/v in sterile dH_2O) in a microwave, cool to 55°C in a waterbath. Warm RPMI + 20% FBS to 55°C in the water bath. Allow at least 30 min for temperature to equilibrate. Mix equal volumes of the two solutions to give 0.5% Agar + 1× RPMI + 10% FBS.
2. Pour 1 ml mixture per well of a 6-well plate and allow to set. The plates can be stored at 4°C for up to 1 week.
3. *Top Agar*. Melt 0.7% Agar in the microwave, cool to 40°C in a water bath (*see* **Note 24**). Also warm RPMI + 20% FBS to the same temperature.
4. Prepare 5,000 cells in 1 ml prewarmed RPMI + 20% FBS medium.
5. Mix gently 1 ml cell solution with 1 ml Agar and pour onto the base agar in a 6-well plate (*see* **Note 25**).
6. Incubate at 37°C in a humidified incubator for 10–14 days.

7. Stain plates with 0.005% Crystal Violet (0.5 ml per well) for >1 h and count colonies under a dissecting microscope.

Sphere-Formation Assays in UltraLow Attachment Plates

1. For LAPC4 xenografts, 1,000–10,000 dissociated cells are plated per well in a 6-well ultralow attachment plate in 2 ml IMDM + 15% FBS media.
2. For human primary prostate tumors, 10,000 dissociated cells are plated per well in a 24-well ultra low attachment plate in 500 μl serum-free medium (*see* **Note 26**).
3. Feed the spheres every 3–4 days after the first 5 days. Extreme care is needed in not disturbing or disrupting the developing spheres. It is recommended to gently add some fresh medium to the culture wells rather than changing medium. If medium must be changed, do it under a dissecting microscope and replenish only half of the medium.

Sphere-Formation Assays on Swiss 3T3 Fibroblast Feeders

1. 1,000–10,000 dissociated LAPC9 cells are plated per well over Swiss 3T3 fibroblast feeders that have been mitotically arrested with mitomycin C (prepared as described in **Subheading "Preparing the Swiss 3T3 Cells for Use as a Feeder Layer"**).
2. Cell are cultured without disturbance in IMDM plus 15% FBS for 4–5 days.
3. Thereafter, feed the spheres every 3–4 days with IMDM + 15% FBS.

Sphere-Formation Assays in Matrigel (See **Note 27**)

1. Suspend 1,000–10,000 dissociated or sorted prostate tumor cells in modified PrEBM media (or other appropriate media) 1:1 with Matrigel in a total volume of 100–200 μl (*see* **Note 28**).
2. Plate the resulting mixture around the rim of a well of a 12/24-well plate and allow to solidify at 37°C for 10 min before adding 0.5 ml/1 ml of culture media (*see* **Note 26**).
3. Replenish with media every 3–4 days.
4. Count colonies 10–14 days after plating.

Passaging Spheres Formed in UltraLow Attachment Plates

1. Collect the spheres by centrifugation at 130 × *g* for 5 min.
2. Carefully discard the supernatant. Alternatively, individual spheres can be directly picked up with a cut pipette tip. Trypsinize spheres for 5 min at 37°C using 0.25 mg/ml trypsin/EDTA solution. Neutralize it with TNS.
3. Spin at 130 × *g* for 5 min, collect pellet. Resuspend pellet in media.
4. Pass cells through a 27-G syringe 5–10 times.
5. Filter through a 40-μm mesh. Wash it with 1 ml of media. Check cells under the microscope. Repeat the filtration step if there are still cell clusters present.

6. Do a viability count (with trypan blue or erythrosine B).
7. Plate slightly less cells than for the primary sphere formation as the secondary spheres are generally more clonogenic.

Passaging Spheres Uncovered from the Matrigel (See **Note 27**)

1. Aspirate carefully the media from the wells.
2. Digest Matrigel by incubating in 500 μl of 1 mg/ml dispase solution in serum free media for 30 min at 37°C.
3. Spin at 130 × *g* for 5 min and carefully aspirate the supernatant.
4. Resuspend the pellet in 1 ml of 200 units/ml type I collagenase solution (prepared in serum free media) for 15 min at 37°C.
5. Spin at 200 × *g* for 5 min and carefully aspirate the supernatant.
6. Resuspend pellet in 500 μl of 0.05% Trypsin/EDTA and incubate at 37°C for 5 min. Add 500 μl of TNS to stop the reaction.
7. Cells are pelleted, resuspended in media, passed through a 27-G syringe and filtered through a 40-μm mesh as in **Subheading "Passaging Spheres Formed in Ultra-Low Attachment Plates," steps 3–6**.
8. Do a viability count and replate desired cell number after mixing 1:1 with Matrigel, as described for primary spheres.

3.4. In Vivo Assays for Prostate CSC Properties

As discussed earlier, the "gold" standard of measuring CSC properties is that the candidate CSC population should possess enhanced capacity, in an experimental system, to initiate tumors that recapitulate patient tumor histology and can be serially transplanted (**Table 1**; *16, 17*). Therefore, tumor experiments must be done to show this critical capacity associated with the presumed CSCs. Nevertheless, "reconstituting" human tumor development in mice, whether from primary tumor pieces or from dissociated cells, is extremely complicated involving numerous variables associated with both (donor) tumors and (recipient) mice (**Tables 2** and **3**). In this subheading, we shall present the basic protocols of injecting/implanting/transplanting human prostate cancer (HPCa) pieces or cells in NOD/SCID mice. In the next subheading, we will present some experimental data to illustrate some important lessons we have learnt in the past few years on assaying prostate CSC activities in immunodeficient mice.

3.4.1. Subcutaneous (s.c.) Implantation/Injection

1. This is the most commonly used route of implantation for tumor cells or pieces.

2. Cells can be cultured or purified from xenograft tumors or from patient tumors. These can be either unsorted cells, marker-sorted cells (on FACS or MACS), or the SP cells. A limiting dilution assay (LDA) can be planned by using cell numbers of 1 to millions. For low cell numbers (1–1,000), a microscopic examination should be routinely conducted to ensure the presence of cells.
3. Depending on the numbers, cells are resuspended in 20–30 μl culture medium and then mixed with different amounts of concentrated Matrigel (*see* **Note 29**).
4. Alternatively, tumor pieces minced into 1–2 mm^3 can be directly implanted. Larger tumor pieces or smaller pieces soaked in Matrigel can also be used. It is imperative that for every primary patient tumor, tumor pieces be implanted in the mice aiming to obtain the 1° xenografts, which can be potentially used later to obtain large numbers of putative CSCs for molecular and biochemical studies.
5. For patient tumor pieces or cells as well as androgen-sensitive cell lines (e.g., LNCaP) or xenografts (LAPC4/9) we routinely use the male NOD/SCID mice (6–8 week old) supplemented with the testosterone (T) pellet. For androgen-independent cell lines such as PC3 and Du145, we also use female NOD/SCID mice in order to save animals (*see* **Note 30**).
6. For cell suspensions, hold the mouse and inject the cell/Matrigel mixture into the flank(s) by using an insulin syringe (*see* **Note 31**).
7. For tumor pieces, anesthetize the mouse with Ketamine/xylazine solution (concentration: 17.16 mg/ml) with a dose of 65 mg/kg. The anesthesia may be supplemented with inhalant isoflurane mixed 50/50 with mineral oil when the mouse responds to squeezing of the hind foot or shows response during the procedure.
8. Shave the dorsal hair of the mouse and prep the skin with Betadine and then 70% ethanol. Make a small incision (0.5 cm) in the dorsal skin of the mouse, right under the head, open a pocket under the skin with the scissors and insert a T pellet. Staple the incision to close.
9. Make one or two small incision(s) in the flank(s) of the mouse, right above the leg(s); open a pocket under the skin with the scissors and insert a small piece of tissue. Close the incisions with staples.
10. Inject the mouse subcutaneously with 0.02 ml of Benamine followed by 1 ml of saline to aid recovery from the anesthesia. Put the mouse cage on a heating pad. Monitor the mouse until it starts to move.

11. After 7 days of surgery, remove the surgical clamps.
12. Tumor development is monitored starting from the second week. Tumorigenicity is measured mainly by tumor incidence (i.e., the number of tumors/number of injections), latency (i.e., time from injection to detection of palpable tumors), and tumor weight. All animals are terminated at 4–6 months after tumor cell injection. Tumors harvested are fixed in formalin and paraffin sections are made for H&E staining or immunohistochemistry.

3.4.2. Dorsal Prostate (DP) Injection

1. Prepare the cells suspension in 10 μl medium (*see* **Note 32**).
2. Mix cell suspension with 10 μl Matrigel in a 100-μl PCR tube to make 50% Matrigel (*see* **Note** 33). All tubes should be kept on ice.
3. Before injection, prewash Hamilton syringe needle with 100% acetone, 70% ethanol, and PBS. Rinse needles thoroughly when switching to different cell types. After injection, wash the needle in the reverse order.
4. Preparation of the surgical area: the surgical area must be aseptically sterilized, have proper lighting, and be equipped with the appropriate surgical equipment. This room is designated for all rodent surgeries. An aseptically cleaned dissecting microscope is used to perform the surgical procedure. Sterile paper towels are placed on the stage of the microscope and the animal is then placed on top of the paper towel, backside down with the surgical area (abdomen) facing upward. The microscope also has built-in lighting. The surgical room is set up with appropriate heating pads that are used during the surgical procedure and sterile blue pads are placed on top of the heating pads. A heating pad, with sterile blue pad on top of it, is also used to place the rodents on during the postoperative procedures and while they are monitored. A sterile blue pad is also placed on the surgery-operating table next to the microscope for placement of the surgical instruments and supplies that are necessary for the surgical procedure.
5. *Preparation for the surgeon.* The surgeon wears a facemask, hair bonnet, surgical gown and clean, sterile footcovers. Before beginning the surgery, the surgeon thoroughly washes hands and puts on a pair of sterile gloves. Once the surgeon is properly gowned and gloved, he/she places his/her sterile instruments on the sterile blue pad next to the microscope as well as sterile sutures and other necessary supplies.
6. Anesthetize NOD/SCID mice (6–8 weeks) by s.c. injection with Ketamine/xylazine solution (concentration: 17.16 mg/ml) with a dose of 65 mg/kg. Ophthalmic ointment is

then placed on the animal's eyes to prevent them from drying out during surgery. Put the mice in supine position.

7. Animals are fully anesthetized when they lose reflexes. A nose cone containing isoflurane is used only if needed during the surgical procedure. During preparation and surgery, the anesthetic depth is determined by the animal's respiratory pattern and by pinching the animal's foot for reflex response.
8. The animal's hair is clipped on the abdomen side starting from the bottom of the rib cage down to just above the penis sheath and from each side of the abdomen. A small patch of hair is also shaved on the back only if a T pellet is to be implanted. This depends on the cell type injected or implanted. The shaved area(s) are then prepped with betadine scrub, rubbing in a circular motion. 70% alcohol is then applied to the surgical area to remove the betadine and loose hair. This process is repeated two or more times. The betadine and alcohol are applied with sterile gauze. Once the animal is prepped for surgery, it is then laid on its back on the sterile paper towels that lay on top of the stage of the microscope and is now ready for the procedure.
9. With a pair of fine forceps, lift an area of skin 2 mm above the prepucial gland, about 1–2 cm above the penis sheath, and about 2–3 cm below the bottom of the rib cage; make a horizontal incision 1 cm in length through the skin layer using surgical scissors. Then make the same type of incision through the muscle layer. Once through the body cavity, the bladder should be visible. With a pair of fine forceps grip the bladder and lift upward then downward out of the body cavity toward the penis sheath. This will expose the two seminal vesicles. While holding onto the bladder with the left hand, gently hold the seminal vesicles, one by one, and pull them out of the body cavity and lay them facing down on the outer surface of the abdomen. With a pair of large forceps in the left hand, gently tilt back the seminal vesicles at the point of insertion, which is near the bladder neck, toward the penis sheath so that the DP lobes are clearly visible. While placing the seminal vesicle in this position, the Hamilton syringe needle, under the microscope, is inserted horizontally into the DP lobe from the right side and the cells are then injected directly into the prostate lobe. After the injection has been performed, pull the needle out of the DP lobe. Carefully lift the seminal vesicles with the fine forceps and insert them back into the body cavity, one at a time, and then the bladder. Once the organs are placed back into the body cavity, the muscle layer is sutured at least three times with 4/6–0 gut sutures with a reverse cutting, 3/8 circle C-13 needle. Once

sutured, pull together the skin with forceps and close the skin with surgical clamps and close the incision completely.

10. Upon completion of the surgery, the animal is placed on a warm heating pad, covered with a blue pad, to aid in regaining its body temperature. At this time, Flunixin meglumime, an analgesic, is administered at.04 cc subcutaneously in the right flank area. One cc of sterile saline is given subcutaneously directly behind the neck. The animal is placed into a clean cage, which is placed on top of a heating pad at the appropriate temperature, lined with a blue pad. Observe the animal until it has regained full sternal recumbancy and is walking around the cage. This usually takes ~30 min to 1 h.
11. After 7 days of surgery, remove the surgical clamps.
12. Tumor development is monitored starting from the second week. Tumorigenicity is measured mainly by tumor incidence (i.e., the number of tumors/number of injections), latency (i.e., time from injection to detection of palpable tumors), and tumor weight. All animals are terminated at 6–9 months after tumor cell injection. Tumors harvested are fixed in formalin and paraffin sections are made for H&E staining or immunohistochemistry.

3.4.3. The TR/KC Assays: Transplantation Under the KC Upon Tissue Recombination (TR) with the Rat Urogenital Sinus Mesenchyme (rUGM)

The powerful technique was pioneered by Dr. G. Cunha *(37)* and has been widely used to demonstrate whether normal mouse *(38)* or human *(39)* prostate epithelial cells have SC properties. The basic principle is rather straightforward: the candidate prostate epithelial cells are mixed either with pieces of rUGM or rUGM cells and then the two parts are "recombined" in a "setting" solution overnight. The tissue recombinants are then transplanted under the mouse KC. Prostate epithelial cells with regenerative (i.e., SC) capacities will form prostatic glands in a few months. The whole procedure of KC grafting can be found in an illustrated tutorial at http://mammary.nih.gov/tools/mousework/Cunha 001/index.html.

Preparing rUGM from E17/E18 Rat Embryos

1. Add 3 ml of Isoflurane to a paper napkin inside a dome chamber and allow the fumes to percolate the air inside chamber for a few minutes.
2. Use timed pregnant rats (generally 18-day pregnant). Handling rat with the tail, place it inside the chamber and immediately close it. It should take ~3–5 min for the rat to fall asleep. If doing more than one rat, add more isoflurane.
3. Place rat on her back and locate heart by palpating with fingers. Insert the syringe containing buthanesia into the heart at an incline. To determine if the needle is inside heart, draw up the needle slightly. If blood is observed, the needle is in

the heart. Continue by slowly injecting the buthanesia. The heart beat should stop almost instantaneously. Follow with cervical dislocation.

4. Place rat with ventral side facing up again and make a horizontal incision in the muscle wall of the abdomen. The muscle will spasm slightly when cut. Locate the beginning of the uterus and sever from the body using scissors until the uterus is completely removed. Place the uterus in a Petri dish on ice. In rare cases, the embryos will still be alive and if so, place them on ice for about 15–30 min.
5. Cut at the placenta to sever each amniotic sac from the uterus and place in a clean Petri dish. After this point, all further dissections need to be done under the microscope.
6. Place a few paper napkins on the microscope stand and adjust the external light source to focus on the embryos.
7. Place an embryo on the microscope such that the ventral side is facing up. Sever the neck region first and then proceed to make relief cuts at each leg. Make a horizontal cut in the abdomen slightly underneath the liver (visible under the skin because of its red color). Ensure that the muscle is completely cut underneath the skin or it will hamper the dissection. Locate the bladder. It usually has an obvious blood vessel running right by it making it easier to identify. Note that the testes have a similar shape but can be distinguished by the thinner, more convoluted blood vessels running on it.
8. Using microdissecting scissors, make a single snip on the right and left sides of the bladder, followed by a cut above the bladder. When making the cut on top, make sure that the scissors are angled such that they are facing slightly up. Make a scooping movement around the bladder while holding the blades of the scissors in a closed position and using forceps, gently pull the urogenital tract out of the rat pup. Place the tract in the section of the dish labeled "T."
9. Once all the tracts have been removed, throw away paper napkins, clean the platform of the microscope, and place a dry (ethanol wiped clean) beveled glass dish on the platform. Use a Pasteur pipette to move a single tract to the glass dish. Note that at this point, the internal light (at the bottom) of the microscope needs to be turned on. Also, pipette just a small amount of medium from the Petri dish. Too much volume would cause the tract the float and make it difficult to pin down.
10. Using the 27-G ½ in. needles, sever the ureters first, followed by the bladder and the urethra. You should see the sinus, which presents as a donut-shaped object. Clean out

extra tissue around the "donut" and using the needle (jab it in), move to the section of the Petri dish labeled "S."

11. When all sinuses have been isolated, proceed to clean out the epithelium. The epithelium is tricky to work with as it is the same color as the mesenchyme and cannot be distinguished from its beginning 2 h after sinus isolation. However, after about an hour in PBS, it should begin to pop out of the "donut hole" and appear as a flat, shiny sheet of large, square epithelial cells. Gently pull out this layer and remove from the dish. The remaining piece of tissue is the mesenchyme and should be placed in the section of the Petri dish labeled "M."
12. When all the mesenchymes have been isolated, proceed to digest into single cells. The tissue can also sit overnight in RPMI + 7% FBS at this point if digestion cannot be completed that day.
13. Digest in a solution of 187 units/ml Collagenase V (Sigma) in RPMI + 7% FBS at 37°C for 1.5–2 h. At the end, wash 3× with media to remove all traces of collagenase. This step is important or the collagenase traces will digest the collagen used to make drops for implanting into animals. Count cells and use in experiments or freeze down for later use.

Preparing PCa Cells/rUGM Recombinants

1. *Make Setting Solution.* Mix 10× EBSS 100 ml, NaHCO3 2.45 g, 1 M NaOH 7.5 ml, and sterile distilled water 42.5 ml. The resulting solution is filtered on Whatman No.1 filter paper and then sterilized by filtration with a 0.20-μm filter. The setting solution is stored in 5-ml aliquots at −20°C.
2. Count epithelial cells and recombine with rUGM cells at desired ratio and keep aside.
3. Add one-fifth volume of sterile 1 M NaOH to ice-cold setting solution and mix on ice.
4. Working on ice, aliquot 1 ml of rat tail collagen (see later) and add 200 μl of setting solution. Gently pipet to mix. The solution will turn yellow. Add ~50 μl more of setting solution dropwise and keep mixing until the color turns bright pink (*see* **Note 34**).
5. Spin the combination of epithelial cells and rUGM at 380 × *g* for 5 min and remove media completely.
6. Add 30 μl of collagen per injection to the cells and mix gently on ice (*see* **Note 35**).
7. Transfer the mixture dots to a 6-well plate; make 1–3 dots per well and incubate at 37°C for 15–30 min for the collagen to gel.

8. Add 2 ml medium slowly and incubate overnight at 37°C for the next-day implantation.

KC Transplantation

1. *Mouse irradiation*. When applicable, the mice are sublethally irradiated with 320 rads either the day before or on the same day as transplantation (*see* **Notes 36** and **37**).
2. The general procedure for mouse anesthetization and surgery are the same as for DP implantation.
3. After anesthetization, the mouse is placed on its side (we always inject the right kidney).
4. A vertical incision is made through the skin along the animal's spine, about 2 cm from the base of the tail to the top curve of the spine.
5. The surgeon's right thumb is placed on the animal's abdomen, applying some pressure and with the left hand the skin is pulled down, so that the kidney is close to the incision and in your view.
6. Once the kidney is found, a 0.5-cm incision is horizontally made on the animal's right side with surgical scissors. With the aid of the surgeon's thumbs and a little pressure on the abdomen, the kidney is exposed and popped out of the body cavity (*see* **Note 38**).
7. A small tear with a pair of sharp fine forceps is made in the kidney capsule. With the left hand, hold and pull up gently, so that the glass pipette can be inserted just beneath the kidney capsule and the top of the kidney. Be careful not to puncture the kidney. Move the tip of the glass pipette in a circular motion, making a small pocket for the recombinant to be placed.
8. Guide the recombinant into the pocket, under the kidney capsule, with the help of the glass pipette. Push the recombinant as far in as possible, so that it stays in place and does not leak out.
9. The kidney is now ready to go back into the body cavity. With a pair of fine forceps, hold onto the muscle layer surrounding the kidney and pull up all around the kidney, being careful not to damage the kidney. The kidney will fall back into place into the body cavity.
10. Once the kidney is in the body cavity, the muscle layer is sutured, using 4–0 gut suture with a reverse cutting, 3/8 circle C-13 needle.
11. Pull the skin layer upward and clamp the skin together using sterile surgical clamps, making sure that all the skin is closed together and no gaps are left.

12. Upon completion of the surgery, the animal is placed on a clean cage over a warm heating pad, covered with a blue pad, to aid in regaining its body temperature. At this time, Banamine, an analgesic, is administered at.04 cc subcutaneously in the right flank. One cc of sterile saline is given subcutaneously directly behind the neck. The animal is then transferred to a clean cage over a heating pad at the appropriate temperature. The animal is monitored until it has awakened from the procedure and is moving around the cage normally.
13. Remove staples. Monitor tumor formation starting from the second week after implantation. Animals are generally terminated in 2–4 months (implants with cultured and xenograft tumor-derived cells are usually terminated earlier).
14. Tumors harvested can be used for secondary tumor transplantation experiments as well as for various IHC and in vitro assays. One critical experiment is to show that any outgrowth has the human origin or at least contains human cells. This is usually accomplished by staining sections using antibodies specifically recognizing human proteins.

3.4.4. AP (Anterior Prostate) Implantation Technique

This is a relatively recently reported technique that has some similarities to the TR/KC assays ***(39, 40)***. In brief, cells are first "recombined" with collagen I prepared from the rat tail and then the tissue recombinants are surgically grafted into the mouse AP tubules.

Preparation of Rat Tail Type I Collagen

1. Tails from mature breeding rats (~250 g) are taken and soaked in 70% ethanol for 45 min.
2. Tails are immediately blotted and stored wrapped in aluminum foil at −20°C until needed.
3. To process, remove rat tails from the freezer and allow them to thaw to RT, soaking again in 70% ethanol for 20 min.
4. Using sterile instruments and aseptic technique, cut the tip of the tail crosswise (about 2 cm from the tip), then split the skin at the tail root and peel it away from the tail.
5. The distal and proximal quarters of the tail are cut away and the remainder is divided into three pieces.
6. Each tendon is separately dissected using a scalpel and the tendons are teased with the blade to separate the fibers.
7. The tendons are weighed (1 g of tendon gives 100 ml collagen solution, each tail yielding 1–1.5 g).
8. Transfer the tendons to a conical flask where they are washed repeatedly with sterile distilled water.
9. When the tendons appear clean, cut each tendon into small pieces and then transfer to 1:500 acetic acid solution.

10. Add 1% Penicillin/Streptomycin and 1% Fungizone. The mixture is left to stir gently at 4°C on a magnetic stirrer for at least a week.
11. The mixture is then decanted into a 50-ml conical tube and centrifuged at 3000 *g* for 15 min.
12. Transfer the supernatant, which is very viscous, and further centrifuge in a Beckman 50.2 Ti rotor at 35,000 × *g* for 1 h.
13. Collect the supernatant in sterilin universal bottles in 15-ml aliquots at 4°C until use.

AP Grafting (Implantation)

1. *Prepare the tissue recombination.* This procedure is the same as described for KC implantation but without adding rat UGM.
2. The general procedure for mouse anesthetization and surgery are the same as for DP implantation.
3. With a pair of fine forceps, lift an area of skin 2 mm above the prepucial gland bulge, about 1–2 cm above the penis sheath, and about 2–3 cm below the bottom of the rib cage; make a horizontal incision 1 cm in length through the skin using surgical scissors.
4. Then make the same type of incision through the muscle layer.
5. Once through the body cavity, the bladder should be visible. With a pair of fine forceps grip the bladder and lift upward then downward out of the body cavity toward the penis sheath.
6. This will expose the two seminal vesicles. While holding onto the bladder with your left hand, gently pull out of the body cavity the animal's left seminal vesicle so that the seminal vesicle is lying downward, facing the animal's penis sheath.
7. A 27-G syringe needle is inserted into the large tubule of the left anterior prostate lobe, from the tip of the prostate lobe to its point of insertion. Make a pocket-like tunnel through the tubule, where the recombinant drop will be placed.
8. The tissue recombinant is inserted into the pocket with a homemade glass pipette. Once the glass pipette tip has guided the recombinant into the AP, near the point of insertion, use forceps to clamp over the pipette as the pipette is pulled out of the prostate lobe. This helps the recombinant stay in place as the pipette is removed. A suture can be used if the recombinant does not stay in place.
9. Carefully lift the seminal vesicle with the fine forceps and insert it back into the body cavity, then the bladder.

10. Once the organs are placed back into the body cavity, the muscle layer is sutured at least three times with 4–0 gut sutures with a reverse cutting, 3/8 circle C-13 needle.
11. Once sutured, with forceps pull together the skin and clamp the skin closed with surgical staples, closing the incision completely.
12. The postoperative procedures are performed as **steps 10** and **11** of **Subheading 3.4.2**.
13. Tumor development is monitored as in **step 12**, **Subheading 3.4.2**.

3.5. Major Lessons Learnt on Reconstituting Human PCa Development in NOD/SCID Mice

3.5.1. Great Influence of Transplantation Sites/Approaches on Tumor Regeneration

1. *Clarification about tumor take vs. tumor growth.* These two terms are often used interchangeably, by mistake. Tumor take refers simply to "grafting," i.e., the initially implanted human tumor pieces (or cells) have stayed or grafted in the mouse but there is no apparent tumor growth. Tumor growth (or regeneration or development) refers to obviously enlarged tumor burden compared to the initial implants.
2. *Matrigel dose-dependently enhances tumor take rate and promotes tumor growth.* In the most commonly used route of implantation, i.e., s.c., we have noticed that with cultured PCa cells, increasing Matrigel concentrations from 25% to 50% dramatically improves tumor take rate as well as tumor growth (unpublished observations).
3. *The s.c. site is the most sensitive and consistent site in reconstituting tumor development of prostate CSC-enriched cells in NOD/SCID mice.* CSC-enriched PCa cell populations, including both SP *(29)* and CD44+ *(30)* cells, when implanted into the 4 different sites, i.e., s.c., DP, AP, and KC using the implantation approaches described earlier, significant differences were observed with respect to tumor development **(Fig. 3)**. With both SP and CD44+ cells purified from LAPC9-GFP cells, the s.c.-implanted cells demonstrated the highest tumor regeneration **(Fig. 3)**. Differences between the SP vs. CD44+ cell population were also observed in terms of tumor regeneration **(Fig. 3)**.
4. *The s.c. site is the most sensitive and consistent site in reconstituting tumor development of primary HPCa pieces or cells in NOD/SCID mice.* In fact, when tumor pieces or single cells from primary HPCa samples (6 Gleason 6 tumors, 8 Gleason 7 tumors, and 4 Gleason 8/9 tumors) were similarly implanted into the s.c., KC, or AP sites, the subcutaneum was also found to be the most sensitive site for tumor regeneration **(Table 4; Fig. 4)**.
5. *The prostate-implanted CSCs show more extensive metastasis.* Interestingly, in the experiments carried out with the SP and

A Tumor development of SP and $CD44^+$ LAPC9-GFP cells implanted at different sites

Cells[a]	Incidence[b]	Latency (d)[c]	Termin. (d)[d]	Weights (g)[e]	Metastasis[f]		
					lung	LN	pancreas
SP (1,000 cells)							
s.c	8/10 (80%)	41-55	67	1.5	−to +	−	−to ±
DP	5/9 (56%)*	58	69	0.8	± to +++	− to +++	−to ±
AP	2/9 (22%)*	N/A	89	0.01	−	−	−
KC	7/13 (54%)*	N/A	82	0.08	−to +	−to +	−to ±
$CD44^+$ (10,000 cells)							
s.c	8/8 (100%)	27	65	1.1	−to +	−to ±	−to ±
DP	2/6 (33%)*	71	85	0.7	N.D	N.D	N.D
AP	7/7 (100%)	N/A	97	2.7	+ to +++	−to +++	± to +++
KC	0/8 (0)*	N/A	117	N/A	−	−	−

[a]The SP and $CD44^+$ LAPC9-GFP cells were acutely purified from xenoggraft tumors and implanted (cell numbers in parentheses) at one of the four anatomical sites indicated. For s.c and DP, cells were implanted in 50% Matrigel. For AP and KC, cells were 'recombined' with rat collagen and rUGM, respectively, prior to transplantation either in the AP tubules or the KC. Male NOD/SCID mice (6-8 weeks old) supplemented with the T pellets were used.
[b]Refers to tumors developed/the number of implantations. The % incidence was indicated in parentheses. *$P < 0.05$ when compared to s.c groups.
[c]Time in days (d) from implantation to palpable tumors. N/A, not applicable (because it was difficult to palpate tumors developed in AP and KC).
[d]Time in days (d) when animals were terminated due to overwhelming tumor burden.
[e]Mean tumor weights in grams (g). Note that both DP and AP tumors might contain some residual host prostate. KC tumor weights were obtained by subtracting the intact kidney (i.e., the untransplanted) from the same host.
[f]Metastasis was determined by examining GFP+ cells in lymph nodes (LN), lung, and pancreas.

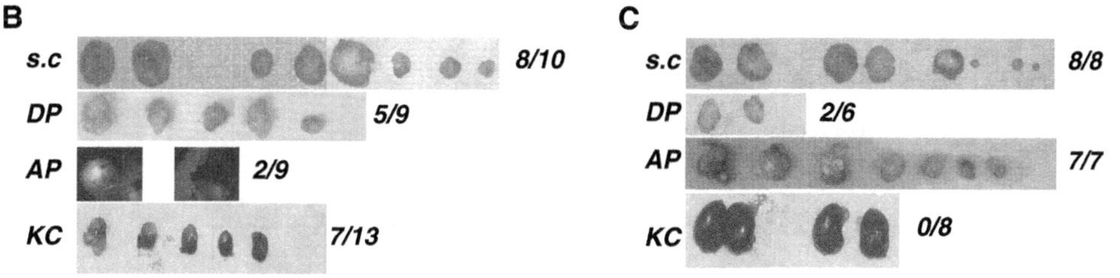

Fig. 3. Comparison of tumor development of prostate CSC populations (i.e., the SP and CD44+ cells acutely purified from the GFP-tagged LAPC9 tumors) implanted at different sites. (**A**) Table presentation of results. (**B**) and (**C**) Tumor images of SP (**B**) and $CD44^+$ (**C**) LAPC9 cells implanted at different sites. In **B**, the two AP tumors were very small and therefore the GFP images were shown. In **C**, none of the eight transplants in four kidneys developed into tumors.

CD44+ LAPC9-GFP cells, when we examined the GFP+ tumor cells that disseminated into the lymph nodes (LN), lung, and pancreas, we observed that cells injected into DP (for SP cells) or AP (for CD44+ cells) showed more extensive dissemination than the s.c.-implanted cells (**Fig. 3A**). These observations are consistent with the long-established concept that the orthotopically implanted tumor cells demonstrate higher metastatic capacity.

3.5.2. The rUGM/KC Assays are not Ideal for Reconstituting Prostate Tumor Development

1. *The rUGM/KC transplantation approach is, surprisingly, not sensitive in reconstituting either tumorigenicity or metastasis.* In the LAPC9 SP and CD44+ cell experiments, the KC-implanted CSC-enriched cells showed lower tumor regeneration and metastasis (**Fig. 3**). Similarly, in primary PCa experiments, the KC-implanted tumor pieces or cells showed lower tumor take and less regeneration (**Fig. 4; Table 4**).

Table 4
Reconstitution of primary human PCa in NOD/SCID mice

Patient	Samples[a] (implant. & harvest time)	Tumor incidence[b] s.c	KC	AP
HPCa2 (Gleason 6)	700k cells (163 d)	0/2	–	–
HPCa9 (Gleason 6)	Tumor piece (196 d)	–	–	0/1
	10k cells (165 d)	–	0/9	–
	100k cells (165 d)	–	0/1	–
HPCa10 (Gleason 6)	Tumor piece (153 d)	–	0/5	0/1
	10k cells (153 d)	–	0/4	–
	50k cells (153 d)	–	0/3	–
HPCa12 (Gleason 6)	Tumor piece (153 d)	0/2	0/6	–
HPCa13 (Gleason 6)	Tumor piece (153 d)	0/2	0/6	–
HPCa16 (Gleason 6)	Tumor piece (165 d)	–	0/7	0/7
	10k cells (141 d)	–	0/6	–
	100k cells (141 d)	–	0/7	–
HPCa4 (Gleason 7)	Tumor piece (121 d)	–	0/8	–
HPCa6 (Gleason 7)	Tumor piece (185 d)	–	0/3	0/4
HPCa7 (Gleason 7)	Tumor piece (180 d)	–	0/3	0/2
HPCa8 (Gleason 7)	Tumor piece (180 d)	–	0/4	0/1
HPCa11 (Gleason 7)	Tumor piece (161 d)	0/4	–	–
	50k cells (161 d)	–	0/1	–
	200k cells (161 d)	–	–	0/1
HPCa14 (Gleason 7)	Tumor piece (160 d)	4/4	0/5	0/4
	10k cells (171 d)	–	0/8	–
	100k cells (171 d)	–	0/8	–
HPCa17 (Gleason 7)	Tumor piece (183 d)	0/4	0/4	0/2
HPCa18 (Gleason 7)	Tumor piece (154 d)	2/4	0/2	0/3
	1 million cells (150 d)	0/4	–	–
HPCa15 (Gleason 8)	Tumor piece (130 d)	4/6	1/9	0/4
HPCa5 (Gleason 9)	Tumor piece (40 d)	–	1/2	–
HPCa19 (Gleason 9)	Tumor piece (152 d)	4/4	–	2/2
HPCa20 (Gleason 9)	Tumor piece (145 d)	2/2	1/6	1/2

[a]All samples (either ~2–3 mm^3 tumor pieces or purified HPCa cells) were implanted into male NOD/SCID mice supplemented with testosterone pellets. Harvest time are in days (d) since implantation.

[b]Tumor incidence refers to obvious tumor growth (not tumor take). For s.c., tumor pieces or tumor cells in Matrigel (50%) were implanted. For KC (kidney capsule) transplantation, tumor pieces or purified cells recombined with rUGM were transplanted. For AP, either tumor pieces or purified tumor cells recombined with collagen (*see* **Subheading 3**) were transplanted.

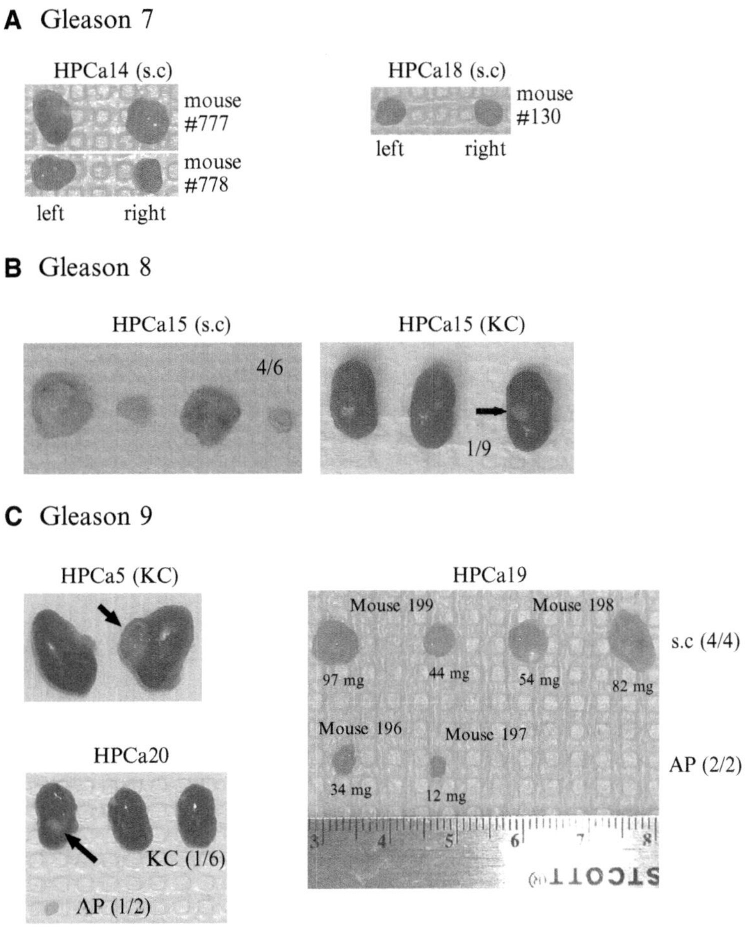

Fig. 4. The s.c. site is also the most sensitive microenvironment to reconstitute tumor development from pieces or cells of the primary PCa samples. Shown are representative images of tumors regenerated from Gleason grade 7 (**A**), 8 (**B**), and 9 (**C**) primary prostate tumor samples. Note obvious differences in the sizes of s.c. vs. AP or KC-regenerated tumors.

2. *In the TR/KC transplantation assays the ratio of rUGM/PCa cells significantly affects tumor development.* To further determine the utility of TR/KC assays in assaying CSC activities, we recombined 100 – 10,000 LAPC4 cells with rUGM pieces and carried out KC transplantation experiments. Increasing tumor development related to the input cell number was observed in 2 months (**Fig. 5A**). Histologically, 1,000 LAPC4 cells recombined with the rUGM formed the expected poorly differentiated carcinoma that stained positive for AR, PSA, and human-specific mitochondria (**Fig. 5B**). We then recombined 100 – 100,000 GFP-tagged Du145 cells, which are negative for AR and PSA,

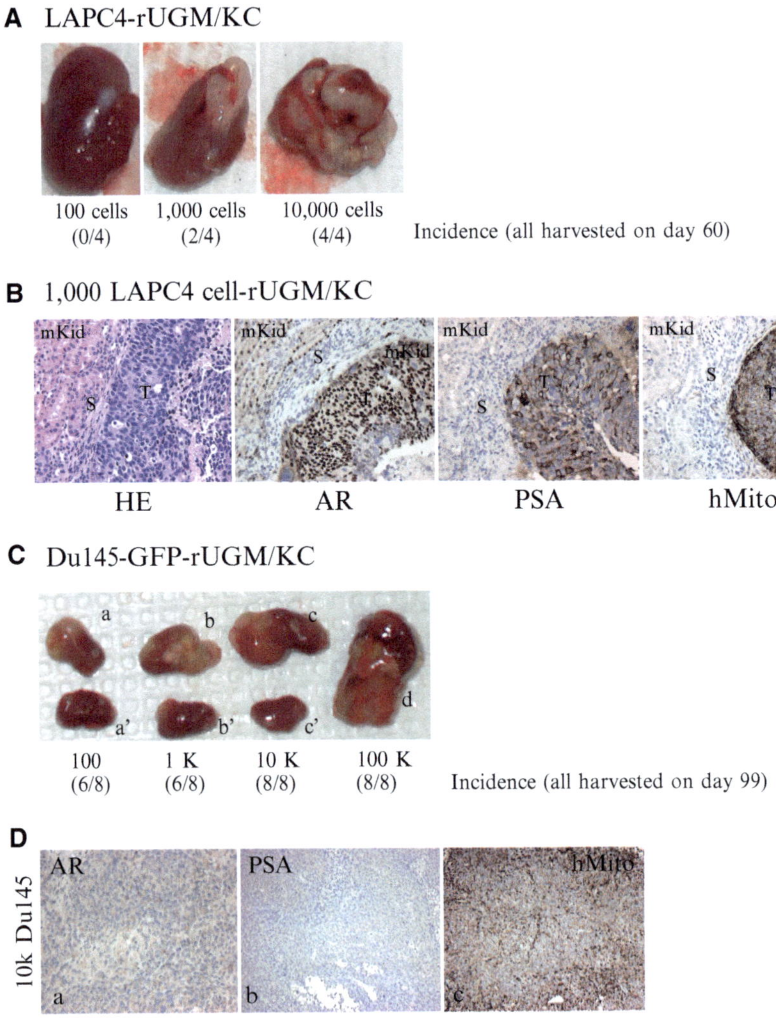

Fig. 5. The TR/KC transplantation assays using either LAPC4 (**A**, **B**) or Du145-GFP (**C**, **D**) cells. (**A**) Images of LAPC4-rUGM KC transplants. (**B**) Images of tissue section staining from 1,000 LAPC4 cell transplants (200×). AR, androgen receptor; PSA, prostate-specific antigen; mKid, mouse kidney; hMito, human mitochondria; S, stroma; T, tumor. (**C**) Gross images of KC transplants derived from 100 to 100,000 Du145-GFP cells. Each kidney initially received four implants (only 1 of the two 100,000 transplants was shown). **a–a′**, **b–b′**, and **c–c**′ refer to the two corresponding kidneys (from two different animals) that received the same numbers of Du145-GFP cells (i.e., 100–10,000). Note the heterogeneity of outgrowths. (**D**) Immunostaining for AR, PSA, and hMito (×100).

with rUGM and carried out KC transplantation experiments. We observed very sensitive reconstitution of tumorigenicity in that 6 of the 8 (75%) recombinants derived from 100 Du145 cells formed tumors within ~3 months (**Fig. 5C**). However, we observed significant variations in the sizes of outgrowths (**Fig. 5C**). Histologically, the outgrowths presented as poorly differentiated carcinomas that stained positive for human mitochondria and negative for PSA and AR (**Fig. 5D**).

To circumvent the variable nature of the rUGM pieces (**Fig. 5**), we isolated rUGM cells (away from the epithelium) and carried out two sets of "ratio" experiments, in which we recombined 1,000 or 10,000 of either Du145-GFP or LAPC4 cells with increasing numbers of rUGM cells. As shown in **Fig. 6** and **Table 5**, at a fixed PCa cell number, increasing rUGM cell number generally led to increased incidence and size of the outgrowths. This was especially prominent with LAPC4 cells–either 1,000 or 10,000 LAPC4 cells in the absence of rUGM formed no tumors (**Fig. 6**; **Table 5**). When 1,000 LAPC4 cells were recombined with 1,000 rUGM

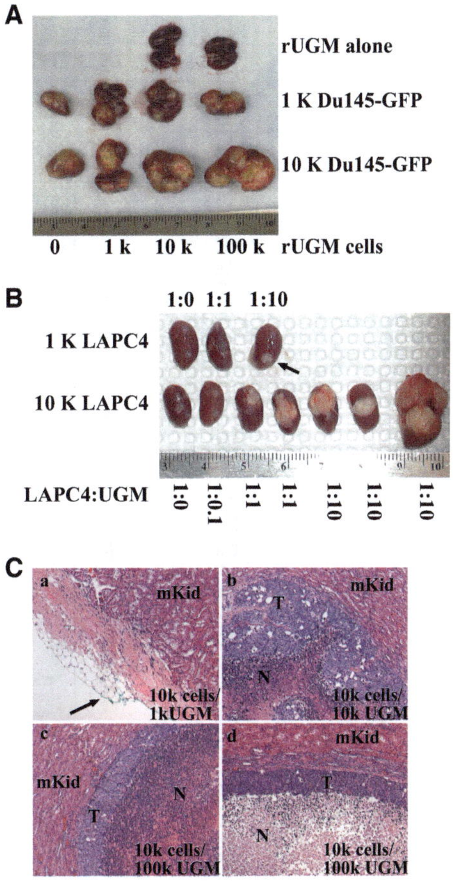

Fig. 6. PCa cell/rUGM ratio critically determines the outcome of tumor reconstitution. (**A**) Du145-GFP cells at 1,000 or 10,000 were recombined with 0–100,000 rUGM cells and the recombinants transplanted under the male NOD/SCID mouse kidney capsule. Three months later the recombinants were harvested. Note that the tumors were greenish due to GFP fluorescence. Some kidneys were bisected in order to better show the growths. (**B**) and (**C**) 1,000 LAPC4 cells were recombined with 0–10,000 rUGM cells or 10,000 LAPC4 cells with 0–100,000 rUGM cells. Three months later the recombinants were harvested (**B**) and tissue sections stained for H-E (C; ×200). In B, the arrow points to a small outgrowth. In C, two tumors (c and d) were shown to illustrate prominent necrosis in these large tumors. In Ca, the arrow points to the site of transplantation identified by the inking marker. *mKid* mouse kidney; *T* tumor; *N* necrotic areas.

Table 5
Tumor development and metastatic profiles of DU145-GFP or LAPC4 cells recombined with rUGM and transplanted under the KC

			Metastasis[b]								
Tumor cells	rUGM cells	Incidence[a]	Femur	Tibia	Lung	Kidney	Liver	Pancreas	Brain	Spleen	LN[c]
DU145-GFP *(all implanted 01/11/06 and harvested 04/12/06)*											
1,000 *cells (animal#)*											
860	0	2/3	–	–	±	–	–	–	–	–	–
841	1,000	3/4	–	–	+	–	+	–	–	–	–
304	10,000	3/4	–	–	–	–	–	–	–	–	–
350	100,000	4/4	–	–	+	–	–	+	–	–	–
10,000 cells											
859	0	2/3	–	–	+	–	–	–	–	–	–
857	1,000	3/4	–	–	+	–	–	–	–	–	–
858	10,000	3/4	–	–	+	–	–	+	–	–	–
305	100,000	4/4	–	–	++	+	±	±	–	+	+++[d]

LAPC4 *(all implanted 04/13/06 and harvested 07/18/06)*			
1,000 cells (animal#)			
776	0	0/3	
625	1,000	0/4	
624	10,000	2/4	
10,000 cells			
623	0	0/3}	N.D
622	1,000	0/3	
621	10,000	1/2	
620	10,000	2/2	
619	100,000	1/1	
618	100,000	1/1	
617	100,000	1/1	

[a]Incidence refers to the number of outgrowth/number of implants

[b]Metastasis to different organs was determined at necropsy by GFP imaging and was confirmed by tissue disaggregation combined with fluorescence microscopy. Signs of – to +++ refer to relative levels of metastasis (i.e., the abundance of GFP^+ cells in the dissociated cells from the whole organ) no GFP^+ cells; ± 1–5 GFP^+ cells; + dozens of GFP^+ cells; ++ hundreds of GFP^+ cells; +++ thousands or tens of thousands of GFP^+ cells. The metastasis for the implanted LAPC4 cells was not determined (N.D) because these cells were not labeled with GFP

[c]Lymph nodes examined including renal, sciatic, mesenteric, inguinal, and pyloric LN

[d]This animal had only renal LN metastasis

cells or when 10,000 LAPC4 cells were recombined with 1,000 rUGM cells, still no tumor formation was observed (**Fig. 6**; **Table 5**). However, when 1,000 LAPC4 cells were mixed with 10,000 rUGM cells or when 10,000 LAPC4 cells were recombined with 10,000 rUGM cells, tumor development was observed. Larger tumors formed when 10,000 LAPC4 cells were recombined with 10 times more (i.e., 100,000) rUGM cells (**Fig. 6B**) and the tumors generally showed increased necrosis (**Fig. 6C**). Transplantation of rUGM alone (at up to 100,000 cells or a whole rUGM) did not result in any outgrowth (n = 16; two kidneys with 4 initial transplants were shown in **Fig. 6a**, top). *These results emphasize the critical importance of the input tumor cell number as well as the ratio of tumor cells vs. rUGM stromal cells when carrying out TR/KC transplantation assays.*

3. *The rUGM/KC transplantation approach has a tendency to regenerate benign outgrowths that do not recapitulate the parental tumor histology.* We purified PCa cells from the Gleason 9 HPCa5 tumor (**Fig. 4C**). The patient tumor was made of mainly undifferentiated carcinoma cells (**Fig. 7A**) that were strongly positive for AR and PSA (not shown). We recombined 3,000 or 30,000 HPCa5 cells with 1 rUGM and transplanted the recombinants under the KC. Although the TRs of 3,000 HPCa 5 cells did not form any outgrowth (n = 6), the one 30,000 rUGM/HPCa5 cell recombinant formed transparent benign growths (**Fig. 7B**). Histologically, the tissue recombinants formed differentiated glandular structures (rather than the undifferentiated carcinoma) with atypia that displayed luminal and basal cell organization with intraluminal secretion and occasional PIN-like papillary protrusions (**Fig. 7C**). The prostatic glands were positive for basal cell markers K5 (**Fig. 7Da**) and p63 (not shown) and luminal marker AR (**Fig. 7Db**). These regenerated glands did not express PSA (**Fig. 7Dc**).
4. *It is often difficult to prove that the epithelial (tumor) cells in the rUGM/TR-derived outgrowths truly have a human origin, especially with the primary prostate tumor cells.* For example, the benign glands derived from the HPCa5 cells did not show obvious human-specific mitochondrial staining (**Fig. 7Dd**), although this marker stained strongly the undifferentiated carcinomas derived from metastatic LAPC4 (**Fig. 5B**) or Du145 (**Fig. 5D**) cells. In our experience, the human-specific mitochondria antibody works well only in highly proliferative cells. Other human-specific antibodies, e.g., the anti-Ku70 antibody, may work better for identifying human cells the rUGM outgrowths.

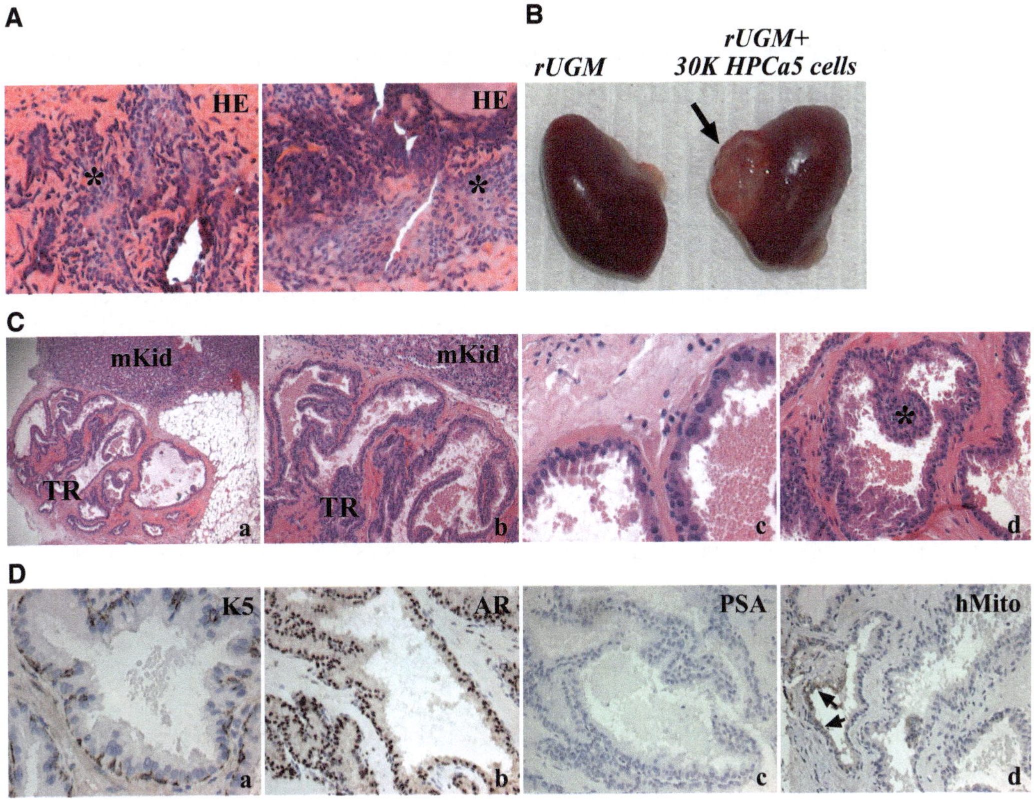

Fig. 7. The rUGM/KC-transplanted HPCa5 patient-derived outgrowth does not recapitulate the original patient tumor histology. (**A**) Representative histological features illustrating two areas of poorly differentiated adenocarcinoma (*asterisks*) from the original patient Gleason grade 9 tumor. (**B**) Images of KC transplants. The *arrow* indicates the benign outgrowth. Note that rUGM alone did not form any outgrowth ($n = 4$). (**C**) Representative HE images illustrating the formation of well-differentiated glandular structures (**a** and **b**) that show typical luminal and basal cell organization with intraluminal secretion (**c**) and occasional PIN-like papillary protrusions (**d**; *asterisk*). (**D**) Representative images stained for K5 (a), AR (b), PSA (c) hMito (d). In (**d**), the *arrows* point to weekly hMito-positive glands. *TR* tissue recombinants; m*Kid* mouse kidney; *hMito* human mitochondria.

3.5.3. In Contrast to other CSCs (Table 1), Unsorted or Marker-Sorted Primary PCa Cells, when Injected/Implanted "Naked," Fail to Initiate Tumor Development in NOD/SCID Mice

1. Since all CSCs so far reported (**Table 1**) and subpopulations of xenograft-derived PCa cells (SP/non-SP, ABCG2+/-. CD44+/–, and α2β1+/–) *(29–32)* were injected "naked" into the mice, we purified, from nine primary prostate tumors, the CD44+/– PCa cells (n = 5; 1 GS6, 3 GS7 and 1 GS9), CD133+/– cells (n = 3; 2 GS6 and 1 GS7), or CD44+ CD133+ and double-negative cells (1 GS6). The purities of marker-positive and marker-negative cells were 82–95% and >98%, respectively. Surprisingly, when 100–10,000 CD44+ and 1,000–500,000 CD44– cells, upon purification, were directly injected s.c. in 50% Matrigel, or first recombined with rUGM and then implanted under the KC, or first recombined with collagen and then implanted in the AP, we did not observe any tumor development (all animals terminated in 4–6 months) with either marker-positive or negative PCa cells out of a total of 242 injections/implantations (**Table 6**).

Table 6
Marker-sorted HPCa cells, either "naked" or recombined with rUGM, fail to initiate tumor development in NOD/SCID mice

HPCa sample	Marker/number	Injection/implantation/ time[a]	Outcome[b]
HPCa4 (GS7)	CD44+/5k (1x); 70k (1x)	TR/KC (091605-011306)	0/2
	CD44–/70k (2x)	TR/KC (091605-011306)	0/2
	CD44+/1k (1x), 10k (2x), 100k (1x)	AP (091605-021006)	0/4
	CD44–/10k (2x), 500k (2x)	AP (091605-021006)	0/4
HPCa6 (GS7)	CD44+/1k (10x), 10k (9x), 100k (5x)	TR/KC (010506-070606)	0/24
	CD44–/70k (3x), 100k (6x), 1 mil (4x), 2 mil (5x)	TR/KC (010506-070606)	0/18
HPCa8 (GS7)	CD44+/100, 1k, 10k, 50k	TR/KC (011306-070606)	0/12
	CD44–/100, 1k, 10k, 50k, 250k	TR/KC (011306-070606)	0/15
	CD44+/2k (2x), 20k (2x)	AP (011306-070606)	0/4
	CD44–/2k (2x), 20k (2x), 250k (2x)	AP (011306-070606)	0/6
HPCa9 (GS6)	CD44+CD133+/1k (2x)	TR/KC (012606-071106)	0/2
	CD44+CD133–/1k (7x), 10k (8x) 100k (2x)	TR/KC (012606-071106)	0/17
HPCa10 (GS6)	CD44+/100 (6x), 1k (6x), 10k (4x)	TR/KC (021006-021106)	0/16
	CD44+/100 (8x), 1k (8x), 10k (4x) 50k (4x)	TR/KC (012606-071106)	0/24
HPCa12 (GS7)	CD133+/100 (10x), 1k (10x), 10k (2x)	TR/KC (030106-080206)	0/22
	CD133–/1k (8x), 10k (8x), 100k (5x), 1 mil (3x)	TR/KC (030106-080206)	0/22
HPCa13 (GS6)	CD133+/5k (1x), 50k (1x)	TR/KC (041206-080206)	0/2
	CD133–/5k (1x), 50k (1x)	TR/KC (041206-080206)	0/2
HPCa16 (GS6)	CD133+/1k (4x), 10k (3x)	TR/KC (070106-121406)	0/7
	CD133–/1k (5x), 10k (5x)	TR/KC (070106-121406)	0/10
HPCa42 (GS9)	CD44+/100 (4x), 1k (6x), 10k (4x)	s.c. (080907-121607)	0/14
	CD44–/1k (8x), 10k (4x), 100k (1x)	s.c. (080907-121607)	0/13
HPCa24 (GS9) - 1° xenograft	CD44+/1k (2x), 5k (1x)	s.c. (052907-092407)	0/3
	CD44–/1k (2x), 5k (2x)	s.c. (052907-092407)	0/2
HPCa25 (GS8) - 1° xenograft	CD44+/5k (2x), 50k (2x)	s.c. (071107-121607)	0/4
	CD44–/5k (2x), 50k (2x), 500k (2x)	s.c. (052907-092407)	0/6

(continued)

Table 6 (continued)

HPCa sample	Marker/number	Injection/implantation/ time[a]	Outcome[b]
HPCa32 (GS8) - 1° xenograft	CD44+/1k (6x), 10k (3x)	s.c. (082907-010908)	0/3
	CD44−/1k (10x), 10k (5x), 100k (1x)	s.c. (082907-010908)	0/16
HPCa33 (GS8) - 1° xenograft	CD44+/1k (5x), 10k (4x), 50k (1x)	s.c. (081507–121607)	0/10
	CD44−/1k (4x), 10k (4x), 100k (3x)	s.c. (081507–121607)	0/8

[a]TR/KC, cells were recombined with rUGM and then transplanted under the kidney capsule (KC); AP, cells were recombined with collagen overnight and then implanted into the anterior prostate (AP); s.c., injected subcutaneously in 50% Matrigel; time refers to date of injection to date of termination (month/day/year). All injections/implantations were carried out in 6–8 week old male NOD/SCID mice supplemented with testosterone pellet.

[b]Number of tumor development/number of injections or implantations.

2. In fact, when PCa cells were purified from eleven HPCa samples (three GS6, four GS7, two GS8, and two GS9) and, without sorting, directly injected or implanted (1,000–1 million cells) in NOD/SCID mice, no tumor regeneration was observed in a total of 94 implants **(Table 7)**.
3. We then purified CD44+/− cells from four first-generation (1°) xenograft tumors derived from four HPCa samples (three GS8 and one GS9) and injected 1k–50k (for CD44+) or 1k–500k (for CD44−) cells s.c. into the NOD/SCID mice. Of a total of 52 injections, we did not observe any tumor development **(Table 6)**. In similar experiments using cultured or xenograft-derived tumor cells, we could observe tumor regeneration from as few as 100 cells *(29–32)*.
4. These results suggest that in contrast to the other reported CSCs, HPCa (stem) cells from primary tumors, when injected "naked" in Matrigel or implanted upon recombination with rUGM, fail to reinitiate tumor development in NOD/SCID mice. This is a critical lesson learnt and an important conclusion drawn from our 3 years of studies.
5. In retrospect, these "negative" data may not be particularly surprising. *First*, HPCa, compared to most other tumors **(Table 1)**, are much more indolent and less aggressive. *Second*, in many of the CSC studies reported **(Table 1)**, including breast cancer, which is perhaps most closely related to

Table 7
Unsorted HPCa cells, either "naked" or recombined with rUGM, fail to initiate tumor development in NOD/SCID mice

HPCa sample	Gleason	Cell number	Injection/implantation/time[a]	Outcome[b]
HPCa2	6	700k	s.c. (032505-090805)	0/2
HPCa3	7	10k	DP (070805-020806)	0/2
HPCa5	9	3k + 30k	TR/KC (101205-022006)	0 for 3k; 1 for 30k[c]
HPCa10	6	1k (8x), 10k (4x), 50k (3x)	TR/KC (0210-06-071106)	0/15
HPCa11	7	200k	AP (022406-080406)	0/1
		50k	TR/KC (022406-080406)	0/1
HPCa14	7	500k	AP (042006-101006)	0/2
		10k (10x), 100k (8x), 500k(4x)	TR/KC (042106-101006)	0/22
HPCa15	8	250k	AP (0526060-121306)	0/3
		250k (2x), 500k(4x)	TR/KC (0526060-121306)	0/4
HPCa16	6	200k	AP (070106-121406)	0/4
		1k (6x), 10k (6x), 100k (7x)	TR/KC (0 070106-121406)	0/19
		1 mil.	s.c. (070106-121406)	0/2
HPCa18	7	500k (2x), 1 mil (1x)	DP (090106-013007)	0/3
		1 mil.	s.c. (090106-013907)	0/4
HPCa21	9	500k	s.c. (1215060041707)	0/2
HPCa37	8	125k	TR/KC (061607-101107)	0/4

[a]s.c., injected subcutaneously in 50% Matrigel; DP, injected in 50% Matrigel in the dorsal prostate; TR/KC, cells were first recombined with rUGM and then transplanted under the kidney capsule; AP, cells were recombined with collagen and implanted in the anterior prostate tubules; time refers to date of injection to date of termination.

[b]Number of tumor development/number of injections or implantations.

[c]This outgrowth does not histologically resemble the original tumor (*see* **Fig. 7**).

PCa, cells from metastases or xenotransplants (rather than from primary patient tumors) were used. *Third*, NOD/SCID mice have some residual NK cell and other immune functions; therefore, in some studies the investigators "precondition" these mice with injections of etoposide *(4)* or anti-NK antibodies *(6)* or using "humanized" *(5)* or irradiated *(6)* mice.

6. These considerations suggest that the recipient mice likely need be preconditioned (e.g., irradiated) for human PCa cells

to engraft. Furthermore, considering the critical importance of stroma cells in PCa development and progression *(41, 42)*, the candidate prostate CSC populations (e.g., CD44+ or CD44+CD133+) may have to be coinjected with carcinoma-associated fibroblasts (CAFs) or human mesenchymal SCs (hMSCs), both of which have been shown to promote tumor development/progression *(43–45)*, into the irradiated mice. Finally, more aggressive prostate tumors, such as those from patients who have failed multiple treatments (hormonal, chemo-, and radiation therapies), should be considered as the source materials to enrich for tumorigenic prostate CSCs. We are currently exploring all these possibilities.

4. Notes

1. Trypsin, which is harmful to cells, especially the primary cells, must be neutralized with TNS or with serum-containing medium (which has antitrypsin activity).
2. The kit is a magnetic labeling system for the depletion of mature hematopoietic cells such as T cells, B cells, monocytes/macrophages, granulocytes, and erythrocytes, as well as some mouse stromal (fibroblasts, smooth muscle, and endothelial) cells. For this purpose, cells are labeled with a cocktail of biotinylated antibodies against a panel of "lineage" antigens (CD5, CD45R, CD11b, Gr-1, and Ter-119) and then incubated with antibiotin microbeads. Lineage marker-positive cells are removed by magnetic columns. Detailed protocols are provided in the manufacturer's instruction manual.
3. Staining buffer can be stored in 4°C for 4 weeks although freshly made solution is preferred.
4. To make a single-cell suspension, pipet the medium up and down with a 1,000-μl pipetman. Epithelial cancer cells, and especially primary cells, have a tendency to aggregate. Therefore, the cell suspension frequently needs to be filtered through a 40-μm nylon mesh.
5. Both Accumax and Histopaque (see later) should be stored at 4°C in the dark. Before the experiment, bring them to room temperature. When using a not-yet-opened bottle of Accumax, thaw the bottle at RT, not in a waterbath. They should be stored cold at all times after opening.
6. The digestion reaction should never be done above 37°C as this temperature inactivates the enzymes present in the Accumax solution.

7. To maximize cell yield, do not discard the chunks. Pass the whole solution through a 40-μm cell strainer with the gentle help of a sterile glass pestle or a plunger of a 5 ml syringe.
8. If a large number of cells is obtained through digestion (e.g., >50 million live cells), then only a portion of that number should be used for further processing as the flow cytometry time and sorting speed will put a limit on analyzing a maximum of ~20 million cells.
9. The volumes given are for up to 5 million cells. When working with higher numbers, scale up the reagent volumes and total volumes. Refer to the detailed protocol from Miltenyi.
10. Volumes given are for up to 10 million cells; scale up for higher numbers.
11. When working with human primary specimens, cell viability is very important. We are still studying the optimal dissociation conditions for our specimens; namely, we have been trying to decrease the digestion time to as short as 2 h instead of overnight while increasing the concentration of the collagenase solution. Other type of collagenases from Sigma possessing higher enzymatic activity can also be tried (e.g., collagenases IV, V). One can also prepare a digestion cocktail, including collagenase and hyaluronidase, which, in our hands, seems to digest the prostate tissues more effectively. For improved cell dissociation, refer to the instructions from Sigma.
12. The Histopaque gradient described for the purification of live cells in prostate xenograft tumors does not work very well with primary prostate samples. A discontinuous Percoll gradient protocol is used instead. Light and Heavy Percoll solutions are prepared on day of use in a sterile environment and stored on ice, as follows (a) *Light Percoll*: 6.8 ml Percoll, 125 μL FBS, 15.7 ml phenol red-colored water, 2.5 ml 10× solution (b) *Heavy Percoll*: 16.225 ml Percoll, 125 μL FBS, 6.275 ml water, 2.5 ml 10× solution. To prepare 500 ml of 10× solution, weigh (a) 38.86 g NaCl (b) 1.94 g KCl (c) 1.42 g $Na_2HPO_4.2H_2O$ (d) 0.78 g $NaH_2PO_4.2H_2O$ (e) 11.92 g HEPES (f) 5.00 g Glucose. Dissolve in 450 ml of distilled water, adjust pH to 7.4 with 5 M NaOH. Adjust volume to 500 ml. Filter sterilize, aliquot in 50-ml tubes and store in the dark at 4°C.
13. Volumes given are for up to 5 million cells; scale up for higher numbers.
14. Volumes given are for up to 5 million cells; scale up for higher numbers.

15. This is a viability dye that only penetrates cells with compromised cell membranes, much like erythrosin B or trypan blue.
16. The sorted cell number reported by the FACS machine is usually inaccurate and it is important to make an actual count as most of the experiments depend on a direct comparison between equal numbers of "+" and "–" cells.
17. During the sort, it is possible that many antigens get knocked off cells because of high-speed collisions making the percentage purity for the "+" population appearing much lower than it actually is. This can be overcome by plating the cells on coverslips for 2–4 h, thereby allowing them to recover antigen expression before restaining them to analyze for purity.
18. Take the example of CD44 staining. Biotin-conjugated antibody also can be used, followed by antibiotin microbeads. Volumes given are for 5 million cells; scale up for higher numbers.
19. Using a lower concentration of cells increases the percent enrichment. For rare populations, the negative fraction is usually quite pure but the positive fraction has to be passed through the column 2–3 times to enrich satisfactorily (>40%).
20. This protocol is modified from Goodell et al *(35)*. HL60-DOX and MCF-7 cell lines can be used as positive control *(29)*.
21. The Hoechst dye concentration is good for up to 10^6 cells/ml; scale it up or down according to cell numbers.
22. All human cell lines and tumor-derived cells should be incubated for 120 min. Incubation time is 90 min for murine cell lines and tumor-derived cells.
23. Use 20 mM for adult mice; dilute to 10 mM for young mice.
24. It is important not to exceed 40°C as higher temperatures will kill the cells.
25. Only do one plate at a time so that agar does not set prematurely.
26. Primary prostate tumor cells can produce spheres in modified PrEBM media, but they seem to lose viability very rapidly. The best sphere-producing culture conditions are currently under study. We have been obtaining better spheres with complete PrEBM conditioned by either CAFs or hMSCs and are now working on a defined medium.
27. This protocol is modified from Xin et al *(39)*.
28. Perform this step on ice.

29. All work involving Matrigel needs be done on ice. The final concentration of Matrigel ranges from 20 to 50% depending on the type of samples. Higher concentration of Matrigel enhances tumor take and growth.
30. We find that androgen-independent PCa cells (e.g., PC3) grow and maintain easily in female NOD/SCID mice.
31. If a testosterone pellet is to be implanted, mice need be anesthetized as described later.
32. The maximum volume is 15 μl if cells are to be injected in 50% Matrigel and 25 μl if cells are to be injected in 25% Matrigel.
33. A 100-μl PCR tube must be used; otherwise, you cannot draw the cells with the Hamilton syringe.
34. The collagen is not ready until the color turns bright pink for a while.
35. The volume for the recombination cannot be over 30 μl; otherwise, the collagen is hard to implant to the kidney capsule.
36. Depending on the purpose of the experiment, mice can be irradiated to promote tumor reconstitution (see earlier discussions).
37. This experimental procedure was described for the injection of human colon cancer cell suspensions under the renal capsule of NOD/SCID mice *(4)*.
38. If the incision is made too large, the kidney will not stay popped out. This incision is critical for the success of the procedure.

Acknowledgments

We thank Mr. Kent Claypool for assistance in FACS, Histology and Animal Facility Cores for technical assistance, and other members (past and present) of the Tang lab for discussion and support. This work was supported in part by grants from NIH (R01-AG023374, R01-ES015888, and R21-ES015893-01A1), American Cancer Society (RSG MGO-105961), Department of Defense (W81XWH-07-1-0616 and PC073751), Prostate Cancer Foundation, and Elsa Pardee Foundation (D.G.T), by DOD-PCRP grant W81XWH-04-1-0867, and by two Center grants (CCSG-5 P30 CA166672 and ES07784). H. Li was supported in part by a predoctoral fellowship from DOD (W81XWH-07-1-0132). L. Patrawala was supported in part by a predoctoral fellowship from Department of Defense (W81XWH-06-1-023). C. Jeter was supported in part by a postdoctoral fellowship from American Urological Association.

References

1. Lapidot T, Sirard C, Vormoor J, et al. (1994) A cell initiating human acute myeloid leukaemia after transplantation into SCID mice. *Nature* **367**, 645–648.
2. Singh SK, Hawkins C, Clarke ID, et al. (2004) Identification of human brain tumor initiating cells. *Nature* **432**, 396–401.
3. Schatton T, Murphy GF, Frank NY, et al. (2008) Identification of cells initiating human melanomas. *Nature* **451**, 345–349.
4. Al-Hajj M, Wicha MS, Benito-Hernandez A, Morrison SJ, Clarke MF. (2003) Prospective identification of tumorigenic breast cancer cells. *Proc Natl Acad Sci USA* **100**, 3983–3988.
5. Ginestier C, Hur MH, Charafe-Jauffret E, et al. (2007) ALDH1 is a marker of normal and malignant human mammary stem cells and a predictor of poor clinical outcome. *Cell Stem Cell* **1**, 555–567.
6. O'Brien CA, Pollett A, Gallinger S, Dick JE. (2007) A human colon cancer cell capable of initiating tumour growth in immunodeficient mice. *Nature* **445**, 106–110.
7. Ricci-Vitiani L, Lombardi DG, Pilozzi E, et al. (2007) Identification and expansion of human colon-cancer-initiating cells. *Nature* **445**, 111–115.
8. Todaro M, Alea MP, Di Stefano AB, et al. (2007) Colon cancer stem cells dictate tumor growth and resist cell death by production of interleukin-4. *Cell Stem Cell* **1**, 389–402.
9. Dalerba P, Dylla SJ, Park IK, et al. (2007) Phenotypic characterization of human colorectal cancer stem cells. *Proc Natl Acad Sci USA* **104**, 10158–10163.
10. Li C, Heidt DG, Dalerba P, et al. (2007) Identification of pancreatic cancer stem cells. *Cancer Res* **67**, 1030–1037.
11. Hermann PC, Huber SL, Herrler T, et al. (2007) Distinct populations of cancer stem cells determine tumor growth and metastatic activity in human pancreatic cancer. *Cell Stem Cell* **1**, 313–323.
12. Yang ZF, Ho DW, Ng MN, et al. (2008) Significance of CD90(+) cancer stem cells in human liver cancer. *Cancer Cell* **13**, 153–166.
13. Eramo A, Lotti F, Sette G, et al. (2008) Identification and expansion of the tumorigenic lung cancer stem cell population. *Cell Death Differ.* **15**, 504–514.
14. Prince ME, Sivanandan R, Kaczorowski A, et al. (2007) Identification of a subpopulation of cells with cancer stem cell properties in head and neck squamous cell carcinoma. *Proc Natl Acad Sci USA* **104**, 973–978.
15. Hope KJ, Jin L, Dick JE. (2004) Acute myeloid leukemia originates from a hierarchy of leukemic stem cell classes that differ in self-renewal capacity. *Nat Immunol* **5**, 738–743.
16. Clarke MF, Dick JE, Dirks PB, et al. (2006) Cancer stem cells – Perspectives on current status and future directions: AACR workshop on cancer stem cells. *Cancer Res* **66**, 9339–9344.
17. Tang DG, Patrawala L, Calhoun T, et al. (2007) Prostate cancer stem/progenitor cells: Identification, characterization, and implications. *Mol Carcinog* **46**, 1–14.
18. Hill, RP (2006) Identifying cancer stem cells in solid tumors: case not proven. *Cancer Res* **66**, 1891–1895.
19. Yu F, Yao H, Zhu P, et al. (2007) let-7 regulates self renewal and tumorigenicity of breast cancer cells. *Cell* **131**, 1109–1123.
20. Bao S, Wu Q, McLendon RE, et al. (2006) Glioma stem cells promote radioresistance by preferential activation of the DNA damage response. *Nature* **444**, 756–760.
21. Wang JCY. (2007) Evaluating therapeutic efficacy against cancer stem cells: New challenges posed by a new paradigm. *Cell Stem Cell* **1**, 497–501.
22. Isaacs, JT (1985) Control of cell proliferation and death in normal and neoplastic prostate: A stem cell model. In: Benign Prostate Hyperplasia. Rogers, CH, and Cunha, GR (Eds). Bethesda. Springer-Verlag, pp85–94.
23. Wang Y, Hayward SW, Cao M, Thayer KA, Cunha GR. (2001) Cell differentiation lineage in the prostate. *Differentiation* **68**, 270–279.
24. Signoretti S, Loda M. (2007) Prostate stem cells: from development to cancer. *Semin Cancer Biol.* **17**, 219–224.

25. Lawson, DA, Witte, ON (2007) Stem cells in prostate cancer initiation and progression. *J Clin Invest* **117**, 2044–2050.
26. Collins AT, Habib FK, Maitland NJ, Neal DE. (2001) Identification and isolation of human prostate epithelial stem cells based on α2β1-integrin expression. *J Cell Sci* **114**, 3865–3872.
27. Richardson GD, Robson CN, Lang SH, et al. (2004) CD133, a novel marker for human prostatic epithelial stem cells. *J Cell Sci* **117**, 3539–3545.
28. Collins AT, Berry PA, Hyde C, Stower MJ, Maitland NJ. (2005) Prospective identification of tumorigenic prostate cancer stem cells. *Cancer Res* **65**, 10946–10951.
29. Patrawala L, Calhoun T, Schneider-Broussard R, Zhou J, Claypool K, Tang DG. (2005) Side population is enriched in tumorigenic, stem-like cancer cells, whereas ABCG2+ and ABCG2- cancer cells are similarly tumorigenic. *Cancer Res* **65**, 6207–6219.
30. Patrawala L, Calhoun T, Schneider-Broussard R, et al. (2006) Highly purified CD44+ prostate cancer cells from xenograft human tumors are enriched in tumorigenic and metastatic progenitor cells. *Oncogene* **25**, 1696–1708.
31. Patrawala L, Calhoun-Davis T, Schneider-Broussard R, Tang DG. (2007) Hierarchical organization of prostate cancer cells in xenograft tumors: the CD44+ alpha2beta1+ cell population is enriched in tumor-initiating cells. *Cancer Res* **67**, 6796–805.
32. Li HW, Chen X, Calhoun-Davis T, Claypool K, Tang DG. (2008) PC3 Human prostate carcinoma cell holoclones contain self-renewing tumor-initiating cells. *Cancer Res* **68**, 1820–1825.
33. Miki J, Furusato B, Li H, et al. (2007) Identification of putative stem cell markers, CD133 and CXCR4, in hTERT-immortalized primary nonmalignant and malignant tumor-derived human prostate epithelial cell lines and in prostate cancer specimens. *Cancer Res* **67**, 3153–3161.
34. Gu G, Yuan J, Wills M, Kasper S. (2007) Prostate cancer cells with stem cell characteristics reconstitute the original human tumor in vivo. *Cancer Res* **67**, 4807–4815.
35. Goodell MA, McKinney-Freeman S, Camargo FD. (2005) Isolation and characterization of side population cells. *Methods Mol Biol* **290**, 343–352.
36. Bhatia B, Tang S, Yang P, et al. (2005) Cell-autonomous induction of functional tumor suppressor 15-lipoxygenase 2 (15-LOX2) contributes to replicative senescence of human prostate progenitor cells. *Oncogene* **24**, 3583–3595.
37. Cunha GR, Lung B. (1978) The possible influence of temporal factors in androgenic responsiveness of urogenital tissue recombinants from wild-type and androgen-insensitive (Tfm) mice. *J Exp Zool* **205**, 181–193.
38. Xin L, Lukacs RU, Lawson DA, Cheng D, Witte ON. (2007) Self-renewal and multilineage differentiation in vitro from murine prostate stem cells. *Stem Cells* **25**, 2760–2769.
39. Wang Y, Revelo MP, Sudilovsky D, et al. (2005) Development and characterization of efficient xenograft models for benign and malignant human prostate tissue. *Prostate* **64**, 149–159.
40. Wang Y, Xue H, Cutz JC, et al. (2005) An orthotopic metastatic prostate cancer model in SCID mice via grafting of a transplantable human prostate tumor line. *Lab Invest* **85**, 1392–1404.
41. Cunha GR, Hayward SW, Wang YZ, Ricke WA. (2003) Role of the stromal microenvironment in carcinogenesis of the prostate. *Int J Cancer* **107**, 1–10.
42. Cher ML, Towler DA, Rafii S, et al. (2006) Cancer interaction with the bone microenvironment: a workshop of the National Institutes of Health Tumor Microenvironment Study Section. *Am J Pathol* **168**, 1405–1412.
43. Olumi AF, Grossfeld GD, Hayward SW, et al. (1999) Carcinoma-associated fibroblasts direct tumor progression of initiated human prostatic epithelium. *Cancer Res* **59**, 5002–5011.
44. Studeny M, Marini FC, Dembinski JL, et al. (2004) Mesenchymal stem cells: potential precursors for tumor stroma and targeted-delivery vehicles for anticancer agents. *J Natl Cancer Inst* **96**, 1593–1603.
45. Karnoub AE, Dash AB, Vo AP, et al. (2007). Mesenchymal stem cells within tumour stroma promote breast cancer metastasis. *Nature* **449**, 557–563.

Chapter 8

Characterization of Nonmalignant and Malignant Prostatic Stem/Progenitor Cells by Hoechst Side Population Method

Murielle Mimeault and Surinder K. Batra

Summary

Recent technical progress in the field of cancer stem/progenitor cell research revealed that these malignant cells may provide critical roles for primary tumor growth, metastases at distant tissues and organs, treatment resistance, and disease relapse. The precise molecular oncogenic events that frequently occur in cancer stem/progenitor cells and their early progenies during the early and late stages of cancer progression as well as their contribution to the treatment resistance and disease recurrence remain poorly defined. This lack of information on the deregulated gene products that may be involved in the malignant transformation of tissue-resident adult stem/progenitor cells into highly tumorigenic and/or migrating cancer stem/progenitor cells emphasizes the urgent need to perform future investigations. Toward this direction, we describe in this book chapter the characterization of nonmalignant and malignant prostatic stem/progenitor cells from well-established cell lines by Hoechst side population method. This novel approach should help to establish novel in vitro and in vivo models of human cancer stem/progenitor cell mimicking more closely the genetic and phenotypic changes occurring during the different stages of prostate carcinogenesis and disease progression in clinical settings. Of therapeutic interest, the identification of new biomarkers and molecular targets specific to these prostatic cancer-initiating cells should also help to develop more effective diagnostic and prognostic tests and chemopreventive and therapeutic treatments for the patients diagnosed at early and late stages of disease progression.

Key words: Prostatic stem/progenitor cells, prostatic cancer stem/progenitor cells, Hoechst side population technique, fluorescence-activated cell sorting

1. Introduction

Technical advancements in the tissue-resident adult stem/progenitor cell biology have allowed researchers to identify certain specific physiological functions of these immature cells and their early progenies endowed with a self-renewal potential.

John S. Yu (ed.), *Cancer Stem Cells*, Methods in Molecular Biology, vol. 568
DOI 10.1007/978-1-59745-280-9_8, © Humana Press, a part of Springer Science+Business Media, LLC 2009

The adult stem/progenitor cells may contribute to the tissue homeostatic maintenance, and more particularly in regenerative process after intense injuries such as chronic inflammatory atrophies and fibrosis *(1–7)*. Furthermore, numerous lines of experimental evidence have revealed that the genetic abnormalities in adult stem/progenitor cells or their early progenies could lead to their malignant transformation into tumorigenic cancer stem/progenitor cells also designated as cancer- or tumor-initiating cells, which can be involved in the etiology and development of cancers *(1–19)*. The accumulation of genetic and/or epigenetic alterations in multipotent and poorly differentiated tumorigenic cancer stem/progenitor cells, and more particularly their acquisition of a migratory phenotype during epithelial-mesenchymal transition, concomitant with the changes in activated stroma, may also lead to their invasion from primary neoplasm and dissemination at distant sites *(5, 7, 15–17, 19)*. Particularly, the sustained activation of numerous tumorigenic signaling cascades such as hedgehog, epidermal growth factor receptor (EGFR), Wnt/β-catenin, and stromal cell-derived factor-1/CXC chemokine receptor-4 (CXCR4) in cancer stem/progenitor cells and/or their early progenies may contribute to their sustained growth, survival, invasion and/or metastatic spread during cancer progression, treatment resistance, and disease relapse *(1–6, 14–17, 19–21)*. Hence, the tumorigenic and migrating cancer stem/progenitor cells possessing aberrant proliferation and differentiation abilities may provide critical roles for the primary tumor formation and metastases at distant tissues/organs by giving rise to the bulk mass of further differentiated cancer cells.

Several techniques have been elaborated for the enrichment or isolation of human adult stem/progenitor stem and their malignant counterpart, cancer stem/progenitor cells with stem cell-like properties from malignant tissues and well-established cell lines. Among the methods frequently used, there are the Hoechst 33342 low-dye side population (SP) method, magnetic-activated cell sorting and fluorescence-activated cell sorting (FACS) using the antibodies directed against the specific stem cell-like surface marker(s) such as CD133, CD44, CD34, CD138, and/or CD20 *(9, 10, 15, 22–38)*. These techniques have notably allowed researchers to isolate a small subpopulation of human cancer stem/progenitor cells comprising about 0.1–3% of total cancer cell mass from leukemia and primary malignant tissues and metastatic neoplasms from patients with skin, brain, gastrointestinal tract, pancreas, liver, breast, prostate, and ovarian cancers as well as the established cell lines *(1, 2, 4, 5, 7–10, 14–18, 22–29, 31–33, 36–45)*. In general, the cancer stem/progenitor cells express several specific stem cell-like markers such as CD133, CD44, ATP-binding cassette (ABC) multidrug transporters, and/or CXCR4 but lack differentiation marker expression

(1, 2, 4, 5, 7–10, 14–18, 22–29, 31–33, 35–37, 39–44). These highly leukemic or tumorigenic cancer stem/progenitor cells were able to give rise in vitro and in vivo to further differentiated tumor cells expressing phenotypes of the original tumor *(8–10, 18, 22–29, 31, 32, 34, 40, 41, 44)*. In addition, it has also been shown that certain established human cancer cell lines may represent a heterogeneous population of cancer cells, and the presence of a small subpopulation of leukemic or tumorigenic cancer stem/progenitor cells expressing stem cell-like markers may be responsible for their capacity to generate a leukemia or form a tumor and metastasize in animal models in vivo with a high incidence *(15, 27, 28, 34, 38, 46, 47)*. Importantly, certain experimental lines of evidence have also indicated that the isolated cancer stem/progenitor cells, which express high levels of antiapoptotic factors, ABC multidrug efflux pumps, and enhanced DNA repair mechanisms could be more resistant than their differentiated progenies to radiation, hormonal, and/or chemotherapeutic treatments *(1, 2, 5, 11, 12, 14–17, 21, 27, 35, 36, 46–50)*. Hence, these immature cancer-initiating cells can then provide major functions in cancer progression to the aggressive and metastatic disease states, and resistance to current clinical treatments and disease relapse.

Although this is an important progress in cancer stem/progenitor cell research, additional studies are necessary in order to more precisely establish the specific deregulated gene products induced via the genetic and/or epigenetic alterations in cancer stem/progenitor cells versus their further differentiated progenies and normal adult stem cells. These differentially expressed gene products in cancer stem/progenitor cells may contribute to their malignant transformation during cancer etiology and progression to locally invasive and metastatic disease stages. In regard with this, we describe here the protocol for the isolation of the human nonmalignant and malignant prostatic stem/progenitor cells from well-established cell lines by the Hoechst SP method. The use of isolated SP and non-SP cell fractions should help to establish novel in vitro and in vivo human cell models relevant to prostate cancer progression as well as identify novel biomarkers and molecular therapeutic targets for improving the efficacy of current diagnostic and prognostic methods and therapies against the aggressive, recurrent, and lethal hormone-refractory prostate cancers.

2. Materials

1. The human PZ-HPV-7 and RWPE-1 nonmalignant prostate epithelial cells and RWPE-2, DU145, and PC3 prostate cancer cell lines were originally purchased from American Type Culture Collection.

2. PZ-HPV-7 and RWPE-1 cells were cultured in keratinocyte serum-free medium supplemented with L-glutamine, bovine pituitary extract, and EGF according to the manufacturer's instructions (GIBCO™, Invitrogen Corp., Carlsbad, CA) (*see* **Note 1**).
3. DU145 and PC3 cells were maintained routinely in RPMI 1640 culture medium containing 10% fetal bovine serum (FBS), 26 mM NaH_2CO_3 of pH 7.4, 1% L-glutamine and antibiotics (100 UI/mL penicillin-100 µg/mL streptomycin).
4. RPMI 1640 medium and all other culture materials were from Life Technologies (Invitrogen Corp., Carlsbad, CA).
5. Dulbecco's phosphate-buffered saline (D-PBS, 1X) solution without calcium and magnesium was purchased from HyClone Laboratories, Inc. (Logan, UT).
6. Trypsin-EDTA solution (1X) consisting of 0.05% Trypsin/ 0.53 mM EDTA in Hank's balanced salt solution (HBSS, 1X) with sodium bicarbonate and without calcium and magnesium was purchased from CELLGRO, Mediatech, Inc. (Herndon, VA).
7. Hoechst 33342, trihydrochloride, trihydrate *FluoroPure™ grade* was purchased from Sigma-Aldrich, Corp. (St. Louis, MO). This product is provided under solid form and must be stored at ≤25°C and protected from light before use.
8. Verapamil hydrochloride, ≥98%, was purchased from Sigma-Aldrich, Corp. (St. Louis, MO). This product is provided under solid form and must be stored in a desiccator and protected from light before use.
9. The analyses and sorting of the SP and non-SP cell fractions were done using a FACS Aria flow cytometer with a DIVA software (Becton Dickinson Biosciences, San Jose, CA).

3. Methods

Among the methods of cell isolation, the Hoechst SP technique is potentially useful for the enrichment of stem/progenitor cells or cancer stem/progenitor cells with the stem cell-like properties from tissue specimens of patients and well-established cell lines, and more particularly in the case where the tissues-specific stem cell markers are not precisely established *(15, 22, 33–39, 45, 46, 48, 49, 51–57)*. This method allows us to obtain a very small fraction of cells designated as "side population," which is generally enriched in primitive and undifferentiated or poorly differentiated progenitor cells endowed with a self-renewal capacity and multilineage differentiation potential and expressing the stemness genes, including ABC multidrug transporters. This technique of cell separation is based on intrinsic high ability of SP cell fraction to actively efflux the fluorescent DNA-binding dye, Hoechst 33342 due to their elevated expression levels of verapamil-sensitive ABC transporters, such as BCRP (breast cancer related protein)/ ABCG2 and/or MDR1/ABCB1/P-gp (multidrug resistance 1 gene encoding P-glycoprotein). In contrast, non-SP cell fraction obtained by this method generally does not express the high levels of stem cell-like markers, including ABC efflux pumps *(15, 18, 22, 33–36, 39, 46, 48, 49, 51–55, 57)*. Hence, the comparative analyses of the genes expression profiles and functional properties observed for SP cell subpopulation with stem cell-like properties versus non-SP cell fraction isolated from total cancer cell mass may help to identify novel potential biomarkers and therapeutic targets. In certain cases, it has notably been reported that the SP cell subpopulation expresses higher levels of antiapoptotic factors and was more resistant to chemotherapeutic treatments than non-SP cell fraction *(15, 22, 34–36, 38, 46, 49, 55)*. In regard with this, we describe here the protocol of Hoechst low-dye SP method for the isolation of the SP and non-SP cell fractions from well-established human prostatic cell lines. Among the tested cell lines, there are human nonmalignant PZ-HPV-7 and RWPE-1 prostatic epithelial cell lines, weakly tumorigenic, and invasive RWPE-2 cells derived from RWPE-1 cells by malignant transformation with K-Ras and tumorigenic and metastatic DU145 and PC3 prostate cancer cell lines established from metastases at brain and bone, respectively *(58–62)*. Of particular technical interest, we also show that the CD133 positive-PC3 stem/progenitor cells detected in total PC3 cell mass are principally enriched in the SP cell fraction detected by this Hoechst SP method.

3.1. Cell Preparation Protocol

1. All the cells were grown in flasks within a chamber with a humidified atmosphere in a 37°C incubator supplied with 5% CO_2.
2. When cells reach about 75% confluence, the culture medium is removed and discarded from the flask (*see* **Note 2**).
3. The cell layer is washed once with 10–15 mL of Ca^{2+}/Mg^{2+}-free D-PBS solution.
4. The cells are trypsinized by using 0.05% Trypsin/0.53 mM EDTA solution and the flask is placed in a 37°C incubator. The cells are observed under a phase-contrast microscope until cell layer is dispersed (usually within 5–10 min).
5. A 10–15 mL of D-PBS solution containing 2% FBS (or 0.1% Soybean Trypsin Inhibitor) is added to flask, which inactivates the trypsin and stops enzyme from working.
6. The cell suspension is aspirated by gently pipetting, transferred to a centrifuge tube, and the cells are spun down at 200–300×*g* for 10 min at room temperature.

7. The cell pellet is gently suspended in 10–15 mL of D-PBS solution and the cell suspension is centrifuged at 200–300 × *g* for 10 min at room temperature.
8. The cell pellet is gently suspended in 10–15 mL of D-PBS solution and the cells are counted. The cell suspension is centrifuged at 200–300 × W*g* for 10 min at room temperature.

3.2. Hoechst Side Population Method Protocol

1. The cell pellet is suspended at a cell concentration of 1×10^6 cells per mL in the culture medium containing 2% FBS and supplemented with a final concentration of 2.0 μg/mL fluorescent Hoechst 33342 dye in the absence or presence of an ABC multidrug-transporter inhibitor, verapamil, at a final concentration of 50 μM (*see* **Note 3**).
2. The cells are incubated in a 37°C water bath for 2 h. The cells are gently agitated every 15 min.
3. After this Hoechst staining period, a 10–15 mL of D-PBS solution is added to the cell suspension and cells are spun down at 200–300 rpm for 10 min at room temperature.
4. The cell pellet is gently suspended in 1 mL of D-PBS solution (*see* **Note 4**).
5. The SP and non-SP cell fractions in the cell suspension are immediately analyzed by FACS or maintained at 4°C until flow cytometry analysis.
6. For FACS analyses, the Hoechst dye is excited with a UV laser at 355 nm and its fluorescence emission measured with both 505 long-pass 670/40 filter (Hoechst Red) and 450/50 filter (Hoechst Blue) using a BD FACS Aria flow cytometer with a DIVA software. The cell debris is excluded from analyses while the viable cells are analyzed and sorted. The examples of the Hoechst dye efflux profiles showing the SP and non-SP cell fractions obtained for the nonmalignant and malignant prostatic cell lines by this method are presented in **Fig. 1**.
7. The SP and non-SP cell fractions are collected by FAC sorting in 1 mL of serum-free culture medium containing exogenous EGF (10 ng/mL) plus fibroblast growth factor at 8 ng/mL and maintained in this serum-free medium before their use.

4. Notes

1. EGF solution added to keratinocyte serum-free medium must not be filtered.
2. The number of the cells detected in the SP cell fraction from well-established cell lines may be significantly influenced by the experimental conditions, including cell density and growth

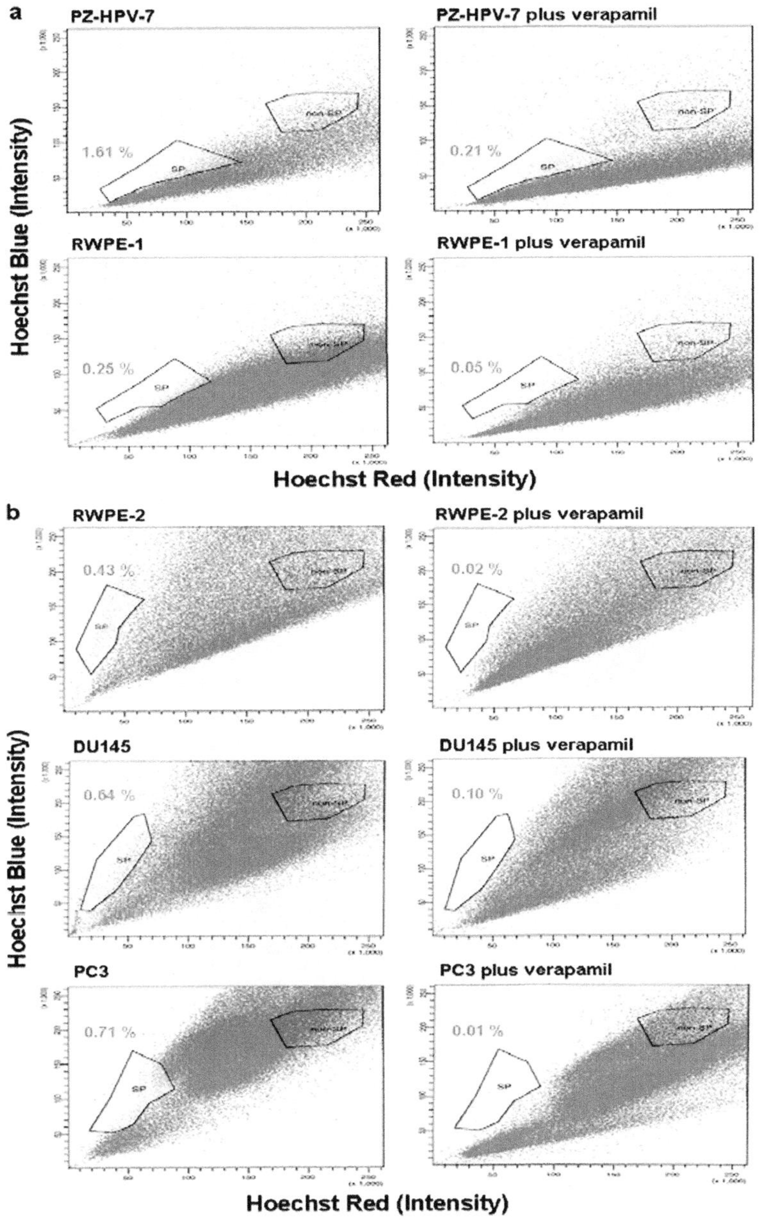

Fig. 1. Fluorescence-activated cell sorting (FACS) analyses of the side population (SP) and non-SP cell fractions detected in nonmalignant and malignant prostatic cell lines. Hoechst dye efflux profiles obtained for (**a**) nonmalignant PZ-HPV-7 and RWPE-1 prostatic cell lines and (**b**) malignant prostatic cell lines, including weakly tumorigenic and invasive RWPE-2 cells and tumorigenic and metastatic DU145 and PC3 cells. The SP cell fraction expressing the stem cell-like markers, including the ABC multidrug efflux transporters and non-SP cell fraction, is indicated. As shown, the cell number in the SP cell fraction was significantly reduced in the presence of ABC efflux pump inhibitor verapamil at 50 μM as compared to untreated cells.

factors used *(15, 35, 39, 54)*. Therefore, it is important to standardize the cell culture conditions used.

3. In the experiments made in the presence of ABC-transporter inhibitor verapamil, an aliquot of the cell suspension sample is preincubated at 37°C for 15 min prior to the addition of fluorescent Hoechst dye. The Hoechst dye is then added to

this cell suspension containing verapamil, and the cells are incubated at 37°C for 2 h with an intermittent agitation.

4. In order to further characterize the expression level of a specific cell surface marker in SP cell subpopulation versus non-SP cell fraction, an additional step may then be performed before the FACS analyses by an incubation of cells, previously stained with the Hoechst dye, with a fluorescent probe-labeled antibody. For instance, the cell pellet may be suspended in D-PBS solution and incubated in presence of phycoerythrin (PE)-conjugated human monoclonal anti-CD133 antibody (293C3, Miltenyi Biotec, Inc.) prior to the FACS analyses. An example of the results obtained is shown in **Fig. 2**. For FACS analyses, the excitation and emission of Hoechst dye is performed as previously described, while the PE is excited with a 488 laser and its fluorescence emission is measured with a 550 long-pass 575/26 filter.

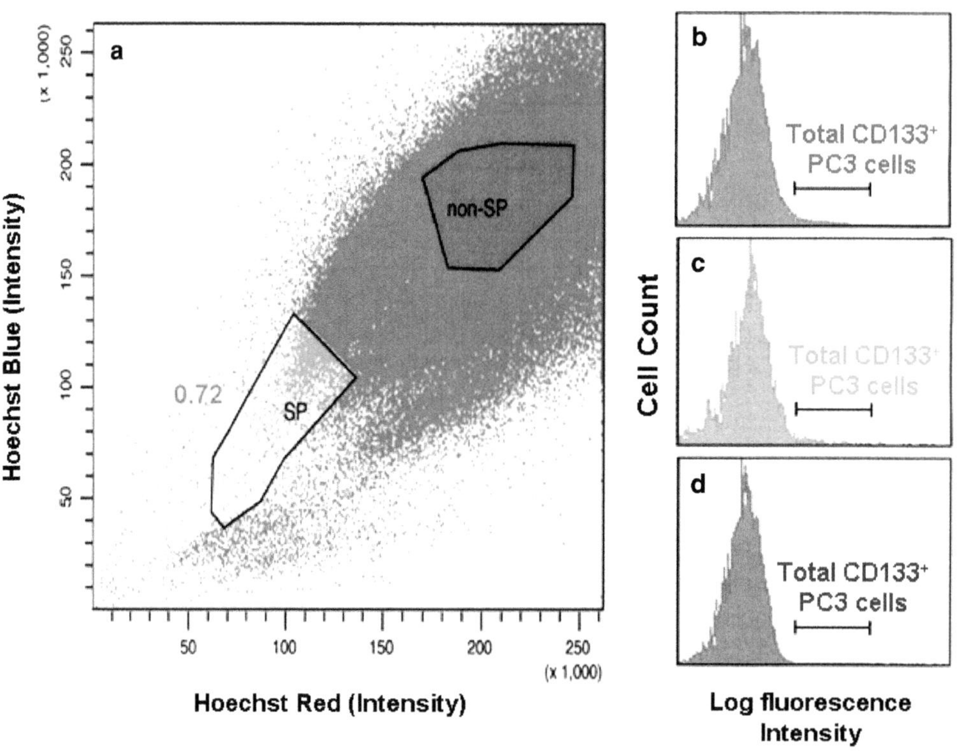

Fig. 2. Hoechst dye efflux profile observed for PC3 cells stained with fluorescent Hoechst dye in the absence or presence of phycoerythrin-labeled anti-CD133 (293C3) antibody (Miltenyi Biotec, Inc.). The Hoechst dye efflux profile showing (**a**) the SP cell fraction, which expresses the stem cell-like markers, including the CD133+, and non-SP cell fraction. The number of total CD133+ PC cells in the total PC3 cells analyzed (**b**) was principally localized in the SP cell fraction (**c**) relative to the non-SP cell subpopulation (**d**).

Acknowledgments

The authors of this manuscript are supported by grants from the U.S. Department of Defense (PC04502) and the National Institutes of Health (CA78590). We thank the Cell Analysis Facility at the University of Nebraska Medical Center for FACS analyses.

References

1. Mimeault, M., Hauke, R., and Batra, S. K. (2007) Stem cells – A revolution in therapeutics – Recent advances on the stem cell biology and their therapeutic applications in regenerative medicine and cancer therapies. *Clin. Pharmacol. Ther.* **82**, 252–264.
2. Mimeault, M., and Batra, S. K. (2008) Stem cell applications in disease research: Recent advances on stem cell and cancer stem cell biology and their therapeutic implications. *In Progress in Stem Cell Applications*, eds. Faraday, A. V., and Dyer, J. T., Hauppauge, NY: NOVA Science.
3. Mimeault, M., and Batra, S. K. (2006) Recent advances on multiple tumorigenic cascades involved in prostatic cancer progression and targeting therapies. *Carcinogenesis* **27**, 1–22.
4. Mimeault, M., and Batra, S. K. (2006) Recent advances on the significance of stem cells in tissue regeneration and cancer therapies. *Stem Cells* **24**, 2319–2345.
5. Mimeault, M., Mehta, P. P., Hauke, R., and Batra, S. K. (2008) Functions of normal and malignant prostatic stem/progenitor cells in tissue regeneration and cancer progression and novel targeting therapies against advanced prostate cancers. *Endocr. Rev.* **29**, 234–252.
6. Mimeault, M., and Batra, S. K. (2008) Recent progress on tissue-resident adult stem cell biology and their therapeutic implications. *Stem Cell Rev.* **4**, 27–49.
7. Mimeault, M., and Batra, S. K. (2008) Targeting of cancer stem/progenitor cells plus stem cell-based therapies: An ultimate hope for treating and curing the aggressive and recurrent cancers. *Panminerva Med.* **50**, 3–18.
8. Bapat, S. A., Mali, A. M., Koppikar, C. B., and Kurrey, N. K. (2005) Stem and progenitor-like cells contribute to the aggressive behavior of human epithelial ovarian cancer. *Cancer Res.* **65**, 3025–3029.
9. Yuan, X., Curtin, J., Xiong, Y., Liu, G., Waschsmann-Hogiu, S., Farkas, D. L., Black, K. L., and Yu, J. S. (2004) Isolation of cancer stem cells from adult glioblastoma multiforme. *Oncogene* **23**, 9392–9400.
10. Galli, R., Binda, E., Orfanelli, U., Cipelletti, B., Gritti, A., DeV, S., Fiocco, R., Foroni, C., Dimeco, F., and Vescovi, A. (2004) Isolation and characterization of tumorigenic, stem-like neural precursors from human glioblastoma. *Cancer Res.* **64**, 7011–7021.
11. Liu, G., Yuan, X., Zeng, Z., Tunici, P., Ng, H., AbdulkadirI, R., Lu, L., Irvin, D., Black, K. L., and Yu, J. S. (2006) Analysis of gene expression and chemoresistance of CD133 + cancer stem cells in glioblastoma. *Mol. Cancer* **5**, 67.
12. Bao, S., Wu, Q., McLendon, R. E., Hao, Y., Shi, Q., Hjelmeland, A. B., Dewhirst, M. W., Bigner, D. D., and Rich, J. N. (2006) Glioma stem cells promote radioresistance by preferential activation of the DNA damage response. *Nature* **444**, 756–760.
13. Haraguchi, N., Utsunomiya, T., Inoue, H., Tanaka, F., Mimori, K., Barnard, G. F., and Mori, M. (2006) Characterization of a side population of cancer cells from human gastrointestinal system. *Stem Cells* **24**, 506–513.
14. Mimeault, M., Hauke, R., and Batra, S. K. (2007) Recent advances on the molecular mechanisms involved in drug-resistance of cancer cells and novel targeting therapies. *Clin. Pharmacol. Ther.* **83,** 673–691.
15. Mimeault, M., Hauke, R., Mehta, P. P., and Batra, S. K. (2007) Recent advances on cancer stem/progenitor cell research: Therapeutic implications for overcoming resistance to the most aggressive cancers. *J. Mol. Cell. Med.* **11**, 981–1011.
16. Mimeault, M., and Batra, S. K. (2007) Interplay of distinct growth factors during epithelial-mesenchymal transition of cancer progenitor cells and molecular targeting as novel cancer therapies. *Ann. Oncol.* **18**, 1605–1619.
17. Mimeault, M., and Batra, S. K. (2007) Functions of tumorigenic and migrating cancer progenitor cells in cancer progression and metastasis and their therapeutic implications. *Cancer Metastasis Rev.* **26**, 203–214.
18. Hermann, P. C., Huber, S. L., Herrler, T., Aicher, A., Ellwart, J. W., Guba, M., Bruns, C. J., and Heeschen, C. (2007) Distinct populations of cancer stem cells determine tumor growth and metastatic activity in human pancreatic cancer. *Cell Stem cell.* **1**, 313–323.

19. Mimeault, M., and Batra, S. K. (2008) Critical roles of tumorigenic and migrating cancer stem/progenitor cells in cancer progression and their therapeutic implications. *In Stem Cells and Cancer*, eds. Bagley, R. G., and Teicher, B. A., Totowa, NJ: HUMANA.
20. Chen, B. Y., Liu, J. Y., Chang, H. H., Chang, C. P., Lo, W. Y., Kuo, W. H., Yang, C. R., and Lin, D. P. (2007) Hedgehog is involved in prostate basal cell hyperplasia formation and its progressing towards tumorige-nesis. *Biochem. Biophys. Res. Commun.* **357**, 1084–1089.
21. Dean, M., Fojo, T., and Bates, S. (2005) Tumour stem cells and drug resistance. *Nat. Rev. Cancer* **5**, 275–284.
22. Wulf, G. G., Wang, R. Y., Kuehnle, I., Weidner, D., Marini, F., Brenner, M. K., Andreeff, M., and Goodell, M. A. (2001) A leukemic stem cell with intrinsic drug efflux capacity in acute myeloid leukemia. *Blood* **98**, 1166–1173.
23. Hope, K. J., Jin, L., and Dick, J. E. (2004) Acute myeloid leukemia originates from a hierarchy of leukemic stem cell classes that differ in self-renewal capacity. *Nat. Immunol.* **5**, 738–743.
24. Hemmati, H. D., Nakano, I., Lazareff, J. A. Masterman-Smith, M., Geschwind, D. H., Bronner-Fraser, M., and Kornblum, H. I. (2003) Cancerous stem cells can arise from pediatric brain tumors. *Proc. Natl. Acad. Sci. U.S.A.* **100**, 15178–15183.
25. Singh, S. K., Hawkins, C., Clarkel, D., Squire, J. A., Bayani, J., Hide, T., Henkelman, R. M., Cusimano, M. D., and Dirks, P. B. (2004) Identification of human brain tumour initiating cells. Nature **432**, 396–401.
26. Al-Hajj, M., Wicha, M. S., Benito-Hernandez, A., Morrison, S. J., and Clarke, M. F. (2003) Prospective identification of tumorigenic breast cancer cells. *Proc. Natl. Acad. Sci. U.S.A.* **100** 3983–3988.
27. Ponti, D., Costa, A., Zaffaroni, N., Pratesi, G., Petrangolini, G., Coradini, D., Pilotti, S., Pierotti, M. A., and Daidone, M. G. (2005) Isolation and *in vitro* propagation of tumorigenic breast cancer cells with stem/progenitor cell properties. *Cancer Res.* **65**, 5506–5511.
28. Patrawala, L., Calhoun, T., Schneider-Broussard, R., Li, H., Bhatia, B., Tang, S., Reilly, J. G., Chandra, D., Zhou, J., Claypool, K., Coghlan, L., and Tang, D. G. (2006) Highly purified CD44+ prostate cancer cells from xenograft human tumors are enriched in tumorigenic and metastatic progenitor cells. *Oncogene* **25**, 1696–1708.
29. Patrawala, L., Calhoun-Davis, T., Schneider-Broussard, R., and Tang, D. G. (2007) Hierarchical organization of prostate cancer cells in xenograft tumors: The CD44+ {alpha}2{beta}1+ cell population is enriched in tumor-initiating cells. *Cancer Res.* **67**, 6796–6805.
30. Tang, D. G., Patrawala, L., Calhoun, T., Bhatia, B., Choy, G., Schneider-Broussard, R., and Jeter, C. (2007) Prostate cancer stem/progenitor cells: Identification, characterization, and implications. *Mol. Carcinog.* **46**, 1–14.
31. Li, C., Heidt, D. G., Dalerba, P., Burant, C. F., Zhang, L., Adsay, V., Wicha, M., Clarke, M. F., and Simeone, D. M. (2007) Identification of pancreatic cancer stem cells. *Cancer Res* 67, 1030–1037.
32. Ricci-Vitiani, L., Lombardi, D. G., Pilozzi, E., Biffoni, M., Todaro, M., Peschle, C., and De, M. R. (2007) Identification and expansion of human colon-cancer-initiating cells. *Nature* **445**, 111–115.
33. Hadnagy, A., Gaboury, L., Beaulieu, R., and Balicki, D. (2006) SP analysis may be used to identify cancer stem cell populations. *Exp. Cell Res.* **312**, 3701–3710.
34. Grichnik, J. M., Burch, J. A., Schulteis, R. D., Shan, S., Liu, J., Darrow, T. L., Vervaert, C. E., and Seigler, H. F. (2006) Melanoma, a tumor based on a mutant stem cell?. *J. Invest. Dermatol.* **126**, 142–153.
35. Hirschmann-Jax, C., Foster, A. E., Wulf, G. G., Nuchtern, J. G., Jax, T. W., Gobel, U., Goodell, M. A., and Brenner, M. K. (2004) A distinct "side population" of cells with high drug efflux capacity in human tumor cells. *Proc. Natl. Acad. Sci. U.S.A.* **101**, 14228–14233.
36. Ho, M. M., Ng, A. V., Lam, S., and Hung, J. Y. (2007) Side population in human lung cancer cell lines and tumors is enriched with stem-like cancer cells. *Cancer Res.* **67**, 4827–4833.
37. Brown, M. D., Gilmore, P. E., Hart, C. A., Samuel, J. D., Ramani V., A., George, N. J., and Clarke, N. W. (2007) Characterization of benign and malignant prostate epithelial Hoechst 33342 side populations. *Prostate* **67**, 1384–1396.
38. Friel, A. M., Sergent, P. A., Patnaude, C., Szotek, P. P., Oliva, E., Scadden, D. T., Seiden, M. V., Foster, R., and Rueda, B. R. (2008) Functional analyses of the cancer stem cell-like properties of human endometrial tumor initiating cells. *Cell Cycle* 7, 242–249.
39. Kondo, T., Setoguchi, T., and Taga, T. (2004) Persistence of a small subpopulation of cancer stem-like cells in the C6 glioma cell line. *Proc. Natl. Acad. Sci. U.S.A.* **101**, 781–786.
40. Prince, M. E., Sivanandan, R., Kaczorowski, A., Wolf, G. T., Kaplan, M. J., Dalerba, P., WeissmanI., L., Clarke, M. F., and Ailles, L. E. (2007) Identification of a subpopulation of cells with cancer stem cell properties in head

and neck squamous cell carcinoma. *Proc. Natl. Acad. Sci. U.S.A.* **104**, 973–978.

41. Fang, D., Nguyen, T. K., Leishear, K., Finko, R., Kulp, A. N., Hotz, S., Van Belle, P. A., Xu, X., Elder, D. E., and Herlyn, M. (2005) A tumorigenic subpopulation with stem cell properties in melanomas. *Cancer Res.* **65**, 9328–9337.
42. Eramo, A., Lotti, F., Sette, G., Pilozzi, E., Biffoni, M., DiV., A., Conticello, C., Ruco, L., Peschle, C., and De Maria, R. (2008) Identification and expansion of the tumorigenic lung cancer stem cell population. *Cell Death. Differ.* **15**, 504–514.
43. Yang, Z. F., Ho, D. W., Ng, M. N., Lau, C. K., Yu, W. C., Ngai, P., Chu, P. W., Lam, C. T., Poon, R. T., and Fan, S. T. (2008) Significance of CD90(+) cancer stem cells in human liver cancer. *Cancer Cell* **13**, 153–166.
44. Maitland, N. J., Bryce, S. D., Stower, M. J., and Collins, A. T. (2006) Prostate cancer stem cells: A target for new therapies. *Ernst Schering Found. Symp. Proc.* **5**, 155–179.
45. Komuro, H., Saihara, R., Shinya, M., Takita, J., Kaneko, S., Kaneko, M., and Hayashi, Y. (2007) Identification of side population cells (stem-like cell population) in pediatric solid tumor cell lines. *J. Pediatr. Surg.* **42**, 2040–2045.
46. Szotek, P. P., Pieretti-Vanmarcke, R., Masiakos, P. T., Dinulescu, D. M., Connolly, D., Foster, R., Dombkowski, D., Preffer, F., Maclaughlin, D. T., and Donahoe, P. K. (2006) Ovarian cancer side population defines cells with stem cell-like characteristics and Mullerian inhibiting substance responsiveness. *Proc. Natl. Acad. Sci. U.S.A.* **103**, 11154–11159.
47. Wei, C., Guomin, W., Yujun, L., and Ruizhe, Q. (2007) Cancer stem-like cells in human prostate carcinoma cells DU145: The seeds of the cell line?. *Cancer Biol. Ther.* **6**, 763–768.
48. Goodell, M. A., Brose, K., Paradis, G., Conner, A. S., and Mulligan, R. C. (1996) Isolation and functional properties of murine hematopoietic stem cells that are replicating *in vivo*. *J. Exp. Med.* **183**, 1797–1806.
49. Haraguchi, N., Inoue, H., Tanaka, F., Mimori, K., Utsunomiya, T., Sasaki, A., and Mori, M. (2006) Cancer stem cells in human gastrointestinal cancers. *Hum. Cell* **19**, 24–29.
50. de Jonge-Peeters, S. D., Kuipers, F., de Vries, E. G., and Vellenga, E. (2007) ABC transporter expression in hematopoietic stem cells and the role in AML drug resistance. *Crit. Rev. Oncol. Hematol.* **62**, 214–226.
51. Goodell, M. A., Rosenzweig, M., Kim, H., Marks, D. F., DeMaria, M., Paradis, G., Grupp, S. A., Sieff, C. A., Mulligan, R. C., and Johnson, R. P. (1997) Dye efflux studies suggest that hematopoietic stem cells expressing low or undetectable levels of CD34 antigen exist in multiple species. *Nat. Med.* **3**, 1337–1345.
52. Scharenberg, C. W., Harkey, M. A., and Torok-Storb, B. (2002) The ABCG2 transporter is an efficient Hoechst 33342 efflux pump and is preferentially expressed by immature human hematopoietic progenitors. *Blood* **99**, 507–512.
53. Seigel, G. M., Campbell, L. M., Narayan, M., and Gonzalez-Fernandez, F. (2005) Cancer stem cell characteristics in retinoblastoma. *Mol. Vis.* **11**, 729–737.
54. Chen, J. S., Pardo, F. S., Wang-Rodriguez, J., Chu, T. S., Lopez, J. P., Aguilera, J., Altuna, X., Weisman, R. A., and Ongkeko, W. M. (2006) EGFR regulates the side population in head and neck squamous cell carcinoma. *Laryngoscope* **116**: 401–406.
55. Chiba, T., Kita, K., Zheng, Y. W., Yokosuka, O., Saisho, H., Iwama, A., Nakauchi, H., and Taniguchi, H. (2006) Side population purified from hepatocellular carcinoma cells harbors cancer stem cell-like properties. *Hepatology* **44**, 240–251.
56. Zhou, J., Zhang, H., Gu, P., Bai, J., Margolick, J. B., and Zhang, Y. (2007) NF-kappaB pathway inhibitors preferentially inhibit breast cancer stem-like cells. *Breast Cancer Res. Treat.* **111**, 419–427.
57. Christgen, M., Ballmaier, M., Bruchhardt, H., von, W. R., Kreipe, H., and Lehmann, U. (2007) Identification of a distinct side population of cancer cells in the Cal-51 human breast carcinoma cell line. *Mol. Cell. Biochem.* **306**, 201–212.
58. Weijerman, P. C., Konig, J. J., Wong, S. T., Niesters, H. G., and Peehl, D. M. (1994) Lipofection-mediated immortalization of human prostatic epithelial cells of normal and malignant origin using human papillomavirus type 18 DNA. *Cancer Res.* **54**, 5579–5583.
59. Bello, D., Webber, M. M., Kleinman, H. K., Wartinger, D. D., and Rhim, J. S. (1997) Androgen responsive adult human prostatic epithelial cell lines immortalized by human papillomavirus 18. *Carcinogenesis* **18**, 1215–1223.
60. Webber, M. M., Bello, D., Kleinman, H. K., and Hoffman, M. P. (1997) Acinar differentiation by non-malignant immortalized human prostatic epithelial cells and its loss by malignant cells. *Carcinogenesis* **18**, 1225–1231.
61. Mickey, D. D., Stone, K. R., Wunderli, H., Mickey, G. H., Vollmer, R. T., and Paulson, D. F. (1977) Heterotransplantation of a human prostatic adenocarcinoma cell line in nude mice. *Cancer Res.* **37**, 4049–4058.
62. Kaighn, M. E., Narayan, K. S., Ohnuki, Y., Lechner, J. F., and Jones, L. W. (1079) Establishment and characterization of a human prostatic carcinoma cell line (PC-3). *Invest. Urol.* **17**, 16–23.

Chapter 9

Anchorage-Independent Growth of Prostate Cancer Stem Cells

Shunyou Wang

Summary

Prostate cancer is a major health concern in the Western world. Prostate cancer stem cells have been implicated to be involved in, if not solely responsible for prostate cancer initiation and relapse after surgical, hormonal, and chemotherapy. Until now, the identity of the presumed prostate cancer stem cell has been illusive and the efforts to characterize such cells have been hampered by the lack of efficient prostate stem cell activity assay. Anchorage-independent growth of brain and mammary stem cells *in vitro* has been proved to be a useful tool to enrich and characterize neural and mammary stem cells. Recently, we have successfully established a prostosphere cultivation method employing strategies similar to neurosphere and mammosphere cultures. Our results demonstrate that prostosphere-forming cells possess stem cell characters and are significantly over-represented in cancer tissues.

Key words: Prostate, Cancer, Stem Cell, Epithelial, Differentiation, Lineage, anchorage independent, Pten, Neurosphere, Mammosphere

1. Introduction

Stem cells are generally rare and present themselves in a specialized microenvironment where they remain quiescent but maintain the capacity to proliferate when necessary *(1, 2)*. This microenvironment, or stem cell niche, includes the extracellular matrix (ECM), stroma, and different cell types that interact with stem cells directly through cell-cell interaction or indirectly through cytokine and growth factor signaling. One challenge to studying stem cell/cancer stem cells (SC/CSC) in vitro is defining conditions to allow their propagation in an undifferentiated state such that enough material can be produced for further studies. To date, the stem cell niche is poorly understood thereby making

John S. Yu (ed.), *Cancer Stem Cells*, Methods in Molecular Biology, vol. 568
DOI 10.1007/978-1-59745-280-9_9, © Humana Press, a part of Springer Science + Business Media, LLC 2009

it difficult to recapitulate it in vitro. Alternatively, SCs can be propagated in an anchorage-independent culture that maintains their undifferentiated property in the case of neural and mammary cells. Brain and mammary SCs form spheroid structures called neurospheres and mammospheres, respectively, when cultured in suspension *(3, 4)*. Neurospheres and mammospheres can be induced in vitro to differentiate into their respective cell lineages present in the original tissues where derived from. When inoculated into cleared mammary fat pad, mammospheres can form well-organized structures that recapitulate the regeneration process of the mammary gland. To explore the possibility of growing prostate SC (PSC) in suspension, we have developed an anchorage-independent culture system and named these cells prostospheres to be analogous with the neurosphere and mammosphere.

Prostate SCs (PSCs) have been prospectively identified in mouse and human tissues. The PSCs can be enriched by surface marker CD133;$\alpha_2\beta_1$ integrin in human *(5)*, and Sca-1 in mouse *(6, 7)*. Collins et al. reported that human prostate cancer cells express CD133;CD44;$\alpha_2\beta_1$ integrin have CSC properties in vitro *(8)*. We have demonstrated that Sca-1$^+$ cells are significantly increased in Pten null mouse prostate cancer *(9, 10)*, and these cells exhibit increased prostosphere-forming activity when examined using the system described here (Wang et al., unpublished data). This in vitro stem cell system could be particularly helpful for efficient screening of SC surface markers and evaluating SC self-renewal pathways.

2. Materials

The following material and methods section will be an in-depth description with respect to mouse prostosphere culture. This method can be easily adapted to human tissue and will be discussed in detail in **Notes 5** and **6**.

2.1. Dissection and Dissociation of Prostate Cancer Tissue

2.1.1. Instruments & Equipment

1. Dissecting microscope (Olympus VMZ, Center Valley, PA)
2. Inverted microscope (Olympus IX70, Center Valley, PA)
3. Scissors and forceps (Fine Science Instrument)
4. Razor blades (Feather Safety Razor Co., Ltd; Hatfield, PA; Cat. 72042-10)
5. Fume hood (Forma Scientific, Waltham, MA. Class II A/B3 Biological Safety Cabinet)
6. Centrifuge (Beckman, Fullerton, CA. CS-6KR Centrifuge)

7. Cell strainer (BD Falcon, Bedford, MA. Cat. 352360 for 100μm; Cat. 352340 for 40 μm;)
8. Hemacytometer (Hausser Scientific, Horsham PA. Cat. 1492)
9. Culture dish (Corning Inc., Corning, NY. Cat. 430167)
10. Incubator (Barnstead International, Dubuque, Iowa. Lab Line 203Q)
11. Rotator (Barnstead International, Dubuque, Iowa. Thermolyne LABQUAKE® TUBE SHAKER/ROTATORS. Cat. C415220)
12. Conical tubes (BD Falcon, Bedford, MA. Cat. 35–2097 for 15 ml; Cat. 352095 for 50 ml; Bedford, MA)
13. CO_2 chamber (Forma Scientific, Waltham, MA. Class II A/B3 Biological Safety Cabinet)

2.1.2. Reagents

1. 70% Alcohol (Lab Chem Inc., Pittsburgh, PA. Cat. LC22210-4)
2. Dulbecco's phosphate-buffered saline (PBS, Invitrogen-Gibco, Carlsbad, California. Cat. 14190-144)
3. Dulbecco's modified Eagle medium [DMEM, Invitrogen-Gibco, Carlsbad, California. Cat. 11965 with High Glucose (4.5 g/L), L-glutamine, without sodium pyruvate]
4. Collagenase (Invitrogen-Gibco, Carlsbad, California. Cat. 17018-029).
5. 0.4% Trypan blue (Sigma, St. Louis, MO. Cat. T8154)

2.2. Cell Culture

1. Dulbecco's modified Eagle Mmedium [DMEM, Invitrogen-Gibco, Carlsbad, California. Cat. 11965 with High Glucose (4.5 g/L), L-glutamine, without sodium pyruvate)
2. Penicillin-Streptomycin-Glutamine (100×) (Invitrogen-Gibco. Cat. 10378-016)
   (a) hEGF (Invitrogen, Cat. 13247-05)
   (b) bFGF (RD Systems, Cat. 133-FB)
   (c) Hydrocortisone (Sigma, Cat. H2270)
   (d) B27 (Invitrogen-Gibco, Cat. 17504-044)
   (e) Heparin (Sigma, Cat. H4784)
   (f) Ultralow Attachment Culture Dish (Corning Inc. Cat. 3471)
   (g) Prostosphere medium (*see* **Table 1**)

2.3. Sphere Count and Size Measurement

1. MCID Software (MicroComputer Imaging Device)
2. Inverted Microscope with CCD camera attached
3. Computer with video capture interface

Table 1
Composition of prostosphere culture medium

Component	Stock	Dilute In	Dilution	Amount for 100 ml	Final Concentration
B27	50×		50	2 ml	1×
EGF	10μg/ml	10 mM Tris; 0.1%BSA	500	0.2 ml	20 ng/ml
bFGF	20μg/ml	10μmM Tris; 0.1%BSA	1,000	0.1 ml	20 ng/ml
Heparin	2 mg/ml		500	0.2 ml	4 μg/ml
Glutamine	200 mM		100	1 ml	2 mM
Penicillin	10,000U/ml		100	1 ml	100U
Streptomycin	10 mg/ml		100	1 ml	100 μg/ml
Insulin	10 mg/ml		2,000	0.05 ml	5μg/ml
Hydrocortisone	5 mg/ml	DMEM	10,000		0.5μg/ml
DMEM	1			94.44 ml	

3. Methods

All animal procedures follow the University of California Animal Research Committee guidelines. Avoid tissue drying out by applying a few drops of PBS or medium on the tissue at anytime during the dissection and subsequent processing. Always use anatomically identical tissue portions when comparing wild-type and mutant mouse prostates. To avoid contamination, all dissecting instruments should be autoclaved and sterile disposable culture plates, pipettes, and tubes be used. Wear gloves at all the time during the process.

3.1. Dissection and Dissociation of Prostate Tissue

1. Kill 4-12-week old mice by CO_2 inhalation. Put mouse in CO_2 chamber for about 5 min or until the mouse stops breathing. Place the mouse on its back on paper towels prewetted with 70% alcohol in dissection hood. Spray the mouse with 70% alcohol.
2. Pull up the abdomen wall with a pair of sterile forceps and open the abdomen cavity by making an incision laterally between the anus and sternum with a pair of scissors. Then cut at the lower abdomen cranially to expose the entire genitourinary (GU) system.

3. The lower GU system includes the bladder, urethra, a pair of seminal vesicles, ureters, ductus of deference, respectively, and three pairs of prostate lobes: ventral (VP), dorsolateral (DLP), and anterial [AP or coagulate gland (CG)]. *See* **Fig. 1** for details.
4. To dissect out the GU system, first cut the urethra at the end closest to the penis to avoid loss of prostate tissue. This end of the urethra can then be used to hold with the tip of a pair of forceps. Pull the whole GU system up gently with the forceps holding the urethra. Cut any tissues connected with the GU system, including the ductus deference, ureters, connective tissue, vessels, and fat tissue.
5. Quickly transfer the GU system to a Petri dish and cover with DMEM.
6. Dissect three pairs of prostate lobes (VP, DLP, and AP) out using two new pairs of sterile fine tip forceps as illustrated in **Fig. 1** under dissecting microscope. (*see* **Note 1**)
7. Transfer all dissected prostate lobes into another dish and add 3 drops of DMEM to keep them moist. Mince the tissues into 1-mm pieces with a sterile blade. The volume of DMEM medium added can be adjusted according to the amount of prostate tissue dissected. If too much medium was added, the tissues will be floating around that will make tissue mincing difficult.
8. Mix all pieces of tissues with 9 ml of DMEM and transfer the mixture to a 15-ml conical tube. Add 1 ml collagenase (8 mg/ml) and mix by inverting the tube several times; then place the tube on a rotator and incubate at 37°C for 2hrs. (*see* **Note 2**)

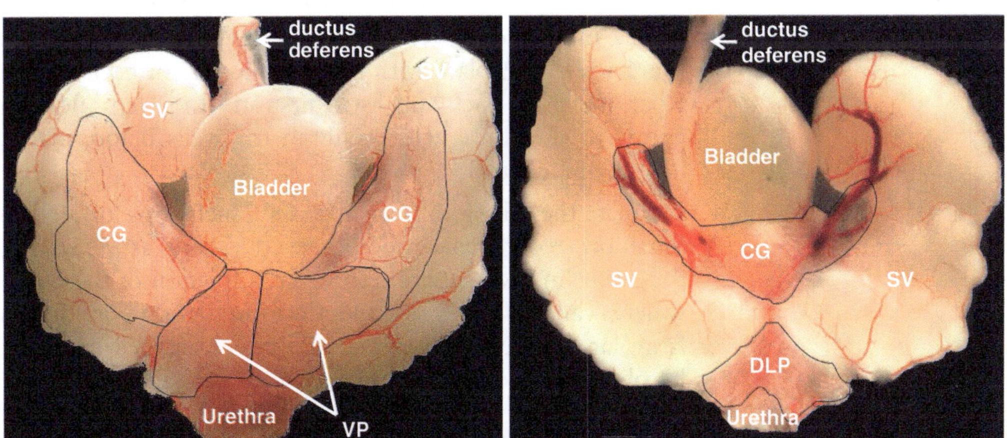

Fig. 1. Ventral and dorsal view of genitourinary (GU) system. *Left panel* ventral view of bladder, seminal vesicle (SV) ductus deferens, urethra, ventral prostate (VP), and anterial prostate or coagulate gland (CG). *Right panel* dorsal view of GU system showing dorsolateral prostate (DLP).

9. After incubation, the incompletely digested tissue pieces can be further dissociated mechanically by pipetting up and down ten times with a 5-ml pipette. (*see* **Note 2**)
10. Put 100-μm and 40-μm strainers on top of individual 50-ml conical tubes. Pass the dissociated cells first through the 100-μm and then 40-μm strainer sequentially to remove incompletely digested tissues and aggregated cell clumps. (*see* **Note 2**)
11. Spin down the flow through from 40-μm strainer at 800 rpm (114× *g*) for 5min to pellet cells.
12. Aspirate supernatant and wash the pellet with 10 ml DMEM once.
13. Aspirate supernatant and resuspend the cell pellet with 10 ml of prostosphere culture medium.
14. Count cell number with hemacytometer by mixing 10 μl of cell suspension with 10 μl of trypan blue. Fill the hemacytometer with 10 μl of the mixture. Count the total cell number (X) in the four corner squares of the grid (1.0 × 1.0 × 0.1 mm each) and calculate the concentration of the cell suspension using the following formula:

    Cell concentration (per ml) = (total cell number X / 4) × dilution ((2)) × 10^4 (*see* **Note 3**)

3.2. Cell Culture

1. Add 1 ml of prostosphere medium to each well of a 6-well Corning ULA plate for 30 minutes. This step will activate the hydrophilic surface of the plate and transform it to a non-adherent hydrophobic surface. Aspirate the medium before adding the cell suspension.
2. Adjust the prepared cell suspension to a concentration of 20,000 cells per ml with prostosphere medium.
3. Seed 2 ml of the cells to each of the medium-treated well of the 6-well ULA plate with three replicates for each sample. Rock the plate back and forth first and then left and right gently to distribute cells evenly in the wells before placing into the incubator.
4. Incubate the cells at 37°C in a 5% CO_2 incubator for 10 days. Small prostospheres can be observed as early as 3 days with 3-10 cells in each sphere. However, a fraction of these early spheres can be short lived, an indication that they are derived from late progenitors. On day 6, significant cell death can be observed and the spheres are more noticeable and larger than at day 3. On day 10, the spheres have reached their maximum size (**Fig. 2**) and begin to age (coated with non-cellular dark clouds when examined under phase contrast microscope). Rock the plate each day to keep the spheres distributed evenly in the well.

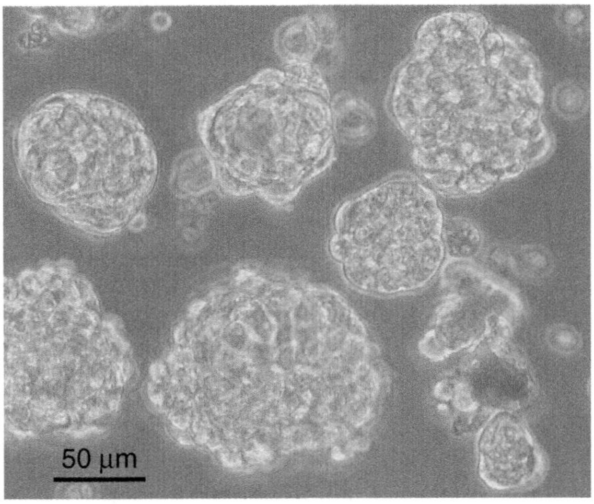

Fig. 2. Representative morphology of 10-day-old prostospheres.

3.3. Sphere Count and Size Measurement

1. On day 10, mix the sphere solution completely by pipetting several times. Take three 100-μl aliquots from each well and transfer to a 96-well plate.
2. Count the number and size of all the spheres under microscope using MCID program as described later.
3. Start MCID program.
4. Calibrate the system by clicking on the ruler icon. From the menu bar, click establish → distance. Check off calibrated, click OK.
5. Set Horizontal pix 65, 100μm; Vertical pix 65, 100μm.
6. Click on sample icon on the menu bar, and then linear length tool
7. Click on camera icon. Adjust light and focus to see spheres clearly on the computer screen, and use 10x object lens. Set the field on the top of the well and start to collect data. (*see* **Note 4**)
8. Press F9 to show the measurement cursor (red arrow). Drag arrow across the diameter of the sphere and then release (confirm that length appears on table).
9. Press F9 again to switch between measuring and view field functions. Hit camera icon for live image. Scan the whole well from top to bottom. (*see* **Note 4**)
10. Calculate the mean and standard deviations for the cell counts. Discount any spheres that are smaller than 20μm or greater than 120μm since they may represent spheres with limited proliferation capacity or spheres that result as aggregates of several individual spheres, respectively.

4. Notes

1. We have used normal mouse prostate to optimize and standardize the prostosphere cultivation conditions. Our results demonstrate that the ideal concentration for clonal prostosphere culture is 20,000 per ml of dissociated mouse prostate cells and that the average number of sphere-forming cells from the prostate of mouse with a genomic background of mixed C57BL/129/BALB/c is about 5 per hundred cells seeded. The average sphere number varies among different mouse strains. Though not all the sphere-forming cells are epithelial stem cells as determined by lineage analysis, almost all the sphere-forming cells are Sca-1 expressing cells (Wang et al, unpublished data). We have previously shown that Sca-1$^+$ population increased dramatically in Pten null mouse prostate cancer versus wild-type tissue *(10)*. Consistent with this observation, the level of prostosphere-forming cells was significantly elevated in Pten null tumor as shown in **Fig. 3**.

   We further examined whether this method could be applied to human prostate stem cells. Dissociated cells from all four human prostate samples with normal histological appearance formed prostospheres even more robustly under the culture conditions described. We also demonstrated that human prostate epithelial cells can be enriched in a short incubation period in prostosphere medium on regular culture plates. The attached epithelial cells formed prostospheres in suspension

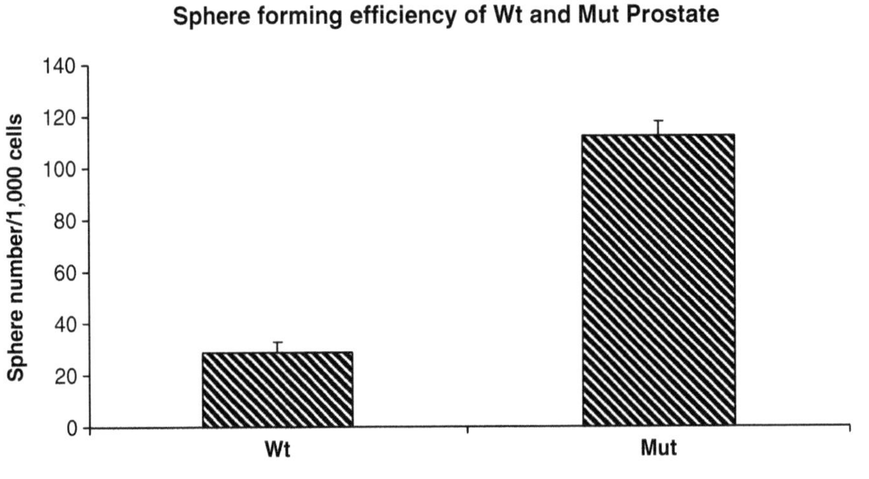

Fig. 3. Enhanced prostosphere-forming capacity in prostate cancer tissues. A representative result of the prostosphere-forming efficiency in wt and Pten null prostate. Prostate from two pairs of genetic and age-matched wt and mut mice was used to examine the amount of prostosphere-forming cells. The fold change was normalized to the number of prostospheres formed by wt mouse prostate in each pair.

culture when detached and displayed epithelial basal cell phenotype (Wang et al., unpublished data).

2. *Dissect mouse prostate:* The GU system is fragile and easy to break. When pulling it up from the pelvic cavity, the forceps can only be put on the urethra or bladder. Pay extra attention not to break the seminal vesicle because the content of it will interfere with the dissection if leaked out. There are two ways to dissect out all the prostate lobes. One is to dissect each lobe separately one after another. The other is to clean up all the non-GU tissue first and then get rid of seminal vesicles, bladder, ductus deferens, and ureters. Lastly, strip the urethra out from prostate.
3. *Dissociate prostate tissue:* Always keep the tissue wet. Even a short period of drying will severely affect the viability of the cells. If there are many incompletely digested tissues after 2-hr collagenase digestion, extend the digestion with fresh collagenase and/or mechanically disrupt the remaining tissues. Collect the remaining tissues in the strainer and incubate them in fresh collagenase solution for 2 more hours at 37°C. Alternatively, place the strainer with remaining tissues in a 35-mm Petri dish covered with 0.2 ml of medium and grind the tissue with the tip of a sterile eppendorf tube.
4. *Cell counting:* To minimize cell aggregation after dissociation, the dissociated cells are resuspended at low density. The few cells loaded onto the hemacytometer may not accurately reflect the actual cell concentration and therefore result in greater variations in the subsequent calculation of sphere-forming efficiency. An alternative cell counting method is to use a microplate. Add 10μl of cell suspension in triplicate and count all the cells in the plate. Also keep in mind to mix the cell suspension well before taking aliquots for counting.
5. *Sphere Counting:* It is easy to distinguish cell aggregates versus spheres under the microscope (**Fig. 2**). Only spheres greater than 20μm (composed of about 10 cells or more) are taken into account. Printing a grid on a transparency sheet and placing it underneath the 96-well plate will help to align each step when moving the microscope stage to scan the entire well.
6. *Human prostate dissociation:* Human prostate has more stroma than mouse prostate. An overnight digestion at 37°C in 0.25-0.5% collagenase is required. To prevent fungal contamination, an appropriate amount of Fungizone can be added to the dissociation solution.
7. *Human prostate epithelial enrichment:* Human prostate epithelial cells can be enriched by incubating the dissociated cells in a regular tissue culture flask in prostosphere medium for 2 days at 37°C. The adherent cells show typical epithelial cell phenotype and can be used for prostosphere assay (Wang S., unpublished data).

Acknowledgments

We thank Dr. Rachel Kim for reading the manuscript and suggestions. This work is supported by grant PC060326 from the Department of Defense

References

1. Reya, T., Morrison, S. J., Clarke, M. F., and Weissman, I. L. (2001) Stem cells, cancer, and cancer stem cells. Nature, **414**, 105–111.
2. Fuchs, E., Tumbar, T., and Guasch, G. (2004) Socializing with the neighbors: stem cells and their niche. Cell, **116**, 769–778.
3. Reynolds, B. A. and Weiss, S. (1996) Clonal and population analyses demonstrate that an EGF-responsive mammalian embryonic CNS precursor is a stem cell. Dev Biol, **175**, 1–13.
4. Dontu, G., Abdallah, W. M., Foley, J. M., Jackson, K. W., Clarke, M. F., Kawamura, M. J., and Wicha, M. S. (2003) In vitro propagation and transcriptional profiling of human mammary stem/progenitor cells. Genes Dev, **17**, 1253–1270.
5. Richardson, G. D., Robson, C. N., Lang, S. H., Neal, D. E., Maitland, N. J., and Collins, A. T. CD133, (2004) a novel marker for human prostatic epithelial stem cells. J Cell Sci, **117**, 3539–3545.
6. Xin, L., Lawson, D. A., and Witte, O. N. (2005) The Sca-1 cell surface marker enriches for a prostate-regenerating cell subpopulation that can initiate prostate tumorigenesis. Proc Natl Acad Sci U S A, **102**, 6942–6947.
7. Burger, P. E., Xiong, X., Coetzee, S., Salm, S. N., Moscatelli, D., Goto, K., and Wilson, E. L. (2005) Sca-1 expression identifies stem cells in the proximal region of prostatic ducts with high capacity to reconstitute prostatic tissue. Proc Natl Acad Sci U S A, **102**, 7180–7185.
8. Collins, A. T., Berry, P. A., Hyde, C., Stower, M. J., and Maitland, N. J. (2005) Prospective identification of tumorigenic prostate cancer stem cells. Cancer Res, **65**, 10946–10951.
9. Wang, S., Gao, J., Lei, Q., Rozengurt, N., Pritchard, C., Jiao, J., Thomas, G. V., Li, G., Roy-Burman, P., Nelson, P. S., Liu, X., and Wu, H. (2003) Prostate-specific deletion of the murine Pten tumor suppressor gene leads to metastatic prostate cancer. Cancer Cell, **4**, 209–221.
10. Wang, S., Garcia, A. J., Wu, M., Lawson, D. A., Witte, O. N., and Wu, H. (2006) Pten deletion leads to the expansion of a prostatic stem/progenitor cell subpopulation and tumor initiation. Proc Natl Acad Sci U S A, **103**, 1480–1485.

Chapter 10

Identification of Human Pancreatic Cancer Stem Cells

Chenwei Li, C. J. Lee, and Diane M. Simeone

Summary

Emerging evidence suggests that malignant tumors are composed of a small subset of distinct cancer cells, termed "cancer stem cells" (typically less than 5% of total cancer cells based on cell surface marker expression), which have great proliferative potential, as well as more differentiated cancer cells, which have very limited proliferative potential. Data have been provided to support the existence of cancer stem cells in several different types of cancer, including human blood, brain, prostate, ovarian, melanoma, colon, and breast cancers. We have recently reported the identification of a subpopulation of pancreatic cancer cells that express the cell surface markers CD44+CD24+ESA+ (0.2–0.8% of all human pancreatic cancer cells) that function as pancreatic cancer stem cells. The CD44+CD24+ESA+ pancreatic cancer cells are highly tumorigenic and possess the stem cell-like properties of self-renewal and the ability to produce differentiated progeny. Pancreatic cancer stem cells also demonstrate upregulation of molecules important in developmental signaling pathways, including sonic hedgehog and the polycomb gene family member Bmi-1. Of clinical importance, cancer stem cells in several tumor types have shown resistance to standard therapies and may play a role in treatment failure or disease recurrence. Identification of pancreatic cancer stem cells and further elucidation of the signaling pathways that regulate their growth and survival may provide novel therapeutic approaches to treat pancreatic cancer, which is notoriously resistant to standard chemotherapy and radiation.

Key words: Pancreatic cancer, stem cells, niche, CD44, ESA, CD24, NOD/SCID, xenografts, tumorigenic potential, self-renewal

1. Introduction

Pancreatic adenocarcinoma is a highly lethal disease which is usually diagnosed in an advanced state when there is no effective cure. It has the worst prognosis of any major malignancy, with a 5-year survival rate as low as 3% and is the fourth most common cause of cancer death per year in the United States *(1)*.

John S. Yu (ed.), *Cancer Stem Cells,* Methods in Molecular Biology, vol. 568
DOI 10.1007/978-1-59745-280-9_10, © Humana Press, a part of Springer Science + Business Media, LLC 2009

Despite advances in surgical and medical therapies, little impact has been made on the mortality rate of this disease.

Recently, there has been a paradigm shift in our understanding of how cancers develop and propagate, termed "the cancer stem cell theory." In the traditional model of how cancer develops and propagates, it is thought that all cells within the cancer are able to proliferate extensively and form new tumors. In the cancer stem cell theory, tumors arise from a small population of cancer cells, usually less than 5% of cells, that have properties of adult stem cells, particularly the ability to self-renew and differentiate into the multiple cell types found in the tumor. Of clinical relevance, this model may explain why standard chemotherapy and ionizing radiation therapy may result in tumor shrinkage, but do not prevent cancer recurrence, since cancer stem cells are resistant to these standard types of therapies. This concept was first proven in the context of acute myeloid leukemia *(2, 3)* and has since been proven in multiple myeloma *(4)* and breast *(5)*, brain *(6)*, prostate *(7)*, ovarian, colon *(8, 9)*, and pancreatic cancer *(10)*.

We recently identified a CD44 + CD24 + ESA + subpopulation of human pancreatic cancer cells as putative pancreatic cancer stem cells by utilizing a xenograft model in which primary human pancreatic adenocarcinomas were implanted in nonobese diabetic/severe combined immunodeficient (NOD/SCID) mice. Using this model system, we sorted primary pancreatic cancer cells based on different cell surface markers and were able to identify a specific subset of pancreatic cancer cells with markedly enhanced tumorigenicity. We also demonstrated that these cells possess the stem cell properties of self-renewal and the ability to produce differentiate progeny in vivo. H&E staining of the tumors generated from the CD44 + CD24 + ESA + pancreatic cancer stem cells were indistinguishable from the patient's primary tumor.

The biologic function of stem cells has been shown to be highly dependent on the local tissue microenvironment, or the niche *(11)*. To validate our findings of enhanced tumorigenicity of CD44 + CD24 + ESA + pancreatic cancer stem cells in a relevant niche, we injected pancreatic cancer cells and triple marker negative cells into the tail of the mouse pancreas. We observed tumor formation with CD44 + CD24 + ESA + cells, but not triple marker negative cells, validating the findings we observed with subcutaneous implantation.

To determine if select developmental signaling pathways important in normal stem cell function might be upregulated in pancreatic cancer stem cells, we measured sonic hedgehog (SHH) and the polycomb gene family member Bmi-1 using quantitative RT-PCR in normal pancreatic epithelial cells, CD44 + CD24 + ESA + pancreatic cancer stem cells, CD44 – CD24 – ESA – cells and bulk pancreatic cancer cells. We found marked upregulation of SHH and Bmi-1 in CD44 + CD24 + ESA + cells compared to the other cell types, verifying activation of developmental signaling pathways in pancreatic cancer stem cells.

2. Materials

2.1. Primary Tumor Specimen Implantation

1. NOD/SCID mice (Jackson Laboratory, Bar Harbor, Maine).
2. RPMI 1640 medium (Gibco). Store at 4°C.
3. Phosphate-buffered saline (PBS) (Gibco). Store at room temperature.
4. Ketamine (Fort Dodge, Fort Dodge IA) and xylazine (Lloyd Shenandoah, IA) are mixed and diluted in PBS and used to anesthetize NOD/SCID mice at the final concentration of 100 mg/kg for ketamine and 5 mg/kg for xylazine. Store at room temperature.
5. Coated Vicryl braided 3–0 suture (Ethicon, Somerville, NJ)

2.2. Preparation of Single-Cell Suspensions of Tumor Cells

1. Medium 199 (with sodium bicarbonate) (Gibco). Store at 4°C.
2. Collagenase IV (Worthington Biochemical, Lakewood, NJ) is dissolved in sterile water and used in Medium 199 for a final concentration of 200 unit/mL. Store in single-use aliquots at –20°C (short term), or at –80°C (long term) (*see* **Note 1**).
3. 40-μm nylon mesh (BD Falcon).
4. Hanks' Balanced Salt Solution (HBSS) supplemented with 20% fetal bovine serum (FBS) (Gibco). Store at 4°C.
5. FBS supplemented with 10% dimethyl sulfoxide (DMSO) is used as a freezing solution (*see* **Note 2**).

2.3. Staining for Flow Cytometry

1. Sterile 5 mL polystyrene round-bottom tubes (12 × 75 mm style) (Becton Dickinson Labware, Franklin Lakes, NJ).
2. Sandoglobin (ZLB Behring LLC, Kankakee, IL). Sandoglobin is a human immunoglobulin used to block nonspecific antibody binding during flow cytometry. Dilute the immunoglobulin to 12% in sterile water and add the preservative sodium azide to a final concentration of 0.02%. Store at 4°C as stock solution.
3. 4′,6-diamidino-2-phenylindole, dihydrochloride (DAPI) (Molecular Probes, Eugene, OR). To make a 5 mg/mL stock solution, dissolve 10 mg (one vial) in 2 mL sterile water. Store at 4°C (short term), or at –20°C (long term).
4. Antibodies: anti-CD44 allophycocyanin and anti-CD24 (PE) (BD Biosciences, San Diego, CA), antimouse H2K (South-

ernBiotech, Birmingham, AL), Streptavidin-PE-Cy5Conjugate (BD PharMingen, Franklin Lakes, NJ) (secondary for antimouse H2K antibody), and antiepithelial-specific antigen (ESA)-FITC (Biomeda, Foster City, CA). Store all antibodies at 4°C.

5. Flow cytometry was performed on both a FACS DiVa and FACS Aria machine (BD Immunocytometry Systems).

2.4. Sorted Cell Implantation into NOD/SCID Mice

1. Matrigel (BD Biosciences). Store in single-use aliquots at –20°C, and thaw on ice.

2.5. Immunohistochemistry

1. Citrate Buffer, pH = 6.0 (Zymed/Invitrogen).
2. Vector Antigen Unmasking Solution (Vector Laboratories, Burlingame, CA).
3. ImmEdge Pen (Vector Laboratories).
4. Dako Cytomation Biotin Blocking System (Dako North America, Carpintria, CA).
5. Histostain Plus AEC Broad Spectrum Kit (Zymed/Invitrogen).
6. Goat polyclonal anti-S100P antibody (R&D Systems).
7. Mouse antistratifin Monoclonal Antibody (ABcam, Cambridge, UK).
8. Clearmount Mounting Solution (Zymed/Invitrogen).

2.6. Real-Time RT-PCR

1. RNeasy Micro Kit (Qiagen) (*see* **Note 3**).
2. Reverse transcription system (Promega).
3. Real-time quantitative RT-PCR primers for SHH were designed and purchased from Applied Biosystems as Assay-on-Demand™ Gene Expression Products (Applied Biosystems).
4. 2X Taqman™ universal PCR master mix (Roche).
5. ABI PRISM 7900HT sequencing detection system (Applied Biosystems).

3. Methods

While it is preferable to sort primary pancreatic cancer cells directly obtained from the patient, it is often difficult to obtain adequate sample size for direct sorting, and thus primary tumors are often established as xenografts in NOD-SCID mice. Xenografts are also used for tumorigenicity studies. Importantly, the validity of using xenografts is supported by previous work showing that pancreatic cancer xenografts retain many of the features of the primary tumor on multiple passaging *(12)*. We have found 100% take of individual primary pancreatic tumors with implantation of 3, 2 × 2 minced pieces placed bilaterally into the midabdomen of four separate NOD/SCID mice. The development of a human primary pancreatic cancer xenograft program is critical before undertaking studies on pancreatic cancer stem cells. NOD/SCID mice are most commonly used to establish to establish xenografts of primary human tumors, but other immunodeficient mouse strains may also be used. Tumor development following implantation of sorted single cells is optimized by additional of matrigel to the cell suspension. The average length of time for development of a 1-cm tumor following implantation of minced pieces of primary tumor is approximately 3–4 months.

Additional technical issues are important to consider in pancreatic cancer stem cell isolation and separation. We developed a specific protocol to prepare single-cell suspensions of pancreatic cancer cells from primary pancreatic cancers and xenografts, since these tumors are particularly fibrotic and need different digestion protocols than those used with other tumor types. Flow cytometry offers a sensitive and specific method to isolate and purify cancer stem cells, but several issues need to be considered. For example, doublets (two cells sticking together) can occasionally sort together and need to be eliminated. Microscopy is needed to show that single cancer cells were indeed isolated. Another factor to consider is that most flow cytometers are typically set up to sort blood cells using small-diameter streams at high pressures, conditions that may not be tolerated by cancer cells found in solid organs. Optimization of the diameters of the liquid stream and sorting pressures needs to be done to ensure a high level of viability of the sorted cells. In addition, the determination of cancer stem cells based on cell surface markers may be highly variable, depending on method of cell preparation, setting of the gates for flow cytometry, and the specific antibody used. For individuals planning on embarking on cancer stem cell studies, training in an experienced stem cell laboratory to master flow cytometry isolation of cancer stem cells is highly recommended *(13)*.

3.1. Primary Tumor Specimen Implantation

1. Suspend samples in sterile RPMI medium 1640 on ice after samples of human pancreatic adenocarcinomas are obtained from operating room (*see* **Note 4**).
2. Remove all extrapancreatic tissue (fat, blood vessels, etc.) using sterile scissors; then mince tumor with a sterile scalpel blade over ice to yield 2 × 2 mm pieces. Wash the tumor pieces with serum-free PBS before implantation.
3. Anesthetize 8-week-old male NOD/SCID mice using an i.p. injection of 100 mg/kg ketamine and 5 mg/kg xylazine. The anesthetics typically take 5–10 min to produce the desired effect (*see* **Note 5**).

4. Shave the abdomen and sterilize the skin with 70% alcohol. Make a 5–10 mm transverse incision in the skin on one side of the midabdomen. Be careful not to penetrate the muscle and the peritoneum. Create a small subcutaneous pocket using gentle blunt dissection of adequate size to accommodate three pieces of minced tumor (2 × 2 mm) that you have prepared.
5. Close the skin incision with absorbable suture. Mice will awaken from the anesthesia after 30–120 min. Check the mice daily. The mice will be monitored weekly for tumor growth for 16–20 weeks.

3.2. Preparation of Single-Cell Suspensions of Tumor Cells

1. Before digestion with collagenase, cut xenograft tumors from mice or primary human tumors into small pieces with sterile scissors in HBSS or RPMI 1640 medium, and then mince the tissue mechanically with scissors until the pieces are 1 mm in size (able to be pipetted without difficulty using 10-mL pipettes). Wash the tissue with HBSS or RPMI 1640 twice, centrifuge samples, and discard the solution (*see* **Note 6**).
2. Resuspend the minced tissue in 20–30 mL medium 199 in a 50-mL centrifuge tube depending on the amount of tissue. Large amounts of tissue may require additional tubes.
3. Add ultrapure collagenase IV in medium 199 at a final concentration of 200 units/mL.
4. Incubate the sample at 37°C with shaking at a speed of 120 RPM for 1.5 h for first step enzymatic dissociation.
5. Pipette the sample for 3 min using a 25-mL pipette to mechanically dissociate the sample and put it back to 37°C incubation.
6. Further mechanically dissociate the sample every 15–20 min by pipetting with a 10-mL or 5-mL pipette until whole tumor is dissociated (*see* **Note 7**).
7. Add medium 199 to a total volume of 50 mL and spin down.
8. Discard the supernatant and resuspend the samples with RPMI containing 20% FBS.
9. Filter through a 40-micron nylon mesh.
10. Wash with HBSS/2% FBS twice and resuspend cells in HBSS/2% FBS.
11. Count cells. Cells are ready for staining.
12. Excess cells may be frozen for future use by placing cells in a solution of 90% FBS with 10% DMSO. Digestion of a 1-cm^3 xenograft tumor will typically result in 10–20 million cells.

3.3. Staining and Flow Cytometry

1. Resuspend each million cells in 100–200 μL of HBSS containing 2% FBS in a 5-mL tube on ice. All solutions should be on ice during the whole process. For frozen samples, wash twice with HBSS containing 2% FBS, and resuspend in HBSS containing 2% FBS. Set the centrifuge at 4°C.
2. Add Sandoglobin solution (1 mg/mL) at a dilution of 1:20 and incubate the sample on ice for 20 min, then wash the sample twice with HBSS/2% FBS, and resuspend it in 2–3 mL HBSS/2% FBS (*see* **Note 8**).
3. Prepare the controls by aliquoting 100 μL samples into five individual tubes and label the tubes DAPI, H2K, CD44, CD24, and ESA, respectively. For primary human pancreatic cancer cells, four controls (DAPI, CD44, CD24, ESA) are needed as H2K is an antibody used to remove mouse-specific cells and is only needed for cells derived from xenografts.
4. Avoid exposure to light from this step forward. Add respective antibodies (except DAPI) to each of the control tubes, and add all antibodies to sample tube at a 1:40 dilution and incubate for 20 min on ice.
5. Wash all samples twice with HBSS/2% FBS.
6. (If H2K was added) Resuspend cells in H2K and sample tubes to the original volume with HBSS/2% FBS and add secondary antibody to the control H2K tube and sample tube at a dilution of 1:200. Incubate samples for 20 min on ice.
7. Spin samples down and wash with HBSS/2% FBS twice. Resuspend cells in HBSS/2% FBS containing (1 μg/mL) DAPI.
8. Run all five controls to adjust the voltage and gates (*see* **Note 9**).
9. Prerun sample and set the gates for sorting, then sort the sample. In all experiments using human xenograft tissue, infiltrating mouse cells should be eliminated by discarding H2K+ (mouse histocompatibility class I) cells during flow cytometry. Dead cells should be eliminated by using the viability dye DAPI. Side scatter and forward scatter profiles should be used to eliminate cell doublets. Examples of sorting for CD44, CD24, and ESA from three individual patient tumors are shown in **Fig. 1**.
10. Reanalyze the cells for purity, which will demonstrate if complete separation of different subpopulations is obtained (*see* **Note 10**).

3.4. Sorted Cell Implantation into NOD/SCID Mice

1. Wash sorted cells with serum-free HBSS twice after flow cytometry (Since the sorted cancer stem cell number is low, the pellet will not visible in most cases). Discard the supernatant very carefully (do not use vacuum-mediated supernatant aspiration).

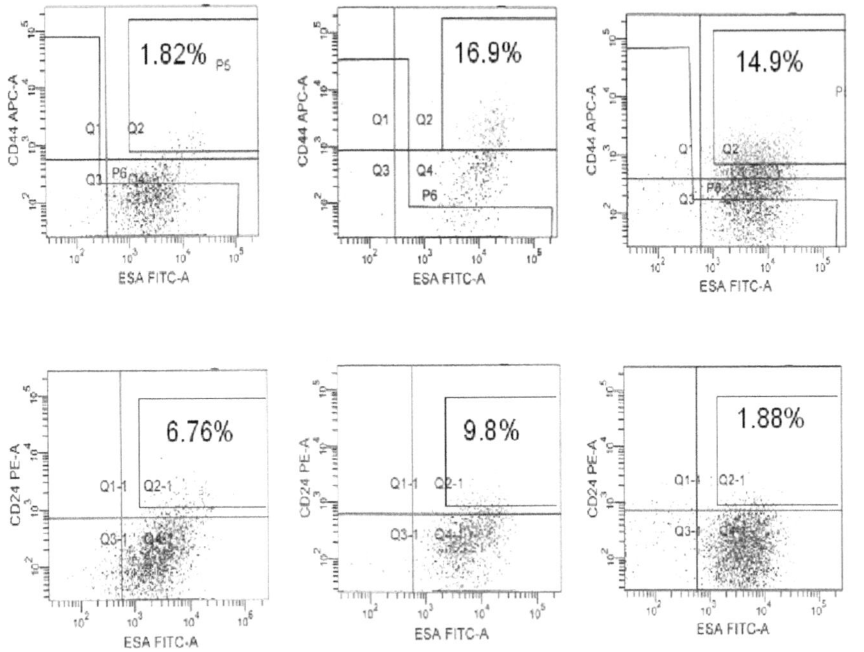

Fig. 1. Isolation of tumorigenic pancreatic cancer cells. Flow cytometry was used to isolate subpopulations of human pancreatic cancer cells which were tested for tumorigenicity in NOD/SCID mice. Cells were stained with antibodies against CD44, CD24, ESA, H2K, and DAPI. Dead cells and mouse cells were eliminated from the analyses. The plots depicted are representative examples of patterns of CD44 and ESA staining (*top three plots*) and CD24 and ESA staining (*bottom three plots*) of viable human pancreatic cancer cells from individual patients, with the frequency of the boxed tumorigenic cancer cell population as a percentage of cancer cells in the specimen shown.

2. Resuspend cells in serum-free RPMI 1640 medium. Count cells using a hemacytometer and dilute cell concentrations accordingly. A certain portion of sorted cells will be nonviable after flow cytometry (up to 10–20%); therefore, viability should be checked before determining cell number.
3. Prethaw the Matrigel on ice and mix sorted cells with Matrigel (1:1 volume).
4. Inject the mixture subcutaneously into the right and left midabdominal areas using a 25 1/2-gauge needle. The injection volume should be no more than 200–300 μL for each side. Monitor tumor growth for 16–20 weeks. An example of tumor formation from injected CD44+CD24+ESA+ pancreatic cancer stem cells is shown in **Fig. 2**.
5. Orthotopic injection. Anesthetize mice with an i.p. injection of 100 mg/kg ketamine and 5 mg/kg xylazine. Further use of isofluorane inhalation anesthetic may be utilized if sedation is not adequate. A 2-cm left subcostal incision is made to enter the peritoneal cavity. Care is taken during the incision to avoid injury to the liver as it typically drapes over midline

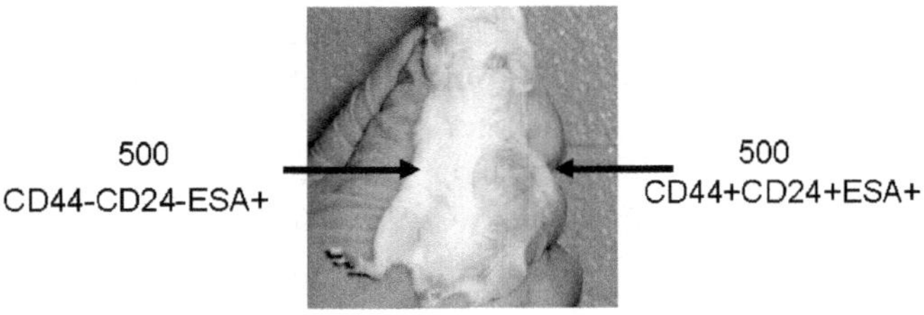

Fig. 2. Tumor formation in NOD/SCID mice injected with highly tumorigenic pancreatic cancer cells. A representative experiment depicting tumor formation in a mouse at the injection site of 500 CD44+CD24+ESA+ cells, with no tumor formation seen at the injection site of 500 CD44−CD24−ESA− cells.

and is located immediately underneath the facial layers as you enter the peritoneum. The stomach is identified and gently eviscerated out of the abdomen. The pancreas will be visible once the stomach is eviscerated and the posterior surface of the stomach is exposed. Identify the spleen and the tail of the pancreas. Gently eviscerate the spleen and the pancreatic tail. Prepare single-cell suspensions of sorted cells at a dilution of 5,000–10,000 sorted cells per 100 μL (resuspend in PBS in a volume of 50–100 μL mixed with 50–100 μL Matrigel) prior to the procedure. Using a 30-gauge needle, gently inject 50–100 μL of the cell/media/matrigel solution into the parenchyma of the distal pancreas. A successful injection into the pancreatic tissue will create a wheal (tissue fluid bubble) without leakage of fluid through the capsule of the pancreas. No more than 200 μL of cell/media/matrigel mixture should be injected at a time. The laparotomy incision should be closed with absorbable suture. Animals are monitored for tumor growth up to 16 weeks. At the time of autopsy, tumor samples should be harvested and processed for histological analysis and cell sorting.

3.5. Immunohistochemistry

1. Fix tissue samples in 10% phosphate-buffered formalin, and embed samples in paraffin.
2. Cut the formalin-fixed, paraffin-embedded blocks into 4-μm-thick sections, mount on poly-l-lysine-coated slides (Sigma), and dry overnight at 37°C.
3. Bake slides at 60°C overnight in a drying oven.
4. To remove paraffin, place slides into xylene and incubate for 10 min (two times) at room temperature.
5. Place slides in a series of ethanol solutions: 100% ethanol for 5 min, followed by 90% ethanol for 5 min, followed by 70% ethanol for 5 min.
6. Rinse in ddH_2O for 5 min (stain with H&E followed standardized protocols).

7. Place slides in a glass staining holder and place into a beaker containing 10 mM citrate buffer of pH = 6.0 (pH may need to be adjusted for different antibodies). Make sure the slides are completely covered with solution. Heat slides at 96°C–98°C for 40 min, and then cool for 20 min.
8. Rinse slides in ddH_2O for 5 min, and then rinse slides in PBS for 5 min.
9. Encircle sections using an ImmEdge Pen. Allow barrier to dry completely.
10. Add 3% hydrogen peroxide and incubate for 10 min. Rinse slides in PBS for 2 min (three times). **Steps 10–17** need to be performed in a humidified chamber at room temperature. Always place enough solution onto the slides to cover the entire tissue section.
11. Add Avidin Blocking Solution onto slides and incubate for 10 min. Rinse slides in PBS for 2 min (three times). Add Biotin Blocking Solution onto slides and incubate for 10 min. Rinse slides in PBS for 2 min (three times).
12. Add Serum Blocking Solution onto slides and incubate for 10 min. Drain and blot off excess solution. Do not rinse.
13. Add primary antibody solution onto slides and incubate for 30–60 min. Rinse slides in PBS for 2 min (three times).
14. Add biotinylated secondary antibody solution onto slides and incubate for 10 min. Rinse slides in PBS for 2 min (three times). Use Ready-to-Use Reagent B (biotinylated secondary antibody).
15. Add Enzyme Conjugate Solution (Ready-to-Use Reagent C) onto slides and incubate for 10 min. Rinse slides in PBS for 2 min (three times).
16. Add AEC Chromogen Solution onto slides and incubate until brown color appears (usually between 2 and 5 min). Do not incubate for more than 5 min. Place slides into ddH_2O for 2 min to stop the reaction. Rinse slides in ddH_2O for 2 min (three times). Use Ready-to-Use Reagent D (AEC Chromogen).
17. For counterstaining, add hematoxylin solution (diluted 1:8 in ddH_2O) onto slides and incubate for 30 s, and then place slide into ddH_2O for 2 min to stop the reaction. Rinse slides in ddH_2O for 2 min (three times).
18. Use a Kimwipe to dry slides. Cover stained tissue section with a few (2–3) drops of Clearmount Mounting Solution. Place slides in a drying oven set at 60°C. Incubate for 30–45 min or until slides are thoroughly dried. Slides are now ready for microscopic evaluation. **Figure 3** depicts

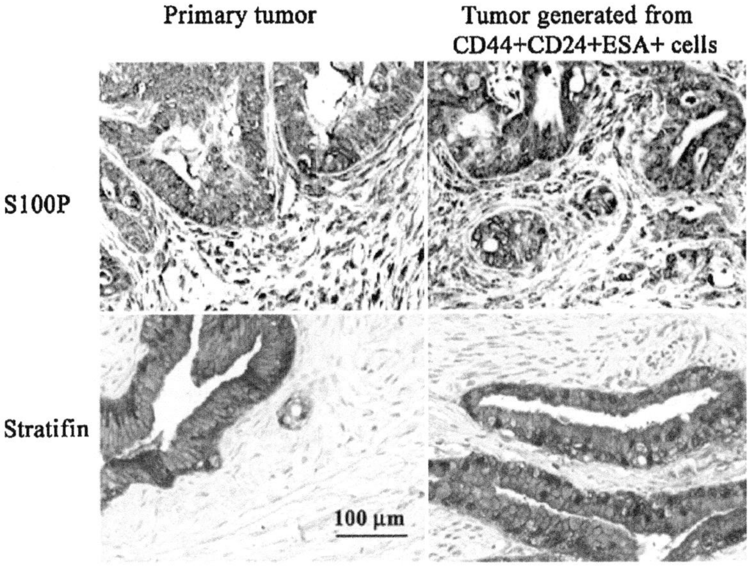

Fig. 3. Expression of differentiation markers in tumors derived from highly tumorigenic pancreatic cancer cells. Tissues were examined for the presence of S100P (*top two panels*) and stratifin (*bottom two panels*) in a primary patient tumor (*left*) and a tumor derived from CD44+CD24+ESA+ cells from the same patient (*right*). Antibody localization was performed using horseradish peroxidase, with dark staining indicating the presence of the specific antigen.

immunohistochemical staining for S100P and stratifin using the described protocol.

3.6. Real-Time RT-PCR

1. Wash the sorted cells with HBSS or PBS twice, and purify total RNA from sorted cells using a RNeasy Micro kit according to the manufacturer's instructions.
2. Synthesize cDNA using equivalent amounts of total RNA (100–500 ng) with random primers in a 20-µL reverse-transcriptase reaction mixture using the reverse transcription system from Promega following the protocol.
3. Prepare the reaction mix for each sample in 20 µL of PCR mixture contained 10 µL of 2X Taqman universal PCR master mix, 1 µL of 20X working stock of expression assay mix containing primers, and 20–50 ng of RNA converted DNA, and load mixture into reading plates provided by ABI.
4. Real-time PCR can be performed using an ABI PRISM 7900HT sequencing detection system. The reactions for each sample should be performed in triplicate. Fluorescence of the PCR products can be detected using same apparatus. The number of cycles required for the amplification plot to reach the threshold limit, the Ct-value, is used for quantification.

4. Notes

1. Collagenase IV from Worthington works best for digesting pancreatic cancer samples into single-cell suspensions. Different lots of the collagenase may have different enzyme activities; therefore, each lot should be tested before usage.
2. FBS with 10% DMSO is recommended for freezing single-cell suspensions for future use. We find that cells remain viable when frozen in this solution for up to 3 years when stored in liquid nitrogen. For short-term freezing, RMPI 1640/20%FBS with 10% DMSO is an alternative solution that also works well.
3. It may be difficult to purify adequate amounts of RNA due to the low number of cancer stem cells present in tumors. We can typically harvest 2,000–20,000 CD44+CD24+ESA+ cells from a 1-cm^2 xenograft tumor, which will yield 5–500 ng of total RNA using an RNeasy micro kit.
4. Primary human pancreatic cancer tissue should be implanted within 24 h of acquisition. Implantation at time points beyond 24 h will result in significantly reduced engraftment rate.
5. The sensitivity of NOD/SCID mice to these anesthetic agents is variable, and anesthetic agents should be carefully administered to avoid overdose.
6. Cellular aggregation can be reduced by adding 100 U/mL DNase to digestion solution.
7. The length of time for digestion is critical. Overdigestion will cause damage to the cells and low viability. Underdigestion will result in a low yield.
8. If necessary, this step can be shortened to 5 min.
9. Before conducting any experiment, a cell line is needed to test all the antibodies to determine the reactivity and specificity of each antibody. We used the Panc-1 cell line to test the CD44, CD24, and ESA antibodies because Panc-1 has a high level of surface expression of these three makers.
10. We highly recommend double sorting to enhance the purity of the sorted cell population desired. However, it is important to realize that with double sorting, one may lose up to 30% of the cells obtained after a single sort.

References

1. Hoyert DL, Heron MP, Murphy SL, Kung HC. (2006) Deaths: final data for 2003. Natl Vital Stat Rep 19: 1–120.
2. Bonnet D, Dick JE. (1997) Human acute myeloid leukemia is organized as a hierarchy that originates from a primitive hematopoietic cell. Nature Med 3: 730–737.
3. Lapidot T, Sirard C, Vormoor J, Murdoch B, Hoang T, Caceres-Cortes J, Minden M, Paterson B, Caligiuri MA, Dick JE. (1994) A cell initiating human acute myeloid leukemia after transplantation into SCID mice. Nature 17: 645–648.
4. Matsui W, Huff CA, Wang Q, et al. (2004) Characterization of clonagenic multiple myeloma cells. Blood 103: 2332–2336.
5. Al-Hajj M, Wicha MS, Benito-Hernandez A, Morrison SJ, Clarke MF. (2003) Prospective identification of tumorigenic breast cancer cells. Proc Natl Acad Sci U S A 100: 3983–3988.
6. Singh SK, Hawkins C, Clarke ID, Squire JA, Bayani J, Hide T, Henkelman RM, Cusimano MD, Dirks PB. (2004) Identification of human brain tumour initiating cells. Nature 432: 396–401.
7. Collins AT, Berry PA, Hyde C, Stower MJ, Maitland NJ. (2005) Prospective identification of tumorigenic prostate cancer stem cells. Cancer Res 65(23): 10946–10951.
8. O'Brien CA, Pollett A, Gallinger S, Dick JE. (2007) A human colon cancer cell capable of initiating tumour growth in immunodeficient mice. Nature 445(7123): 106–110.
9. Ricci-Vitiani L, Lombardi DG, Pilozzi E, Biffoni M, Todaro M, Peschle C, De Maria R. (2007) Identification and expansion of human colon-cancer-initiating cells. Nature 445(7123):111–115.
10. Li C, Heidt DG, Dalerba P, Burant CF, Zhang L, Adsay V, Wicha M, Clarke MF, Simeone DM. (2007) Identification of pancreatic cancer stem cells. Cancer Res 67(3): 1030–1037.
11. Li L, Neaves WB. (2006) Normal stem cells and cancer stem cells: the niche matters. Cancer Res 66: 4553–4557.
12. Hahn SA, Seymour AB, Hoque AT, et. al. (1995) Allelotype of pancreatic adenocarcinoma using xenograft enrichment. Cancer Res 55: 4670–4675.
13. Clarke MF, Dick JE, Dirks PB, Eaves CJ, Jamieson CH, Jones DL, Visvader J, Weissman IL, Wahl GM. (2006) Cancer stem cells – Perspectives on current status and future directions: AACR workshop on cancer stem cells. Cancer Res 66(19): 9339–9344.

Chapter 11

Cancer Stem Cells in Head and Neck Squamous Cell Carcinoma

Laurie Ailles and Mark Prince

Summary

The cancer stem cell hypothesis states that within a tumor only a subset of cells, the "cancer stem cells" (CSC), are capable of initiating and propagating the disease. In various cancers such cells have been identified and prospectively isolated based on the presence of specific cell surface antigens. In head and neck squamous cell carcinomas, we have shown that the CSCs are contained within the CD44+ subset of tumor cells. This subset contains cells capable of initiating tumor growth in mice that recapitulates the original tumor heterogeneity. Furthermore, they have a primitive cellular morphology, express high levels of nuclear BMI1, and are arrayed in characteristic tumor microdomains. The methods used to purify this subset of tumorigenic cells, to characterize their gene expression profiles, and to identify their physical location within the context of the whole tumor are described here.

Key words: Head and neck squamous cell carcinoma, cancer stem cells, fluorescence-activated cell sorting, gene expression analysis, immunostaining

1. Introduction

Head and neck cancer is a malignancy that affects approximately 40,000 new patients in the United States each year. Despite advances in therapy, which have improved quality of life, survival rates have remained static for many years. It is therefore essential that we develop a deeper understanding of the biology of this disease in order to develop more effective therapies. The majority of head and neck tumors are squamous cell carcinomas, derived from the stratified squamous epithelium that lines the mouth, larynx, pharynx, and upper esophagus. Normal stratified squamous epithelium consists of layers of cells, the innermost of which is the basal layer overlying the basement membrane. The latter contains

John S. Yu (ed.), *Cancer Stem Cells,* Methods in Molecular Biology, vol. 568
DOI 10.1007/978-1-59745-280-9_11, © Humana Press, a part of Springer Science+Business Media, LLC 2009

stem and progenitor cells that proliferate and continuously give rise to new cells that differentiate as they move to the surface where they are sloughed off and constantly replaced. The basal cells must include a population of self-renewing stem cells in order to maintain the continuous production of mature squamous cells throughout life and to regenerate the epithelium after injury.

We have recently reported the existence of a subpopulation of cancer stem cells (CSCs) in head and neck squamous cell carcinoma (HNSCC), as demonstrated by the ability to isolate a subpopulation of cells from patient HNSCCs that have the exclusive ability to give rise to new tumors in an immunocompromised mouse model *(1)*. The remaining bulk of cells in the tumor do not have this ability. Furthermore, this tumorigenic subpopulation of cancer cells, which are defined by the lack of expression of markers for nontumor cells (immune cells, endothelial cells, stroma) and expression of the cell surface protein CD44, possess many properties attributed to stem cells: they generate differentiated progeny, they self-renew, they express immature cell markers and genes, and they do not express differentiation markers. Furthermore, they have a basaloid phenotype similar to normal squamous epithelial stem cells, so that in well-to-moderately differentiated tumors, the physical location of the CSC is in the "basal layer," adjacent to tumor stroma.

The identification, isolation, and characterization of HNSCC CSC will dramatically impact our understanding of basic carcinogenic processes in HNSCC and have far-reaching implications for clinical management and treatment of this disease. Specifically, early and rapid diagnostic techniques focusing on stem cell populations can be developed to improve current strategies for screening, prevention, and disease monitoring. New molecular targets unique to CSCs can lead to novel chemotherapeutic and biologic therapies with improved efficacy, decreased potential for tumor resistance, and lower risk of toxicity to normal tissues.

In this chapter we describe the methods used for isolating a CSC-enriched population of cells from HNSCC samples based on CD44 expression, and techniques for further characterization of these cells, including flow cytometry, quantitative RT-PCR, and immunostaining.

2. Materials

2.1. Tissue Harvest and Transport

1. Container for tumor transport (e.g., 50-mL conical tube).
2. Media: DMEM-F12 with 10% fetal calf serum, 100 U/mL Penicillin, and 100 μg/mL streptomycin.

2.2. Tumor Propagation in Mice

1. Immunocompromised mice.
2. 6-cm tissue culture dish.
3. Media (see earlier).
4. Sterile scalpel.
5. Hank's balanced salt solution (HBSS).
6. Ketamine and Xylazine.
7. Saline: 0.7% NaCl.
8. 1-cc syringe.
9. Scale to weigh mice.
10. Shaver.
11. Betadine.
12. Sterile gauze.
13. Sterile field and surgical gloves.
14. Sterile surgical instruments – scissors, forceps, trocar.
15. Liquid skin sealant or wound clip applier.
16. Matrigel™ (BD Biosciences).
17. 0.5-mL syringe with 27- or 30-gauge needle.
18. Isoflurane and machine for isoflurane delivery (optional).

2.3. Tissue Processing to Generate a Single-Cell Suspension

1. 6- or 10-cm tissue culture dish (size dependent on the size of the specimen).
2. Sterile scalpel or razor blade.
3. HBSS with 2% heat-inactivated calf serum (HICS).
4. 50 mL conical centrifuge tubes.
5. Dissociation solution: DMEM-F12 with 300 U/mL collagenase IV, 100 U/mL hyaluronidase, and 125 U/mL DNAse.
6. 40 micron nylon mesh sieve.
7. Red blood cell lysis buffer: 150 mM NH_4Cl, 10 mM $KHCO_3$, 0.1 mM EDTA in H_2O.
8. Trypan blue.
9. Hemacytometer.

2.4. Flow Cytometry: Analysis and Cell Sorting

1. 5 mL polystyrene tubes.
2. HBSS with 2% HICS.
3. Antibodies: mouse-antihuman CD2, CD3, CD10, CD16, CD18, CD31, CD64, CD140b (Lineage cocktail; BD Pharmingen), mouse-antihuman CD44-PE (clone 515; BD Pharmingen), anti-$H2K^d$ (BD Pharmingen), goat-anti-mouse-IgG-Tricolor (Caltag), any other antibodies you may

care to include in your analysis, conjugated to fluorochromes that can be recognized by your FACS machine and do not overlap with the colors used for the detection of the lineage and CD44 antibodies (like FITC, PE-Cy7, etc.).

4. Isotype controls: unconjugated mouse IgG, mouse IgG-PE, etc. You should purchase isotype controls for any other antibodies you include in your analysis; the isotype of the control should match the isotype of the specific antibody used; the fluorochrome conjugate of the control should match the fluorochrome conjugate of the specific antibody used.
5. 1 mg/mL mouse IgG for blocking.
6. 1 mg/mL Propidium Iodide in HBSS.
7. CompBeads (BD Pharmingen), antimouse.
8. Flow cytometer with cell sorting capability.

2.5. Gene Expression Analysis by qRT-PCR

1. Trizol.
2. Linear acrylamide.
3. Chloroform.
4. Isopropanol.
5. Ethanol.
6. RNAse-free water.
7. Superscript III RT-PCR kit from Invitrogen, or similar product from alternative source.
8. TaqMan® Gene Expression Assays for genes of interest, plus at least two TaqMan® endogenous controls (Applied Biosystems Inc.).
9. TaqMan® universal PCR master mix (Applied Biosystems).
10. Real-time PCR machine.
11. Software for data analysis.

2.6. Immunohistochemistry and Immunofluorescence

1. Embedding medium for frozen tissue specimens.
2. Cryomolds, 15 mm × 15 mm × 5 mm.
3. Cryostat.
4. Glass slides (positively charged).
5. Acetone.
6. Phosphate-buffered saline (PBS) with 0.1% Tween-20.
7. Bovine serum albumin.
8. Sodium azide.
9. Hydrogen peroxide.
10. Antibodies: mouse antihuman CD44-biotin (clone G44–26 from BD Pharmingen), biotinylated isotype control, streptavidin-horseradish peroxidase (HRP), mouse anti-Bmi1

(clone 1.T.21 from Abcam), unconjugated mouse IgG1 isotype control, goat-antimouse-IgG-Alexa594, Streptavidin-Alexa488.

11. Liquid-repellent slide marker pen.
12. Humidified chamber.
13. 3,3′-diaminobenzidine tetrahydrochloride (DAB) substrate kit from Vector labs, or equivalent from an alternative source.
14. Solutions for hematoxylin counterstaining: hematoxylin 7211, clarifier, and bluing solution from Richard-Allan Scientific, or similar products from an alternative source.
15. Ethanol.
16. Xylene.
17. Toluene or xylene-based mounting media (e.g., Permount or Histomount).
18. 10 mg/mL of Hoechst 33342.
19. Aqueous mounting media designed to reduce fluorochrome quenching (e.g., Fluoromount).

3. Methods

When a tumor specimen is obtained, the size of the specimen will vary considerably. In some cases, the specimen will be large enough to carry out multiple different protocols, while in other cases, the sample may be very small and then it is necessary to prioritize the procedures to be performed. Our priorities are as follows:

1. The first priority is to propagate the tumor in mice. If the specimen is very small, this may be the only thing you can do. Once you have the tumor growing in mice, you can then proceed with additional experiments from the xenografted tumors.
2. A fraction of the specimen is frozen in embedding medium for tissue sectioning and immunostaining.
3. If the sample is large enough, an additional portion is dissociated into a single-cell suspension, which can then be injected into mice in known cell doses, and can be stained for flow cytometry and cell sorting.

All these procedures are described in detail below.

3.1. Tissue Harvest and Transport

Approval must be obtained from the institution's Internal Review Boards to conduct the research using human tissue and the animal model. Animal experiments must be performed in accordance with federal and institutional regulations and guidelines for the ethical care and use of animals. The risks and potential complications of tissue collection should be discussed and how the tissue will be utilized should be explained to subjects prior to obtaining written consent. Obtaining the assistance of a knowledgeable head and neck surgeon is a vital part of specimen collection.

Samples of HNSCC can be collected in either the clinic or the operating room and under local anesthesia or general anesthesia. It is absolutely critical that the tissue samples that are collected for research purposes do not interfere with the subject's treatment planning or therapy. If a tissue sample is collected at the same time that a biopsy is being performed for diagnostic purposes, the research tissue should be held "in quarantine" until the diagnostic biopsy is confirmed pathologically. In this way if the diagnostic biopsy is inadequate or nondiagnostic the research tissue could be sent to the pathologist and potentially used to obtain a diagnosis. When tissue samples are obtained during tumor resections, where the diagnosis is already established, care should be taken to avoid interfering with pathologic analysis of tumor margins.

Ideally tumor samples are collected while the tumor is still well vascularized. This improves the viability of the cells contained in the specimen and is particularly important if RNA is to be isolated from a fresh tissue sample. Collecting tissue from a vascularized tumor is a not a concern when the specimen is collected during a biopsy procedure as the tumor remains vascularized throughout the procedure. When the specimen will be obtained during a tumor resection, however, collecting tumor samples prior to the commencement of the tumor resection provides the best opportunity to reduce warm ischemia time for the tissue sample. In this situation the specimen can be collected at the beginning of the case, as if a biopsy was being performed, prior to the commencement of the tumor resection. If the specimen is collected in this fashion care must be taken to avoid interfering with tumor margins and the tumor resection. If for some reason the tumor resection must be performed prior to tissue samples being collected the sample should be harvested as soon as possible after the tumor is excised. Despite the disadvantage of potentially reduced cell viability, one advantage of collecting samples from a tumor that has already been resected is that it may be possible to obtain a larger specimen.

In most instances the viability of the tumor sample needs to be maintained to perform subsequent experiments. If indicated, a portion of the sample can be flash frozen in liquid nitrogen or preserved in formalin and paraffin embedded. To ensure maximum viability the specimen should be placed directly into cell culture media, such as DMEM supplemented with 10% fetal bovine serum, and transported on ice. The routine addition of antibiotics to the media is recommended as the majority of the tumor specimens are obtained from nonsterile mucosal surfaces in the upper aerodigestive tracts.

3.2. Tumor Propagation in Mice

Tumors can be propagated either by implantation of small pieces of tumor under the skin or by subcutaneous injection of a single-cell suspension (*see* **Note 1**). The method for creating a single-cell suspension is described in **Subheading 3.3**. Specimens to be implanted should be processed using sterile instruments and sterile technique.

For implantations of small pieces of tumor:

1. Place the specimen in a tissue culture dish and chop with a scalpel blade to yield approximately 1–2 mm^3 sized pieces.
2. Wash the tumor pieces with serum-free HBSS before implantation. Keep the tumor pieces in media, on ice, until implanted. The implantation should be performed as quickly as possible to avoid cell death.
3. Weigh immunocompromised mice (*see* **Note 2**) and calculate the appropriate dose of Ketamine/Xylazine (Ketamine: 80–120 mg/kg and xylazine: 10 mg/kg for induction, diluted in saline) or other suitable anesthetic. Inject anesthetic intraperitoneally.
4. Once the mouse appears asleep, perform the toe pinch reflex test to ensure complete anesthesia. That is, squeeze the mouse's hind foot between forceps or fingernails. If the anesthetic is sufficient, the mouse will not move; if not there will be a twitch response and you should wait a few more minutes, and then test again.
5. Tumor pieces can be implanted subcutaneously into the mouse flank or at the base of the neck. It is possible to grow larger tumors in the flank. Shave the implantation site and swab with betadine or other disinfectant. Make a 2-mm skin incision. While holding the incision open with forceps, use a trocar to implant and advance a small tumor piece subcutaneously 5–10 mm from the incision.
6. Seal the incision with a liquid skin sealant or a wound clip.
7. The mice must be kept warm postoperatively and should be observed until they fully recover from the anesthetic. They should be caged separately from other nonoperated mice to ensure they are not injured. The mice should be monitored at least once or twice a week for tumor growth. Once tumor growth is visible they need to be monitored more frequently to be certain they stay healthy and active.

Two implantations can be performed on each mouse, one on each side. Even with a small tumor specimen, ten small tumor pieces can usually be obtained and thus five mice implanted.

Single-cell suspensions are injected subcutaneously into the flank or base of neck, as follows:

1. Suspend the desired number of cells in HBSS/2%HICS to give 100 μL per injection. To ensure tumor growth, a minimum of 1 million cells should be injected per injection site.

2. Add 100 μL of Matrigel solution to form a final volume of 200 μL (*see* **Note 3**).
3. If desired, anesthetize the mice by inhaled isoflurane. For this method of anesthesia, an appropriate machine that mixes isoflurane with 100% oxygen and delivers the isoflurane at a flow rate of 2 L/min should be used. When removed from the isoflurane chamber or nose cone, mice will wake up within 1–2 min; thus, the injections should be performed quickly. Alternatively, those who have animal handling experience can perform subcutaneous injections without any anesthetic.
4. Clean the injection site with an ethanol swab. Using a small gauge needle (25–30 gauge) inject the mice subcutaneously at the base of the neck or the flank with the single-cell suspension. The needle should be advanced 5–10 mm subcutaneously to prevent extrusion of the sample from the injection site.

Once the tumor reaches a size of 1–1.5 cm in diameter, the mouse should be euthanized and the tumors removed and processed as described in the following paragraphs.

3.3. Tissue Processing to Generate a Single-Cell Suspension

The same method is used whether dissociating a human specimen directly from the clinic, or a tumor that has been propagated in a mouse.

1. Using sterile technique finely mince the specimen with a scalpel blade or razor blade. This is done in a 6- or 10-cm tissue culture dish, depending on the size of the specimen.
2. Add 5–10 mL of DMEM F-12 containing 300 U/mL collagenase IV, 100 U/mL hyaluronidase, and 125 U/mL of DNAse to the dish and disperse the minced tumor. Incubate at 37°C.
3. Pipette up and down with a 10-mL pipette to dissociate cells every 10–15 min until the tumor is well dissociated (*see* **Note 4**). The specimen should not be left in the enzyme solution for longer than necessary to achieve cellular dissociation (*see* **Note 5**).
4. Filter the cells through a 40-μm nylon sieve into a 50-mL conical tube. Wash the cells twice by adding 20 mL of HBSS/2% HICS and centrifuging for 5 min at 350 × g.
5. If the cell pellet appears red due to the presence of red blood cells, resuspend in 1–2 mL of red cell lysis buffer, incubate on ice for 10 min, and wash with HBSS/2% HICS again.
6. Resuspend the cells in the desired final volume of HBSS/2% HICS and perform a viable cell count with trypan blue (mix 10 μL of the cell suspension with 40 μL of trypan blue, and place 10 μL on a hemacytometer. Viable cells are those that exclude trypan blue).

At this point the cells are ready for injection of single-cell suspensions into mice (as described earlier) or staining for flow cytometry (as described in the following paragraphs).

3.4. Flow Cytometry: Analysis and Cell Sorting

3.4.1. Staining of Tumor Cells for FACS

The tumor cells should be suspended in HBSS/2% HICS at a concentration of 1 million cells per 100 μL in a 5-mL polystyrene tube. Two small aliquots of tumor cells (0.5–1 million cells) should be set aside for controls for FACS. One should be labeled "Unstained Control" and the other "Isotype Control." The remaining cells (the number depends on how many cells are obtained upon tumor processing, and can range from a few million to tens of millions) can be stained to use for analysis of cell surface molecules and cell sorting. The steps for staining are as follows:

1. For specimens derived directly from human patients, add lineage cocktail antibodies (unconjugated), at a dilution of 1:50 each (e.g., for 10 million cells, resuspend in 1 mL and add 20 μL of each antibody; *see* **Note 6**). The lineage cocktail contains antibodies to CD2, CD3, CD10, CD16, CD18, CD31, CD64, and CD140b. In the case of a tumor that has been grown in a mouse, the lineage cocktail is replaced by the appropriate anti-H2K antibody (anti-H2K^d for NOD/SCID mice or RagγDKO mice on the Balb/C background, anti-H2K^b for RagγDKO mice on the C57/B6 background) at a dilution of 1:400. Incubate on ice for 15 min (*see* **Note 7**).
2. Fill tube with HBSS/2% HICS and centrifuge for 5 min at 350×*g*.
3. Resuspend cell pellet in the same volume and add goat-anti-mouse-Tricolor, 1:50; incubate on ice for 15 min.
4. Fill tube with HBSS/2% HICS and centrifuge for 5 min at 350×*g*.
5. Resuspend cell pellet in half the original volume and add mouse IgG, 1:50; incubate on ice for 5 min. Do not wash.
6. Bring volume up to original volume and add anti-CD44-PE at a 1:50 dilution, and other directly conjugated antibodies or biotinylated antibodies that may be of interest to you; incubate on ice for 15 min.
7. Fill tube with HBSS/2% HICS and centrifuge for 5 min at 350×*g*.
8. If you have used a biotinylated antibody in **step 6**, add Streptavidin conjugated to an appropriate fluorochrome; incubate on ice for 15 min. Otherwise proceed to **step 10**.
9. Fill tube with HBSS/2% HICS and centrifuge for 5 min at 350×*g*.

10. Resuspend the cell pellet in propidium iodide diluted 1:1000 in HBSS/2%HICS, to give a final concentration of 10 million cells/mL in 1 μg/mL propidium iodide (*see* **Note 8**). If you have less than 3 million cells, resuspend in 300 μL anyway.

While performing these staining steps, in parallel stain the isotype control sample in exactly the same way, substituting an isotype control antibody for the lineage, CD44, etc. antibodies at an equivalent concentration. The unstained control tube can remain on ice until ready to run on the flow cytometer. It is essential that this sample remain completely unstained, including NOT stained with the propidium iodide viability stain.

3.4.2. Preparation of Compensation Controls for FACS

Each color used for flow cytometry must have a single-color control tube in order to allow for compensation of the machine. As it is not possible to perform routine compensations on tumor cells, it is convenient to use "CompBeads" from Becton Dickinson, which are beads that have been coated with antimouse IgG.

1. Mix together one to two drops of blank beads and one to two drops of coated beads in a 5-mL polystyrene tube. Then aliquot 20 μL of the bead mixture to give one tube for each color used, plus one tube for "no stain."
2. Add 180 μL of HBSS/2% HICS to each tube. Add the same antibodies used for staining, at the same dilution (e.g., for CD44-PE, add 4 μL of antibody to the 200 μL of beads in the tube). In the case of directly conjugated antibodies, the compensation control is ready to use (no washing is required). In the case of unconjugated primary antibodies (e.g., anti-H2K^d or lineage) or biotinylated antibodies, the CompBeads can be incubated and washed at the same time and in the same way as the tumor cells. After the incubation with the primary antibody, the beads can be resuspended in 200 μL of HBSS/2%HICS and the secondary antibody, either goat-antimouse-IgG-Tricolor or Streptavidin, added at the appropriate dilution and the tube is now ready to use (no wash necessary).

3.4.3. Flow Cytometry

Flow cytometry will vary considerably depending on the type of machine used and the experimental goals. In most cases, a flow cytometry technician will operate the machine for you.

1. Set the voltages using the "Unstained Control" sample. This is necessary, because tumor cells autofluoresce. Therefore the voltages for all the parameters used must be set to position unstained cells in the first decade prior to running compensation controls. In addition, the FSC and SSC voltages should be set to obtain good positioning of the bulk of the cells in the center of the FSC vs SSC plots.

2. Run the compensation controls. If using FACS Diva software, the controls are each acquired and saved, and then the "Calculate Compensations" function is used. Otherwise compensate manually. It is not necessary to compensate for propidium iodide, as only the negative population will be used.

3. Run the isotype control sample. For this sample, only dead cells will be stained and any positive signal in channels other than propidium iodide represent background levels of staining. The propidium iodide viability stain can be viewed in the same channel as the lineage-Tricolor, and thus both lineage + and dead cells can be gated out simultaneously. Alternatively, on a more advanced machine with appropriate filters, these two parameters can be viewed separately and gated out accordingly. After gating on the propidium iodide negative cells on the isotype control, display the gated cells on an FSC-Area vs FSC-Width plot, and gate on the "singlet" population, as shown in **Fig. 1** (this is only possible with FACS Diva software. Skip this step if you do not have FACS Diva software). Then display this population to show PE (the CD44 channel) vs another parameter, where the latter can be any channel of your choice, such as a channel in which there is no stain,

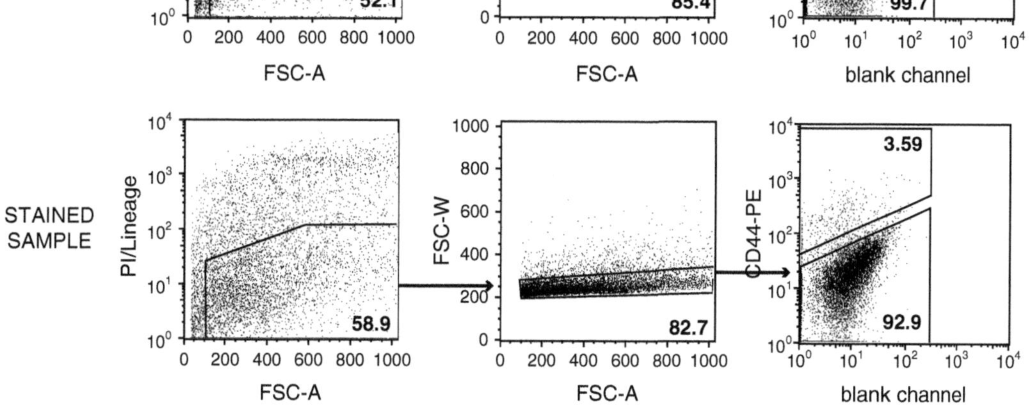

Fig. 1. Gating strategy for sorting CD44+ cells from HNSCC. First, the isotype control is run and a gate is drawn on the viable, lineage population. If using FACSDiva software, the viable, lineage population can be displayed on an FSC-Area vs FSC-Width plot, and the doublets gated out as shown. This population can then be displayed on a CD44-PE vs a blank channel plot, and the gates drawn as shown. The gate for CD44+ cells should be drawn on the isotype control such that <0.1% of the cells are within the gate. The same gates can then be applied to the stained sample.

or a channel in which there is another antibody of interest (e.g., antibody X-FITC), to determine if there is costaining with other markers. For the isotype conrol, all viable cells will be displayed in this plot and should be negative for both parameters shown. One can then set the gates for where the CD44– and CD44+ populations will be (*see* **Fig. 1**).

4. Run the stained sample. The FSC vs propidium iodide/ Tricolor plot will show the Lineage + and dead cells on the Tricolor axis. The percentage of CD44+ cells in the viable, Lin– population can be determined, and the CD44+ and CD44– cells (and any other populations of interest) can be sorted (*see* **Fig. 1**). The cells should be sorted into microfuge tubes containing 200 μL of cold HBSS/2% HICS.
5. After the first round of sorting, the sorted cells should be analyzed for purity. If the populations are <95% pure, they should be run through the sorter a second time.
6. Purified subsets of cells can then be assayed for tumorigenicity by transplantation into mice in the same way as described for single-cell suspensions of whole tumors. Alternatively, they can be used for other biological assays, such as qRT-PCR, to look for differential gene expression.

3.5. Gene Expression Analysis by qRT-PCR

To identify genes that are differentially expressed in subpopulations of cells from HNSCC, RNA is extracted from purified subsets of cells and quantitative RT-PCR performed.

3.5.1. RNA Extractions

1. Either sort cells directly into 1 mL of Trizol, or pellet them by centrifuging at 350 ×*g* and resuspend them in 1 mL of Trizol, and incubate on ice for 10 min.
2. If there are fewer than 5 × 10^5 cells, add 2 μL of 5 mg/mL linear acrylamide.
3. Add 200 μL of chloroform and shake vigorously for 15 s, and then incubate on ice for 10 min.
4. Centrifuge at 14,000 ×*g* for 15 min at 4°C.
5. Add 2 μL of linear acrylamide to a fresh tube. Transfer the upper aqueous layer to this tube, add 500 μL of isopropanol, shake vigorously for 15 s, and incubate at –20°C overnight.
6. Centrifuge at 14,000 ×*g* for 20 min at 4°C.
7. Aspirate supernatant, wash pellet gently with ice cold 75% ethanol, and centrifuge at 5,000 ×*g* for 10 min at 4°C.
8. Aspirate supernatant and briefly air dry the pellet.
9. Dissolve in 8 μL of RNAse-free water per 20,000 cells and transfer to a 200-μL microfuge tube; if you had less than 20,000 cells dissolve in 8 μL. If you had more than 20,000 cells, scale up appropriately, and aliquot RNA into 8-μL aliquots prior to freezing.

10. At this point, you may either freeze the RNA on dry ice and store at −80°C, or proceed with reverse transcription.

3.5.2. Quantitative RT-PCR

1. To 8 μL of RNA (20,000 cells-worth of RNA), add 1 μL of 10× DNAse buffer and 1 μL of DNAse. Remember to have an aliquot for the no reverse-transcriptase control.
2. Incubate at 37°C for 20 min, then 65°C for 10 min, and then transfer to ice.
3. To each tube add 1 μL of random hexamers and 1 μL of 10 mM dNTP and incubate at 65°C for 5 min, then ice for 1 min.
4. Make a mixture containing 2 μL of 10× RT buffer, 4 μL of 25 mM $MgCl_2$, 2 μL of 0.1 M DTT, and 1 μL of RNAseOUT.
5. Add 9 μL of this mixture to the DNAse-treated RNA, incubate at 25°C for 2 min, then add 1 μL of Superscript III (50 U), except for the no RT control.
6. Incubate at 25°C for 10 min, 42°C for 50 min, and 70°C for 15 min; then place on ice.
7. Add 1 μL of RNAse H and incubate at 37°C for 20 min.
8. Dilute cDNA to give 200 cell equivalents per μL (for example, if you started with 20,000 cells, dilute up to 100 μL) with water (*see* **Note 9**).
9. For PCR, order TaqMan Gene Expression Assays for the genes of interest and TaqMan Universal PCR Master Mix from Applied Biosystems. Include primers for at least two housekeeping genes, such as β-actin, GAPDH, HPRT, etc..
10. For each reaction, combine 1 μL of cDNA (200 cell equivalents) and 8 μL of water in one tube, and 10 μL of 2× master mix plus 1 μL of the TaqMan primer/probe in another. Sufficient mixture should be made to perform each reaction in triplicate. Aliquot 9 μL of cDNA mixture into wells of a 96-well optical reaction plate, or other appropriate optical reaction tubes. Add 11 μL of primer mix to each well or tube. Remember to include housekeeping gene controls and a no reverse-transcriptase control for each sample. In addition, a sample in which expression of the gene of interest is known to occur should be included (e.g., embryonic stem cells for stem cell-related genes), both to serve as a positive control for PCR, as well as a calibrator sample. Spin plate or tubes briefly to get all liquid to the bottom.
11. Run PCR on a real-time PCR machine as follows: 50°C for 2 min, 95°C for 10 min, then 40 cycles of 95°C for 15 s, 60°C for 1 min. Analyze data using software provided by your PCR machine manufacturer (if using an ABI machine, ABI Prism software).

12. Expression levels of given genes of interest should be determined relative to a housekeeping gene. The data can then be normalized to the calibrator sample (e.g., embryonic stem cells). This is done by the relative quantitation (RQ) method. First, the "threshold" of the reaction is determined, which can be done automatically by the software program, or manually. The threshold is the level above baseline that is sufficiently low to be within the exponential growth region of the amplification curve. The C_t is the cycle at which the amplification curve crosses the threshold. The δC_t is the difference between the C_ts of 2 different genes. If $\delta C_t = n$, then the difference in expression between the two genes is 2^n-fold. The housekeeping gene which gives the same C_t in all samples should be chosen. The expression levels of genes of interest relative to the housekeeping gene can then be calculated and normalized to the positive control calibrator sample.

3.6. Immunohistochemistry and Immunofluorescence

Immunostaining of sections of HNSCC specimens can be performed to determine the physical location of cells expressing proteins of interest within a tumor sample. Costaining can determine whether two different proteins are expressed in the same cell. Well- and moderately differentiated tumors have a distinct tumor architecture, in which the "basaloid" regions lie adjacent to tumor stroma, and it is in these regions that the CD44+ cells with nuclear Bmi1 reside.

When a specimen is received from the clinic, a portion should be preserved for sectioning and staining as follows:

1. Squeeze some embedding medium for frozen tissue into a cryomold.
2. Lay tissue piece into mold, and cover with more embedding medium and freeze on dry ice. Store at –80°C.
3. Cut 7-micron sections at –20°C on a cryostat, using positively charged glass slides to pick up the sections.
4. Fix slides in ice cold acetone for 5 min, and allow to air dry. Slides can now be stored at –80°C.

3.6.1. Immunohistochemistry for CD44

1. Rinse slides in PBS/0.1% Tween to remove embedding medium. Be sure to have an extra section for staining with an isotype control antibody.
2. Prepare a solution of PBS with 0.5% BSA, 0.1% sodium azide, 0.3% hydrogen peroxide, and 5% mouse serum. Prepare enough to have 100–200 μL per section.
3. Dry around sections with a kimwipe and draw a circle around the sections with a liquid-repellent slide marker pen.
4. Place slides in a humidified chamber (*see* **Note 10**) and pipette 100–200 μL onto each section and incubate at room temperature for 30 min.

5. While blocking, prepare solution of CD44-biotin diluted 1:50 in PBS/0.1% Tween/0.5% BSA. Make enough to have 100–200 μL per section. In parallel, prepare a biotinylated isotype control at the same concentration.
6. Aspirate blocking solution from sections. Without washing, pipette antibody solution onto sections (either CD44-biotin, or isotype-biotin). Incubate at room temperature for 1 h.
7. Wash slides three times in PBS/0.1% Tween, for 5 min each.
8. Prepare solution of Streptavidin-HRP diluted 1:500 in PBS/0.1% Tween.
9. Dry around sections with a kimwipe, ensure circle previously drawn with water-repellent pen is still intact and reinforce if necessary. Lay slides in humid chamber and pipette Streptavidin-HRP solution onto sections. Incubate at room temperature for 30 min.
10. Wash slides three times in PBS/0.1% Tween for 5 min each.
11. Prepare DAB solution using DAB substrate kit or alternative from another supplier (follow instructions supplied by the manufacturer).
12. Dry slides as before, pipette DAB solution onto sections, and incubate at room temperature for 2–10 min. Watch slides, and when brown color begins to appear, stop reaction by transferring slides to water. Incubate in water for 5 min.
13. Counterstain and mount slides as follows:
    a) Dip in hematoxylin for 1 min.
    b) Rinse in tap water for 30 s.
    c) 5 dips in clarifier.
    d) tap water for 30 s.
    e) 5 dips in bluing solution.
    f) tap water for 1 min.
    g) 10 dips in 95% ethanol, twice.
    h) 10 dips in xylene, twice.
    i) coverslip with mounting media such as Permount or Histomount.
14. Allow slides to dry for several hours or overnight before viewing on microscope.

    An example of CD44 staining on HNSCC is shown in **Fig. 2**.

3.6.2. Immunofluorescence for CD44 and Bmi1

1. Rinse slides in PBS/0.1% Tween to remove embedding medium.
2. Prepare blocking solution: PBS/0.1% Tween/0.5% BSA/5% goat serum. Prepare enough to have 100–200 μL per section.

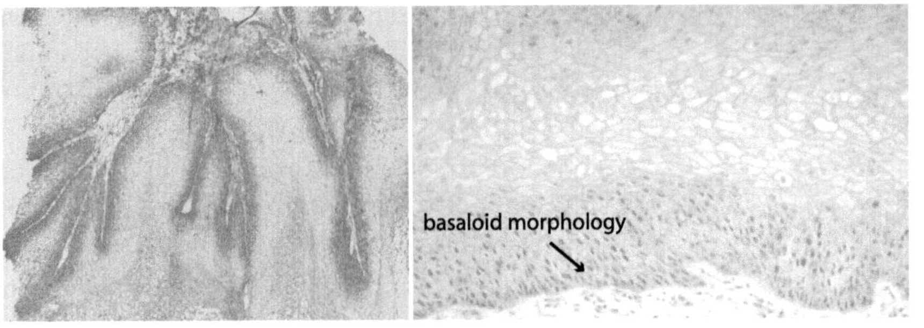

Fig. 2. Immunohistochemistry for CD44 on a well-differentiated HNSCC specimen. On the *left*, a low-power image illustrates the regional staining pattern for CD44. On the *right*, a higher magnification allows observation of the morphology of the CD44+ cells. The CD44+ cells are situated in close proximity to tumor stroma and have a basaloid morphology.

3. Dry around sections with a kimwipe and draw a circle around the sections with a water-repellent pen. Place slide in a humidified chamber and pipet blocking solution onto sections. Incubate at room temperature for 1 h. Remember to have at least one section for the isotype controls.
4. Prepare Bmi1 antibody, and mouse IgG isotype control, each diluted 1:100 in blocking solution. Prepare enough to have 100–200 μL per section.
5. Aspirate blocking solution from slides (do not wash slides) and pipette Bmi1 antibody or isotype control onto sections. Close humidified chamber and incubate at 4°C overnight.
6. Wash slides three times in PBS/0.1% Tween for 5 min each.
7. Dry around sections with kimwipe, ensure circle previously drawn with water-repellent pen is still intact, and reinforce if necessary. Lay slides in humid chamber and pipette a solution of goat-antimouse conjugated to Alexa-594, diluted 1:200 in PBS/0.1% Tween onto sections. Incubate at room temperature for 1 h. Slides should be protected from light from now on (humid chamber can wrapped in aluminum foil).
8. Wash slides three times in PBS/0.1% Tween for 5 min each.
9. Dry slides as before and add solution of mouse IgG diluted 1:50 in PBS/0.5% BSA/0.1% Tween. Incubate for 30 min at room temperature.
10. Aspirate blocking solution from slides (do not wash slides) and add 1:50 CD44-biotin or isotype control-biotin in PBS/0.5% BSA/0.1% Tween. Incubate for 1 h at room temperature.
11. Wash slides three times in PBS/0.1% Tween for 5 min each.
12. Dry slides as before, add a solution of Avidin-Alexa-488 diluted 1:200 in PBS/0.1% Tween, and incubate for 1 h at room temperature.

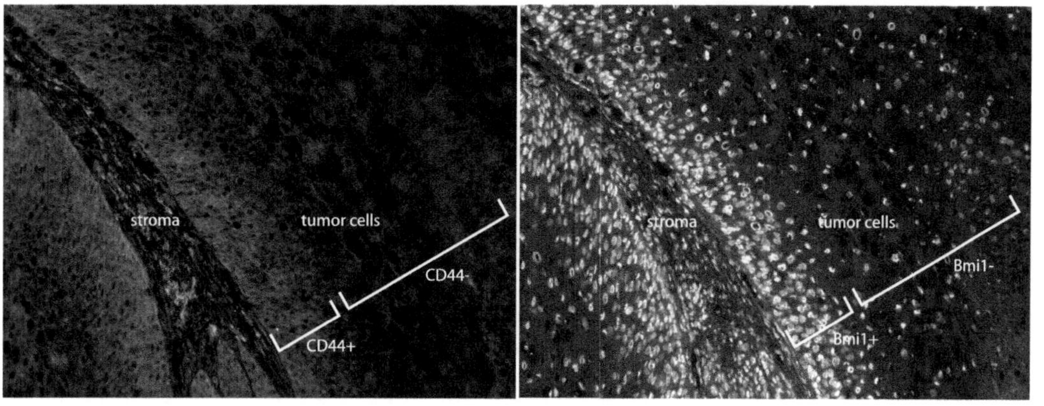

Fig. 3. Immunofluorescent costaining for CD44 (*left*) and Bmi1 (*right*). Nuclear Bmi1 is present in the CD44+ population, with the cells adjacent to the stroma being the most densely stained for Bmi1.

13. Wash slides three times in PBS/0.1% Tween for 5 min each.
14. Immerse slides in PBS/0.1% Tween with 10 μg/mL Hoechst 33342 for 1 min.
15. Rinse slides briefly in PBS/0.1% Tween, and coverslip with Fluoromount, or other equivalent aqueous mounting media. Keep slides in the dark.
16. Allow slides to dry for 10 min before viewing on fluorescence microscope.
17. An example of CD44 and Bmi1 costaining on an HNSCC specimen is shown in **Fig. 3**.

4. Notes

1. The xenografted tumor will generally appear sooner from a tumor piece than from a cell suspension. However, sometimes tumors derived from implanted pieces will ulcerate and/or form cysts, thus decreasing the number of viable cells that can be obtained. Preliminary data suggest that injection of cell suspensions decreases the occurrence of this phenomenon.
2. NOD-SCID *(2)* or RagγDKO *(3)* mice are good recipients for obtaining high take rates in our hands. Other strains of immunocompromised mice may also be appropriate, but must be tested. Ideally, strains of mice that are deficient in T, B, and NK cell function have the highest probability of achieving a high frequency of tumor engraftment.

3. Matrigel solidifies at room temperature. When the Matrigel is received, it should be thawed on ice, aliquotted using a syringe with a large gauge (e.g., 18 gauge) needle, and refrozen. When ready to use for an experiment, the desired volume should again be thawed on ice and measured using a 1-cc syringe with a large gauge needle. Once added to the cells, they should be mixed by gently vortexing or tapping the tube, and should always be maintained on ice. The injections should be performed quickly to prevent the Matrigel from solidifying before the injections are complete. Prechilling the needles on ice should help to prevent premature solidifying of the Matrigel/cell solution.
4. If it is initially not possible to get the tumor slurry to go into a 10-mL pipette, you can start with a 1-mL pipette tip, which has had the tip cut off to make the hole bigger. Be sure to use barrier tips if you do this, as sometimes the chunky mixture can be suddenly sucked up into the tip and contaminate your pipettor.
5. If you want to more carefully monitor cell dissociation, you can at regular intervals, remove a 10-μL aliquot from the plate and place it directly onto a hemacytometer for observation under the microscope.
6. All antibody dilutions have been determined for the particular antibodies described here. If antibodies are purchased from a different company, or different clones are used, they should be titrated to determine their optimal concentration for staining.
7. A more simplified version of the lineage cocktail can be used that contains antibodies to human CD45 (a pan-hematopoietic marker) and CD31 (an endothelial cell marker). In this case, biotinylated antibodies can be used, thus simplifying the staining procedure significantly to the following steps:
   (a) Blocking step with 1:50 mouse IgG.
   (b) Cocktail of antibodies, including CD45-bio, CD31-bio, CD44-PE, other fluorescently conjugated antibodies of interest to you.
   (c) Avidin-PECy5 or other color of your choice.

However, there may be some cell types that are not excluded with this cocktail, in particular fibroblasts. In the case of tumors grown in mice, a biotinylated H2K antibody can be used instead of the unconjugated one, again simplifying the procedure. However, be aware that in both cases, this precludes using a biotinylated antibody for staining for alternative markers that you may be interested in looking at.

8. An alternative viability stain is 1 μg/mL of DAPI, if you have the appropriate laser on your FACS machine to detect it (either a UV or violet laser). This has the added advantage of minimal overlap with other fluorochromes typically used.

9. This protocol is designed assuming that small numbers of cells and thus small quantities of RNA will be obtained. In this situation, it is not practical to O.D. the RNA or cDNA to determine the quantity. Instead, RT-PCR reactions are set up with "cell equivalents" of cDNA. Two hundred cell equivalents should generally be sufficient to perform a single PCR reaction, but if larger numbers of cells are obtained, it may be desirable to use more cDNA for the PCR reactions to allow better detection.
10. A humidified chamber can be made by putting damp paper towels into a plastic box or Tupperware container. Place some pipettes in the box to raise the slides from the paper towels. If doing immunofluorescent staining, the box can be covered in aluminum foil to block out light that may bleach the fluorochromes.

References

1. Prince, M. E., Sivanandan, R., Kaczorowski, A., Wolf, G. T., Kaplan, M. J., Dalerba, P., Weissman, I. L., Clarke, M. F., and Ailles, L. E. (2007) Identification of a subpopulation of cells with cancer stem cell properties in head and neck squamous cell carcinoma. *Proc Natl Acad Sci U S A* **104**, 973–8.
2. Shultz, L. D., Schweitzer, P. A., Christianson, S. W., Gott, B., Schweitzer, I. B., Tennent, B., McKenna, S., Mobraaten, L., Rajan, T. V., Greiner, D. L., and et al. (1995) Multiple defects in innate and adaptive immunologic function in NOD/LtSz-scid mice. *J Immunol* **154**, 180–91.
3. Goldman, J. P., Blundell, M. P., Lopes, L., Kinnon, C., Di Santo, J. P., and Thrasher, A. J. (1998) Enhanced human cell engraftment in mice deficient in RAG2 and the common cytokine receptor gamma chain. *Br J Haematol* **103**, 335–42.

Chapter 12

Pituitary Adenoma Stem Cells

Patrizia Tunici and John S. Yu

Summary

The identification of a subpopulation of brain tumor cells with potent tumorigenic capacity strengthens the cancer stem cell hypothesis of the origin of the tumors that has recently attracted the attention of many researchers. Reports have been published on the identification of tumor cells with stem cells characteristics in different types of tumors (acute myelogenic leukemia, breast cancer, prostate cancer, bone sarcomas, liver cancer, and melanomas). We and other groups have previously reported the isolation of cancer stem cells from adult glioblastoma multiforme. These cells express stem cell markers, and when differentiated they express glial and neuronal markers. In vivo they give a tumor that recapitulates the characteristics of the tumor in the patient. More recently we have isolated tumor stem-like cells also from benign tumors like pituitary adenomas. Cells derived from pituitary adenomas are able to grow as floating aggregates resembling the neurospheres (typical of normal stem cells) in a medium supplemented by growth factors (EGF and bFGF). The immunocytochemical analysis revealed that pituitary tumor stem-like cells are positives for nestin and, when grown for ten days in differentiation medium they express GFAP, BIII tubulin, and S-100. In vitro tumor stem-like cells derived from a patient with a somatotroph adenoma showed high production of growth hormone and prolactin, while cells derived from the same patient but grown in presence of fetal bovine serum showed no production of hormones.

Key words: Pituitary adenoma, Tumor stem cells, Epidermal growth factor, Basic fibroblast growth factor, Growth hormone, Prolactin

1. Introduction

Pituitary adenomas typically are benign neoplasm of epithelial origin and are classified on the basis of plasma hormone levels or immunohistochemical staining in (1) prolactinomas that are the most common and cause amenorrhea, galactorrhea, and infertility in

John S. Yu (ed.), *Cancer Stem Cells*, Methods in Molecular Biology, vol. 568
DOI 10.1007/978-1-59745-280-9_12, © Humana Press, a part of Springer Science+Business Media, LLC 2009

females and hypogonadism in male; (2) somatotrophic adenomas that secrete an excess of growth hormone and cause gigantism in children and acromegaly in adults.'(3) adrenocorticotropic hormone (ACTH)-secreting adenomas that produce Cushing's disease; 4. gonadrotophic (secreting luteinizing hormone and follicle-stimulating hormone) and (5) thyrotropic adenomas are rare and the latter cause hyperthyroidism *(1, 2)*. Some the pituitary adenomas do not secrete hormones and are classified as null cell adenomas *(3)*. The classification of pituitary adenomas depends also on the size of the tumors that can be greater than 10 mm of diameter (macroadenomas) or less than10 mm of diameter (microadenomas) *(4)*.

Recently it was reported that brain tumors contain a subpopulation of tumor stem-like cells and that these cells are responsible for the origin and tumorigenicity of the tumors *(5–9)*. No evidence was reported before for the presence of this subpopulation of tumor stem-like cells also in benign tumors like pituitary adenomas. Here we report protocols and methods to isolate tumor stem-like cells also from pituitary tumors. Specimens of pituitary adenomas were placed in culture conditions that favor the growth of neurospheres, and in vitro experiments were performed to confirm the stem cell properties of these pituitary tumor stem-like cells.

2. Materials

2.1. Cell Culture

1. Spheres tumor cells were cultured in Dulbecco's Modified Eagle's Medium (DMEM)/ F12 Nutrient mixture (1:1, Gibco/BRL) supplemented with B27 1× (Gibco/BRL), penicillin/streptomycin (200 U/ml; Gibco/BRL), fungizone (250 ng/ml, Gibco/BRL), EGF (20 ng/ml; Peprotech), and bFGF (20 ng/ml; Peprotech)
2. Adherent cells were cultured in DMEM/F12 (1:1), 10% Fetal Bovine Serum (Omega Scientific), penicillin/streptomycin (200 U/ml), and Glutamax 1× (Gibco/BRL).
3. *Dissection Medium.* Hank's Balanced Salt Solution 1× (Gibco/BRL), 0.4% Glucose, 1 mM $MgCl_2$, 30 mg/L Catalase, 6 mg/L Deferoxamine, 25 mg/L *N*-acetyl-cysteine, 1.25 mg/L superoxide dismutase, 250 ng/ml fungizone (Gibco/BRL), 110 mg/L sodium pyruvate, 10 mM Hepes buffer
4. *Trypsin solution.* Trypsin/0.25% EDTA solution 1X (Gibco/BRL)

2.2. Immunocytochemistry

1. *PBS 1X/4% paraformaldehyde.* paraformaldehyde was weighed carefully under a chemical hood, and PBS 1× heated at 50°C was carefully added and the solution was let stir until complete dissolving of paraformaldehyde.
2. 0.1% Triton X-100 was freshly prepared in PBS 1×.
3. 10% goat serum was freshly prepared in PBS 1×.
4. *Primary antibodies.* GFAP (Dako, Denmark), betaIII Tubulin (Covance, Berkeley, CA), nestin (Chemicon, Temecula, CA), and CD133 (Miltenyi Biotec, Auburn, CA)
5. *Secondary antibodies.* Cy3-conjugated Affinipure goat anti-rabbit IgG (Jackson ImmunoResearch, Bar Harbor, MD), FITC-conjugated Affinipure goat antimouse IgG (Jackson ImmunoResearch, Bar Harbor, MD)
6. *Mounting medium containing DAPI (Vector Laboratories).* one drop on the top of each slide.

2.3. ELISA Assays

1. ELISA kits for GH, PRL, FSH, LH, TSH were from Anogen, for ACTH from R&D Systems.

3. Methods

3.1. Preparation of Cancer Stem-Like Neurospheres from Pituitary Adenomas

1. Tumor samples from eight pituitary adenoma patients (four pituitary macroadenomas null-cell, two somatotroph (growth hormone producing) pituitary adenomas and two pituitary macroadenomas with acromegaly) were collected within half an hour from the surgical resection as approved by the Institutional Review Board at Cedars Sinai Medical Center.
2. Under a biohazard hood tissues were transferred onto a 50-ml centrifuge tube and weighed using a standard laboratory balance.
3. Part of the tissue was cut and transferred onto labeled cryovials and stored at –80°C for further assay.
4. The remaining specimen was transferred onto a 100-mm cell culture dish and washed three times with 10 ml of PBS 1X to eliminate blood eventually present.
5. Tissue was then transferred onto a new 100-mm cell culture dish and dissected in a *Dissection medium* containing catalase, deferoxamine, N-Acetyl cysteine, and superoxide dismutase (*see Subheading* 2 for details) using sterile scissors, razors, and forceps. Tissue was cut in very small pieces.

6. The dissected tissue was then transferred onto a 50-ml centrifuge tube and centrifuged at 70 x *g* for 1 min and 30 sec in a bench centrifuge.
7. After discarding the supernatant the dissected tissue was resuspended in 10 ml of a *Trypsin solution* and digested at 37°C for at least 10 min (*see* **Note 1**) with constant agitation.
8. Trypsin was then neutralized by adding 30 ml of PBS 1× in the tube. The mixture was then centrifuged at 70 × *g* for 1 min and 30 sec in a bench centrifuge.
9. The digested tissue was resuspended in 10 ml of PBS 1× and triturated by passing it in a tissue sieve (0.38 mm, 40 mesh) and recovering the cells in a 100-mm dish.
10. The recovered cells were then passed through a 70-μm cell strainer (BD Biosciences) and collected in a 50-ml centrifuge tube.
11. Cells were centrifuged at 168 × *g* for 5 min and counted using a hemocytometer chamber.
12. Cells were plated at the density of $1x10^5$ cells/ml in a medium containing DMEM/F12 (1:1) (Gibco), 10% FBS (Omega Scientific), penicillin/streptomycin (200 U/ml; Gibco), and Glutamax 1X (Gibco).
13. After 24–48 h medium was changed with a medium containing DMEM/F12 (1:1) (Gibco), B27 1X (Gibco), penicillin/streptomycin (200 U/ml; Gibco), fungizone (250 ng/ml), EGF (20 ng/ml), and bFGF (20 ng/ml) (*see* **Note 2**).
14. For each cell line daughter cells growing adherent in 10% FBS containing medium were prepared.

3.2. Immunocytochemistry Analysis of Cancer Stem-Like Neurospheres Derived from Pituitary Tumors

1. Neurospheres from pituitary tumors were plated onto polylysine-coated chamber slides at a density of $5x10^3$ cells/ml in the complete medium with or without growth factors and let grow for 10 days to allow cells to differentiate.
2. Cells were fixed in PBS 1×/ 4% paraformaldehyde for 20 min at room temperature
3. Cells were washed with PBS 1× for three times.
4. Cells were then permeabilized in 0.1% Triton X-100 for 15 min at room temperature and blocked with 10% goat serum in PBS 1× for 30 min at room temperature.
5. Cells were then incubated with primary antibodies: GFAP (1:200), betaIII Tubulin (1:400), nestin (1:50), and CD133 (1:200) (*see* **Note 3**).
6. After incubation with FITC or Cy3-conjugated secondary antibodies (1:300) slides were counterstained with a mounting medium containing DAPI before examination by fluorescence microscopy.

3.3. Determination of Pituitary Hormone Production from Cancer Stem-Like Neurospheres Derived from Pituitary Adenomas

1. To quantify the pituitary hormone production from the tumor stem-like cells derived from the pituitary tumors, $5x10^5$ cells from each cell line were plated in five wells of a six-well plate.
2. For each cell line the daughter adherent cells were plated at the same density in parallel in other plates.
3. Conditioned medium from each cell line was collected 2, 4, 8, 16, and 24 h after the plating and kept at –80°C until use.
4. ELISA kits were used for the quantification of the pituitary hormones and the suggested protocols were followed (*see* **Note 4**).

3.4. Results and Conclusions

Using the cell culture conditions reported, we were able to separate from the pituitary adenoma cell population, cells that respond to EGF and bFGF (namely tumor stem-like cells) from the more differentiated tumor daughter cells. These stem-like cells grow *in vitro* as floating neurospheres and, when dissociated, they are able to re-form spheres. Immunocytochemical staining reveals that pituitary tumor stem-like cells are strongly positive for nestin (a neural stem cell marker)(**Fig. 1**) and that, under differentiation conditions, cells express markers of astrocytes and neurons, demonstrating a capacity of these cells to differentiate in multilineage

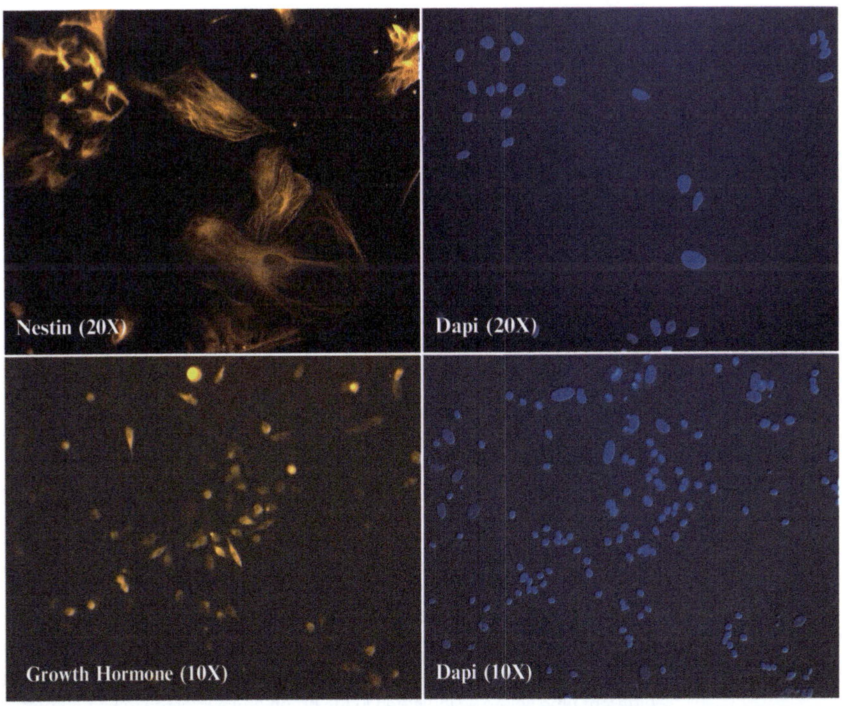

Fig. 1. Immunocytochemistry of pituitary adenoma cells. Pituitary adenoma stem-like cells were stained with nestin and growth hormone antibodies. In the left the correspondent DAPI staining are reported. Pictures are reported at the indicated magnifications. Note the staining of neurofilaments inside the cytoplasm typical of nestin labeling.

direction, a property that is typical of stem cells. In addition they showed staining also for S-100, a protein reported to be present in stellate cells of the pars distalis and tuberalis, in the marginal cells and in pituicytes of the neural lobe of pituitary *(10–11)*. Indeed the correspondent adherent daughter cells have a typical fibroblast-like phenotype.

In one tumor stem-like cell line derived from a patient with a somatotroph adenoma, we found a higher production of prolactin and growth hormone in the spheres cells while the adherent counterpart were not hormone producing (**Fig. 2**).

The protocols and methods presented in this chapter suggest that it is possible to obtain a subpopulation of tumor stem-like cells also in benign tumors like pituitary adenoma, and that these cells possess stem cells characteristics.

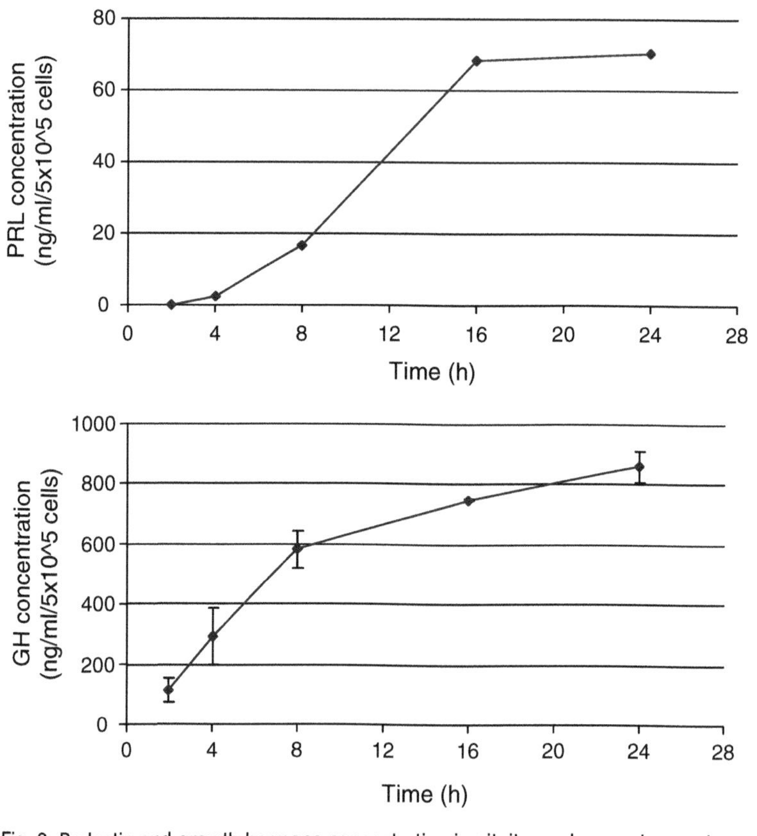

Fig. 2. Prolactin and growth hormone concentration in pituitary adenoma tumor stem-like cells. Tumor stem-like cells and daughter adherent cells from pituitary adenomas were plated at the density of $5x10^5$ cells/ml. The conditioned medium was collected at different time points as indicated in the graph and subjected to ELISA immunoassay. The results are the average ± standard deviation of two independent experiments and all values are calculated by subtracting the blank and the value of unconditioned medium. PRL = prolactin; GH = growth hormone.

4. Notes

1. The incubation time with trypsin solution strictly depends on specimen size. If sample is up to 1 g the incubation time has to be increased to 20–25 min. The agitation of the sample is critical for homogeneous digestion.
2. Cells derived from pituitary adenoma growing as spheres were refed every two days with fresh medium containing the growth factors.
3. Antibodies' dilutions were performed in fresh 10% goat serum in PBS 1X at the indicated concentrations in a final volume of 200 μl for each slide. Slides were put on a rotatory shaker during the incubation time.
4. ELISA assays were performed following the manufacturer-suggested protocols. Since each protocol differs in one or more step it is not possible to describe a standard protocol. In each experiment a standard curve was used as reference.

References

1. Laws, E.R., Jane, J.A. Jr. (2001) Pituitary tumors – long term outcomes and expectations. *Clin. Neurosurg.* **48**, 306–319.
2. Levy, A. (2004) Pituitary disease: presentation, diagnosis, and management. J. Neurol. Neurosurg. Psychiatry 75, 47–52.
3. Greenman, Y. and Melmed, S. (1996) Diagnosis and management of nonfunctioning pituitary tumors. Annu. Rev. Med. 47, 95–106.
4. Melmed, S. (2003) Mechanisms for pituitary tumorigenesis: the plastic pituitary. J. Clin. Invest. 112, 1603–18.
5. Singh, S.K., Clarke, I.D., Terasaki, M., Bonn, V.E., Hawkins, C., Squire, J., Dirks, P.B. (2003) Identification of a cancer stem cell in human brain tumors. Cancer Res. 63, 5821–5828.
6. Singh, S.K., Hawkins, C., Clarke, I.D., Squire, J.A., Bayani, J., Hide, T., Henkelman, R.M., Cusimano, M.D., Dirks, P.B. (2004) Identification of human brain tumour initiating cells. *Nature* **432**, 396–401.
7. Hemmati, H.D., Nakano, I., Lazareff, J.A., Masterman-Smith, M., Geschwind, D.H., Bronner-Fraser, M., Kornblum, H.I. (2003) Cancerous stem cells can arise from pediatric brain tumors. *Proc. Natl. Acad. Sci. USA* **100**, 15178–15183.
8. Polyak, K. and Hahn, W.C. (2005) Roots and stems: stem cells in cancer. *Nature Medicine* **11**, 296–300.
9. Yuan, X., Curtin, J., Xiong, Y., Liu, G., Waschsmann-Hogiu, S., Farkas, D.L., Black, K.L., Yu, J.S. (2004) Isolation of cancer stem cells from adult glioblastoma multiforme. *Oncogene* **23**, 9392–9400.
10. Horvath, E., Kovacs, K. (2002) Folliculo-stellate cells of the human pituitary: a type of adult stem cell? *Ultrastruct. Pathol.* **26**, 219–28.
11. Giometto, B., Miotto, D., Botteri, M., Alessio, L., Scanarini, M., An, S.F., Tavolato, B. (1997) Folliculo-stellate cells of human pituitary adenomas: immunohistochemical study of the monocyte/macrophage phenotype expression. *Neuroendocrinology* **65**, 47–52.

Chapter 13

Genome-Wide DNA Methylation Profiling: The mDIP-Chip Technology

Yin Shen, Shaun D. Fouse, and Guoping Fan

Summary

Aberrant DNA methylation is one of the major characteristics of tumor cells in addition to genetic and other epigenetic alterations. Evidence shows that both regional hypermethylation and global hypomethylation can occur in cancer cells. Increased DNA methylation can be found at select tumor-suppressor gene promoters, causing the silencing of these genes in tumorigenic cells. At the same time, a global decrease in DNA methylation is frequently observed in cancer cells, which may contribute to genome instability. Unlike genetic mutations, hypermethylation at tumor-suppressor gene promoters can be reversed with epigenetic therapy by using DNA demethylating agents.

To better understand the mechanisms of cancer initiation and progression, and to better assess the effects of epigenetic therapy, a reliable high-throughput method for genome-wide DNA methylation analysis is needed. Recently, the process of coupling methylated DNA immunoprecipitation (mDIP) with microarray hybridization has been proven to be a successful strategy to map genome-wide DNA methylation patterns in different cell types.

Key words: DNA methylation, CpG island, Immunoprecipitation, Microarray, Cancer

1. Introduction

DNA methylation refers to the addition of a methyl group to the 5th carbon of the cytosine pyrimidine ring. DNA methylation is universal in vertebrates and is part of the epigenetic code, which refers to heritable change in gene function without a change in DNA sequence. DNA methylation typically occurs at CpG dinucleotides and is catalyzed by a family of DNA methyltransferases (DNMTs) that include de novo DNA methyltransferases (DNMT3A and DNMT3B) and the maintenance DNA

John S. Yu (ed.), *Cancer Stem Cells,* Methods in Molecular Biology, vol. 568
DOI 10.1007/978-1-59745-280-9_13, © Humana Press, a part of Springer Science+Business Media, LLC 2009

methyltransferase (DNMT1) *(1, 2)*. Approximately 70% of CpGs in the mammalian genome are methylated, most of which are in heterochromatic regions containing various repetitive elements. However, CpG islands, which are defined as stretches of DNA containing a high concentration of CpG dinucleotides and frequently locating in the gene promoter region, are mostly unmethylated in normal cells and tissues.

It is well established that cancer cells exhibit abnormal DNA methylation patterns compared to normal cells. Using classic methylation assays such as Southern blot analysis and bisulfite genomic sequencing, we know that the global level of DNA methylation, especially in repetitive regions, is reduced in cancer cells. Concurrently with global hypomethylation, silencing of tumor-suppressor genes by DNA hypermethylation in the promoter CpG islands is frequently observed in tumor cells *(3–5)*. In fact, recent data suggest that genetic mutations in tumor-suppressor genes may only count for a small fraction of cancer cases and a majority of cancers could be attributed to epigenetic alterations *(6, 7)*. These epigenetic alterations include DNA methylation changes, chromatin alterations, and loss of imprinting *(8)*.

While determining the methylation pattern of a particular locus is well established with the bisulfite sequencing method, the tools for mapping genome-wide DNA methylation are just emerging *(9–14)*. We describe here a powerful method that can profile levels of DNA methylation at the genome-wide scale in gene promoters, CpG islands, introns and exons, and intergenic regions. This method first enriches methylated DNA fragments through immunoprecipitation of the methylated DNA with a monoclonal antibody against 5-methyl-cytosine *(11, 15)*. By coupling methylated DNA immunoprecipitation (mDIP) with DNA chip technology (mDIP-Chip), we can obtain a picture of genome-wide DNA methylation patterns. This method has been successfully used to describe promoter DNA methylation patterns in human normal and cancer cell lines, as well as the entire genome methylation in plant *Arabidopsis thaliana* *(11, 13, 14, 16)*.

2. Materials

2.1. DNA Extraction from Cells

1. *Trypsin–EDTA*. 0.25% Trypsin, 2.21 mM EDTA in HBSS (Fisher).
2. *Cell lysis buffer*. 100 mM Tris–HCl, 200 mM NaCl, 5 mM EDTA of pH 8.0, and 0.2% SDS (w/v) (*see* **Note 1**).

3. *Proteinase K (Fisher, molecular biology grade) solution (10 mg/mL).* Proteinase K is dissolved in H_2O to 20 mg/mL, and then an equal volume of 100% glycerol is added. $CaCl_2$ is added to 0.1 mM to stabilize the solution. The Proteinase K solution is aliquoted and stored at –20°C.
4. DNase-free RNase A (Sigma) is dissolved in H_2O at a concentration of 10 mg/mL, aliquoted, and stored at –20°C.
5. Ethyl Alcohol, 100%, molecular biology grade.
6. *TE buffer (pH 8.0).* 10 mM Tris–Cl (pH 8.0) and 1 mM EDTA (pH 8.0).

2.2. Immunoprecipitation

1. Mouse monoclonal antibody against 5-methylcytidine (Eurogentec, # MMS-900P-B) at a concentration of 1 μg/mL in 20 μL aliquots, stored at –20°C (*see* **Note 2**).
2. Dynabeads (M-280 sheep antimouse IgG, Invitrogen).
3. *1× PBS, 0.1% BSA.* Weigh 0.5 g of BSA and dissolve in 50 mL of 1× PBS, filter sterile, and store at 4°C.
4. *IP buffer.* For 10× stock, 100 mM Tris–Cl of pH 7.5, 500 mM NaCl, 10 mM EDTA. For making 1× IP buffer, 10× IP buffer is diluted with H_2O, store this buffer at 4°C.
5. *Low-salt wash solution.* Add 0.3 mL 5 M NaCl to 9.7 mL 1× IP buffer, store at 4°C.
6. *High-salt wash solution.* Add 0.6 mL 5 M NaCl to 9.4 mL 1× IP buffer, store at 4°C.
7. *Elute buffer with 1.5% SDS.* 10% SDS is added to 1× IP buffer so that the final concentration of SDS is 1.5%.
8. *Elute buffer with 0.5% SDS.* 10% SDS is added to 1× IP buffer so that the final concentration of SDS is 0.5%.
9. *Elute buffer with 0.1% SDS.* 10% SDS is added to 1× IP buffer so that the final concentration of SDS is 0.1%.
10. Phenol/Chloroform/Isoamyl Alcohol (25:24:1 mixture) (Fisher).

2.3. DNA Labeling and Hybridization to Microarray

1. BioPrime® Array CGH Genomic Labeling System (Invitrogen).
2. Cy3- dUTP and Cy5-dUTP (1 mM) (Perkin Elmer).
3. Nanodrop ND-1000 spectrophotometer (Nanodrop).
4. *Human Cot-1 DNA (Invitrogen).* 1 mg/mL store at –20°C.
5. *Yeast tRNA (Invitrogen).* Reconstitute in RNase-free H_2O to 5 mg/mL in 1 mL aliquots, store at –20°C.
6. Agilent 2× hybridization buffer (Agilent).
7. Agilent hybridization oven.

8. *20× SSC.* Dissolve 175.3 g NaCl, 88.2 g sodium citrate in 800 mL of H_2O. Adjust pH to 7.0 with a few drops of 14N HCl, and then bring the volume to 1 L with H_2O. Autoclave.
9. *Wash buffer 1.* Mix 25 mL of 20× SSC, 0.5 mL of 10% of Triton-X 102 and 475 mL of H_2O, filter sterile.
10. *Wash buffer 2.* Mix 5 mL of 20× SSC, 0.5 mL of 10% Triton-X 102 and 495 mL of H_2O, filter sterile.
11. Acetonitrile (Fisher).
12. Agilent Stabilization and Drying solution (Agilent).

2.4. Slide Scanning and Data Extraction

1. Agilent scanner (Carousel, 48-position).
2. Feature extraction software (Agilent).

2.5. Data Verification

1. *Bgl*II (New England Biolab).
2. Sodium Bisulfite (Sigma).
3. Hydroquinone (Sigma).
4. *6.3 M Sodium Hydroxide.* Dissolve 2.52 g of NaOH in 10 mL of H_2O.
5. Wizard® DNA Clean-up System (Promega).
6. 10 M Ammonium Acetate pH 7.3.
7. *Glycogen (Invitrogen).* 20 μg/μL store at –20°C.

3. Methods

3.1. DNA Extraction and Preparation for mDIP

1. Cultured cells are harvested by Trypsin–EDTA digestion and spun down at 140 × *g* for 5 min. Supernatant is aspirated and the cell pellet (5 × 10^6 cells) is resuspended with 500 μL of DNA lysis buffer. 5 μL of proteinase K solution is added to a final concentration of 100 μg/mL. This step is also used for DNA extraction from dissected tissues (*see* **Note 3**).
2. The mixed cell lysis solution is incubated at 37°C or 55°C oven with agitation for overnight digestion.
3. For tissue samples, spin the cell lysis mixture at 16,100 × *g* for 15 min. Transfer the supernatant to a clean tube. This is the fraction you will use to extract DNA for the next step.
4. RNAse A solution is added to a final concentration of 20 μg/mL and the cell lysis is incubated for another 1–2 h at 37°C with agitation.
5. Add equal volume of Phenol/Chloroform/Isoamyl Alcohol (500 μL) to the mixed cell lysis solution. Shake the tube vigorously for 15 s (*see* **Note 4**).

6. Centrifuge the mixture at 16,100 × *g* for 10 min.
7. Transfer the supernatant with pipet to a clean Eppendorf tube (*see* **Note 5**).
8. Add 1/10 volume of 3 M NaOAc and 2 volumes of 100% EtOH. Mix well by inverting the tube several times (*see* **Note 6**).
9. Put the tube at –80°C for 30 min to help precipitate the DNA.
10. Spin down DNA at 16,100 × *g* at 4°C for 20 min.
11. Aspirate the supernatant and wash the DNA pallet with 70% of EtOH. Briefly vortex the tube so that the pellet is just lifted off the bottom of tube.
12. Centrifuge at 16,100 × *g* for 5 min and aspirate the supernatant.
13. Dissolve DNA in TE.
14. Check DNA concentration with a UV spectrometer. Also check the DNA quality by running 1 μL of DNA in a 1% agarose gel (*see* **Note 7**).

3.2. mDIP

1. Sonicate 5 μg (in 150 μL volume) genomic DNA to 300–400 bp for 20–25 s on ice with Branson sonifier at setting 3.
2. Run 5 μL on 2% agarose gel with a DNA ladder to check the size of sonication of DNA (*see* **Note 8**).
3. Measure the DNA concentration with UV spectrometer.
4. Mix 2 μg of sonicated DNA, 60 μL of 10× IP buffer, 20 μL of monoclonal 5′m-C antibody (–20°C freezer, 1 vial per sample), and H_2O to 600 μL.
5. Rotate the DNA and antibody mixture at 4°C overnight.
6. Pipet 100 μL of Dynabeads for each sample into an Eppendorf tube. The Dynabeads need to be washed to get rid of nonspecific binding. The washing procedure is facilitated by using a magnet Dynal Invitrogen.
7. Wash 100 μL of Dynabeads (M-280 sheep antimouse IgG, Invitrogen) twice with PBS, 0.1% BSA, and pipet off the supernatant.
8. Wash the beads with 1× IP twice.
9. Resuspend the beads with 100 μL of 1× IP plus 2 μg of sonicated salmon sperm DNA. Rotate at 4°C for 1–2 h.
10. Wash the beads twice with 1× PB, 0.1% BSA.
11. Wash the beads twice with 1× IP.
12. Resuspend the beads in 100 μL of 1x IP and add to the antibody–DNA mixture from the previous day.

13. Rotate at 4°C overnight.
14. Collect the supernatant as the unbound fraction.
15. Wash beads 4 times with 1× IP buffer and collect the supernatant as W1, W2, W3, and W4. Allow 3–5 min rotation at 4°C during each wash.
16. Wash beads once with IP containing 150 mM NaCl. Allow 3–5 min of rotation at 4°C. Collect the supernatant as low-salt washing fraction.
17. Wash beads once with IP containing 300 mM NaCl. Allow 3–5 min of rotation at 4°C. Collect the supernatant as high-salt washing fraction.
18. Freshly prepare elution buffer.
19. Elute in 200 μL IP with 1.5% SDS, vortex for 1 min, collect the buffer as elution fraction.
20. Elute in 200 μL IP with 0.5% SDS, vortex for 1 min, collect the buffer, and combine the elution from **step 19**.
21. Elute in 200 μL IP with 0.1% SDS, vortex for 1 min, collect the buffer, and combine the elution from **steps 19** and **20**.
22. Phenol/chloroform clean half of the unbound, W1, W2, W3, W4, low-salt, high-salt, and elution fractions. Resuspend the DNA from each fraction in 80 μL of H_2O.
23. Test the pull down efficiency of the mDIP procedure. Use 5 μL of DNA from each of the fraction for real-time PCR (*see* **Note 9**).
24. If each fraction of the DNA is proportional to what is expected, clean up the other half of the elute fraction and combine with the previous half elution fraction.
25. Check the Elute DNA concentration with the Nanodrop (*see* **Note 10**).

3.3. DNA Labeling, DNA Hybridization, and Feature Extraction

In our efforts to assay genome-wide DNA methylation patterns, we have successfully used Agilent oligo-microarrays that cover 27,800 CpG islands and 18,000 promoter regions. To perform DNA microarray hybridization, we have followed Agilent ChIP-on-chip protocol (http://www.Agilent.com) with BioPrime® array CGH genomic Labeling Kit from Invitrogen. Nevertheless, readers should be able to obtain satisfactory results with other commercially available DNA microarrays by following the specific instructions from oligo-microarray manufacturers. The following are our adopted procedures for Agilent oligo-microarrays.

1. In two separate Eppendorf tubes, mix 500 ng of input DNA or Elute DNA with 35 μL of 2.5× random primer solution and H_2O to 75 μL.
2. Mix the solution well by either pipeting or vortexing.

3. Place the tube on 95°C heat block and incubate for 5 min.
4. Immediately transfer the tubes on ice and cool for 5 min.
5. Add 8.2 μL of 10× dUTP, 1.5 μL of exo-Klenow, 1.5 μL of either Cy3- or Cy5-dUTP (1 mM), and 1.8 μL of H_2O to the tubes (*see* **Note 11**).
6. Mix the reaction by briefly vortexing and give the tube a quick spin down.
7. Incubate the reaction in a 37°C water bath for 4–5 h in dark.
8. Add 9 μL stop buffer to the reaction and mix. Samples are ready for clean up.
9. Clean the labeled probes using BioPrime® array CGH purification module.
10. Add 400 μL of Purification Buffer A to each reaction tube and mix them by vortexing for 30 s.
11. Transfer the samples to the purification columns in the 2-mL collection tubes.
12. Spin the column at 8,000 × *g* for 1 min at room temperature.
13. Discard flow through and add 600 μL of Purification Buffer B to the column and spin at 8,000 × *g* for 1 min at room temperature.
14. Discard the flow through and add 200 μL of Purification Buffer B to the column and spin at 8,000 × *g* for 1 min at room temperature.
15. Place the column in a new Eppendorf tube.
16. Add 50 μL of sterile water to the column and incubate 1 min at room temperature.
17. Spin the column at 8,000 × *g* for 1 min at room temperature.
18. Keep the elute on ice in dark.
19. Measure the quantity and quality of labeled DNA samples using Nanodrop (*see* **Note 12**).
20. Mix 5 μg of labeled Input DNA with 5 μg of labeled Elute DNA in an Eppendorf tube, and bring the volume to 130 μL with H_2O (*see* **Note 13**).
21. Add 50 μL of 10 × Agilent control target, 50 μL of human cot-1 DNA, and 20 μL of yeast tRNA to the tube (*see* **Note 14**).
22. Add 250 μL of 2× Agilent hybridization buffer, and mix well by pipeting.
23. Heat the mixture at 95°C for 3 min.
24. Immediately transfer the tube to a 37°C H_2O bath and incubate for 30 min.

25. Take out the tube and centrifuge at 16,100 × *g* for 1 min.
26. Load a gasket slide in the Agilent SureHyb Chamber base with the label facing up.
27. Pipet 490 μL and slowly dispense the hybridization mixture onto the gasket slide (*see* **Note 15**).
28. Place a microarray on the gasket slide so that the active side is facing down and in contact with the hybridization buffer.
29. Place the SureHyb Chamber cover onto the sandwiched slide and slide on the clamp assembly. Tighten the clamp by hand.
30. Rotate the assembled chamber so that the slide is wet and make sure there are no stationary bubbles.
31. Hybridize the array in the hybridization oven for 40 h at 65°C with rotation speed at 10 rpm (*see* **Note 16**).
32. Prewarm wash buffer 2 in a 31°C oven overnight.
33. Disassemble the array, and place the array/gasket slide sandwich into a slide wash container with wash buffer 1. Separate the gasket slide from the array slide with a pair of forceps.
34. Transfer the slide into a new container with wash buffer 1. Make sure there is a stir bar in the container and wash with the rotation speed set at medium low.
35. Repeat **steps 2** and **3** for the other slides. You can wash four slides at a time.
36. Wash the slides in wash buffer 1 at room temperature for 5 min.
37. Wash the slides in wash buffer 2 at 31°C for 5 min with stirring at medium low.
38. Slowly move the slides out of wash buffer 2. The slides should be dry.
39. Wash the slides in acetonitrile for 1 min.
40. Dip the slides one at a time into Agilent Stabilization and Drying solution, slowly pulling the slide out (*see* **Note 17**).
41. Warm up the Agilent array scanner by turning it on 20 min before the scanning.
42. Put the washed slide into slide holder so that the numeric barcode is visible.
43. Scan for two colors at 5 μM resolution (*see* **Note 18**).
44. Import the scanned image into Agilent Feature Extraction software.
45. Select the appropriate protocol and grid file and start the feature extraction.
46. Save the Feature Extraction data to the computer.

3.4. Data Analysis and Verification

The methods for analyzing the mDIP data are still evolving. The array you use determines the way you calculate which genes are enriched for methylation. Here we provide ways to analyze both Promoter and CpG island microarrays. The power of your array data will depend on the quality of the hybridizations.

1. Open the Feature Extraction.txt file in your favorite math program (Excel, R, MatLab). The file can be imported into Excel as a tab-delimited file.
2. To compare the enrichment of methylation at a particular locus, sort by Log ratio (*see* **Note 19**). Higher log ratios indicate more enrichment of methylated DNA.
3. If using the CpG island arrays, you can compare the log ratios over an entire CpG island (*see* **Note 20**).
4. When performing replicate arrays, you can compare the average Log Ratio for each probe or set of probes for a CpG island.
5. To further increase you confidence of methylated probes, you can compute a t-test probability for each probe or set of probes. To compute a t-test, use the rProcessedSignal and gProcessedSignal columns from the Feature Extraction .txt file.
6. Pick genes that contain statistically significant probes to confirm. Currently, there is no definitive log ratio cut off for methylated probes. This has to be determined experimentally for each sample. We suggest confirming genes at either end of the log ratio spectrum as well as genes in between to find the threshold of methylation in your array experiments.
7. To confirm a gene's methylation status, use bisulfite sequencing *(17)*.
8. Digest DNA overnight in a 50-μL reaction with a restriction enzyme (*see* **Note 21**).
9. Run 5 μL of digested DNA on a 1% gel and check for a small shift in the genomic DNA band.
10. Heat 20 μL of digested DNA at 97°C for 5 min, and chill on ice for 5 min.
11. Add 1 μL of freshly prepared 6.3 M NaOH to DNA; vortex and centrifuge.
12. If not already, transfer DNA to a 0.2-mL PCR tube.
13. Incubate DNA at 39°C for 30 min.
14. Prepare sodium bisulfite solution by adding 4.05 g $Na_2S_2O_5$ to 8 mL of H_2O. Cover the tube with foil and allowing it to dissolve by incubate the tube at 55°C for 30 min.

15. Prepare hydroquinone solution by adding 0.11 g of hydroquinone to 5 mL of H_2O. Cover the tube with foil and allow it to dissolve by incubating the tube at 55°C for 30 min.
16. Once sodium bisulfite solution has dissolved, adjust pH with 333 μL of 6.3 M NaOH and then add 300 μL of the hydroquinone solution (*see* **Note 22**).
17. After 30 min at 39°C, add 208 μL of the sodium bisulfite solution to the DNA and mix by pipetting.
18. Incubate in a PCR machine with the following cycles
    (a) 55°C for 3 h
    (b) 95°C for 5 min
    (c) Goto **Step 1** 4 times
19. Add 1 mL of Promega's Wizard Clean-Up resin to sample. Desalt and elute according to the manufacturer's protocol.
20. Add 2.5 μL of 6.3 M NaOH to sample and incubate at 37° for 15 min.
21. Spin down briefly.
22. Add 22.5 μL of 10 M Ammonium Acetate of pH 7.3, 0.5 μL of glycogen, and 225 μL of 100% EtOH. Vortex to mix.
23. Place in –80°C for 30 min to help precipitate the DNA.
24. Spin down at 16,100 × *g* for 30 min at 4°C.
25. Resuspend the DNA in 30 μL of TE buffer and store at –20°C.
26. For Bisulfite Sequencing analysis, you can generate primers using MethPrimer (*see* **Note 23**).
27. Use 1/10th of your bisulfite treated DNA for each PCR.
28. After amplification, extract the PCR product by gel purification.
29. Sequence the PCR product using the reverse primer (*see* **Note 24**).

4. Notes

1. Unless stated otherwise, all water used to make the solutions is MilliQ water with a resistivity of 18.2 MΩ-cm.
2. Antibody should avoid repeated freeze and thaw cycles. Aliquot the antibody upon arrival so that one tube is good for one immunoprecipitation reaction for one sample.
3. For tissue samples, usually 500 μL of lysis buffer is enough for a tissue size less than 100 μL in volume.
4. This procedure should be done in a fume hood to prevent inhalation of phenol.

5. Phenol waste should be disposed of in a designated phenol waster container in a fume hood.
6. During DNA extraction procedure, vortexing should be prevented as this could cause shearing of genomic DNA.
7. The ratio of absorbance at 260/280 for DNA should be 1.8–2.0. For good quality of DNA, a high molecular weight band should be seen from the gel.
8. Usually, sonication should yield a DNA smear of 300–400 bp on the agarose gel. Make sure the sonication is complete so that there is no DNA larger than 500 bp.
9. Real-time PCR is performed on all the fractions of mDIP to test the efficacy of the procedure. Loci are picked either for methylated controls or for unmethylated controls. For instance some tumor-suppressor gene promoters can serve as positive control loci, while housekeeping gene promoters should be negative controls. Primers are designed within these loci. The Ct values were converted to a starting quantity and then all amounts were used to determine the total percentage of DNA in each fraction. Only mDIP procedures that resulted in high binding of methylated controls and low binding of unmethylated controls were used for microarray analysis. Quantitative PCR was done on a MyIQ Thermocycler (Biorad) using the Sybr Green Supermix (BioRad). A representative figure of real-time result is shown for an efficient mDIP procedure **(Fig. 1)**.
10. Typically, one IP procedure should yield 500–700 ng of DNA.
11. Typically Input DNA is labeled with Cy3-dUTP, while Eluted DNA is labeled with Cy5-dUTP.
12. Using the Microarray measure function of Nanodrop for this step. The total amount of label DNA should be >5 μg per reaction (>100 ng/μL). Cy3 label efficiency should be >3.5 pmol/μL and Cy5 labeling efficiency should be >2.5 pmol/μL. If the labeled samples do not reach these criteria, repeat the labeling procedure.
13. For Agilent 244K format CpG island array or promoter array, 5 μg of label probes for each channel is required to ensure good signal intensity.
14. Alternatively, Agilent 10× blocking reagent can be used to replace control targets and yeast tRNA. If used, bring volume in **Subheading 3.4, step 1** to 150 μL with H_2O.
15. Make sure that the hybridization mixture is added to the middle of the gasket slide.
16. Check the chambers after 20 min to ensure they are not leaking.

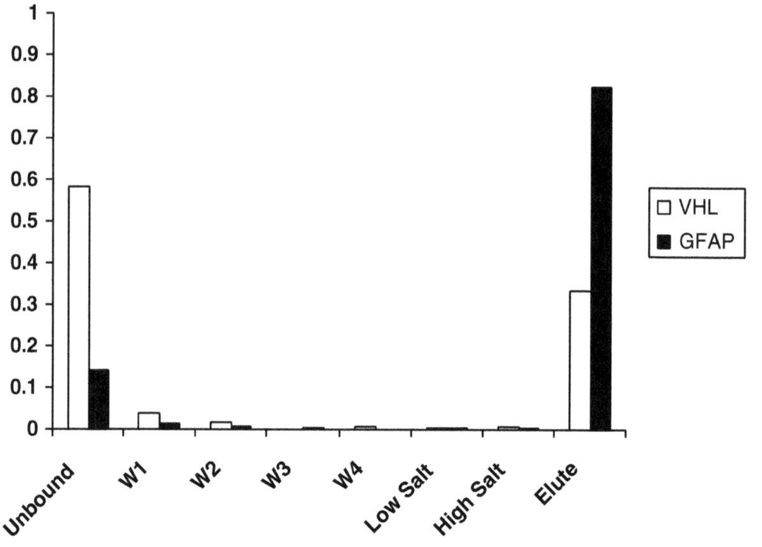

Fig. 1. Real-time PCR confirmation of a successful mDIP experiment using normal leukocytes' genomic DNA. As shown in figure, tumor-suppressor gene VHL promoter is hypermethylated in normal cells (*18*) and consistent with a low pull down efficiency. On the other hand, Gial specific gene GFAP is hypermethylated in normal leukocytes (our unpublished data) and showed a high efficiency of pull down. For more resolution of your sequence of interest, you can clone your PCR product into a Topo vector and generate sequences from at least ten clones.

17. Agilent Stabilization and Drying solution can easily form precipitation at room temperature. Incubate the solution overnight in a 37°C water bath to redissolve the precipitates before use. The acetonitrile and Stabilization and Drying solution should be set up in a fume hood.
18. For Agilent 244K format, we scan at 5 μm resolution. However, for other array format, the scanning resolution can vary.
19. Log Ratio in Feature Extraction is presented as Log10 ofrProcessedSignal over gProcessedSignal.
20. It is known that aberrant CpG island methylation spreads from its starting point throughout the whole CpG island *(19)*. Thus, one can gain great significance from comparing all the probes within one CpG island.
21. It is important to pick a Restriction Enzyme that will not digest the area of interest. A typical enzyme used for this procedure is *Bgl*II.
22. Make the sodium bisulfite and hydroquinone solution fresh each time.
23. The web address for MethPrimer is http://www.urogene.org/methprimer/index1.htmL.

24. In general, reverse primers gives better sequencing results than forward primers with PCR products from Bisulfite treated DNA samples. This is because the forward strand of PCR products from bisulfite converted DNA is T rich, while the reverse strand is A rich.

Acknowledgments

We thank Dr. Howard Cedar at Hebrew University and Drs. Xiaoyu Zhang and Steve Jacobsen at UCLA for sharing with us their technical notes. Funding support for this work is from National Institutes of Health.

References

1. Robertson, K.D. and Wolffe, A.P. (2000) DNA methylation in health and disease. *Nature reviews*, 1, 11–9.
2. Bestor, T.H. (2000) The DNA methyltransferases of mammals. *Human molecular genetics*, 9, 2395–402.
3. Jones, P.A. and Laird, P.W. (1999) Cancer epigenetics comes of age. *Nature genetics*, 21, 163–7.
4. Costello, J.F., Fruhwald, M.C., Smiraglia, D.J., Rush, L.J., Robertson, G.P., Gao, X., Wright, F.A., Feramisco, J.D., Peltomaki, P., Lang, J.C., et al. (2000) Aberrant CpG-island methylation has non-random and tumour-type-specific patterns. *Nature genetics*, 24, 132–8.
5. Feinberg, A.P. and Vogelstein, B. (1983) Hypomethylation distinguishes genes of some human cancers from their normal counterparts. *Nature*, 301, 89–92.
6. Claus, E.B., Petruzella, S., Matloff, E. and Carter, D. (2005) Prevalence of BRCA1 and BRCA2 mutations in women diagnosed with ductal carcinoma in situ. *The Journal of the American Medical Association*, 293, 964–9.
7. Aaltonen, L.A., Salovaara, R., Kristo, P., Canzian, F., Hemminki, A., Peltomaki, P., Chadwick, R.B., Kaariainen, H., Eskelinen, M., Jarvinen, H., et al. (1998) Incidence of hereditary nonpolyposis colorectal cancer and the feasibility of molecular screening for the disease. *The New England journal of medicine*, 338, 1481–7.
8. Feinberg, A.P., Ohlsson, R. and Henikoff, S. (2006) The epigenetic progenitor origin of human cancer. *Nature reviews*, 7, 21–33.
9. Eckhardt, F., Lewin, J., Cortese, R., Rakyan, V.K., Attwood, J., Burger, M., Burton, J., Cox, T.V., Davies, R., Down, T.A., et al. (2006) DNA methylation profiling of human chromosomes 6, 20 and 22. *Nature genetics*, 38, 1378–85.
10. Meissner, A., Gnirke, A., Bell, G.W., Ramsahoye, B., Lander, E.S. and Jaenisch, R. (2005) Reduced representation bisulfite sequencing for comparative high-resolution DNA methylation analysis. *Nucleic acids research*, 33, 5868–77.
11. Weber, M., Davies, J.J., Wittig, D., Oakeley, E.J., Haase, M., Lam, W.L. and Schubeler, D. (2005) Chromosome-wide and promoter-specific analyses identify sites of differential DNA methylation in normal and transformed human cells. *Nature genetics*, 37, 853–62.
12. Zhang, X., Clarenz, O., Cokus, S., Bernatavichute, Y.V., Pellegrini, M., Goodrich, J. and Jacobsen, S.E. (2007) Whole-genome analysis of histone H3 lysine 27 trimethylation in Arabidopsis. *PLoS biology*, 5, e129.
13. Zilberman, D., Gehring, M., Tran, R.K., Ballinger, T. and Henikoff, S. (2007) Genome-wide analysis of Arabidopsis thaliana DNA methylation uncovers an interdependence between methylation and transcription. *Nature genetics*, 39, 61–9.
14. Zhang, X., Yazaki, J., Sundaresan, A., Cokus, S., Chan, S.W., Chen, H., Henderson, I.R., Shinn, P., Pellegrini, M., Jacobsen, S.E., et al. (2006) Genome-wide high-resolution mapping and functional analysis of DNA methylation in arabidopsis. *Cell*, 126, 1189–201.

15 Keshet, I., Schlesinger, Y., Farkash, S., Rand, E., Hecht, M., Segal, E., Pikarski, E., Young, R.A., Niveleau, A., Cedar, H., et al. (2006) Evidence for an instructive mechanism of de novo methylation in cancer cells. *Nature genetics*, 38, 149–53.

16. Weber, M., Hellmann, I., Stadler, M.B., Ramos, L., Paabo, S., Rebhan, M. and Schubeler, D. (2007) Distribution, silencing potential and evolutionary impact of promoter DNA methylation in the human genome. *Nature genetics*, 39, 457–66.

17. Clark, S.J., Harrison, J., Paul, C.L. and Frommer, M. (1994) High sensitivity mapping of methylated cytosines. *Nucleic acids research*, 22, 2990–7.

18. Herman, J.G., Latif, F., Weng, Y., Lerman, M.I., Zbar, B., Liu, S., Samid, D., Duan, D.S., Gnarra, J.R., Linehan, W.M., et al. (1994) Silencing of the VHL tumor-suppressor gene by DNA methylation in renal carcinoma. *Proceedings of the National Academy of Sciences of the United States of America*, 91, 9700–4.

19. Cameron, E.E., Baylin, S.B. and Herman, J.G. (1999) p15(INK4B) CpG island methylation in primary acute leukemia is heterogeneous and suggests density as a critical factor for transcriptional silencing. *Blood*, 94, 2445–51.

Chapter 14

The Contribution of Niche-Derived Factors to the Regulation of Cancer Cells

Julie B. Sneddon

Summary

In normal adult tissues, paracrine signals that derive from the stem cell niche, or microenvironment, play an important role in regulating the critical balance between activity and quiescence of stem cells. Similarly, evidence has emerged to support the hypothesis that signals derived from the microenvironment regulate cancer cells in an analogous manner. We recently reported that in basal cell carcinoma of the skin and in diverse other solid tumors, fibroblasts that comprise the tumor cell niche are, indeed, molecularly distinct from those that comprise the normal stroma. In particular, we found evidence suggesting that expression of secreted BMP antagonists by tumor-associated stromal cells may promote self-renewal of tumor stem cells *in vivo*. This chapter describes methods for identifying and evaluating the molecular signals that derive from fibroblasts in human tumors.

Key words: Tumor stroma, Tumor fibroblast, Cancer cell niche

1. Introduction

In normal adult tissues, stem cells depend on a constellation of both cell-intrinsic and cell-extrinsic factors for proper, homeostatic tissue maintenance *(1, 2)*. Paracrine signals that derive from the stem cell niche, or microenvironment, play an important role in regulating the critical balance between activity and quiescence. In the case of cancer, it has been proposed that tumors arise exclusively from a rare population of cells with stem cell properties, termed cancer stem cells *(3–5)*, and that interactions with the cancer niche may be similarly important in regulating tumorigenesis.

John S. Yu (ed.), *Cancer Stem Cells,* Methods in Molecular Biology, vol. 568
DOI 10.1007/978-1-59745-280-9_14, © Humana Press, a part of Springer Science+Business Media, LLC 2009

In the last several years, increasing evidence has emerged for the hypothesis that signals derived from the microenvironment serve to regulate cancer cells. Specialized endothelial microenvironments in bone marrow, for instance, are required for the homing and engraftment of both normal hematopoietic stem cells and leukemic cells *(6)*. Brain cancer stem cells have been demonstrated to live in a vascular niche that promotes their long-term growth and self-renewal *(7)*. Intriguingly, disrupting this niche impairs brain cancer stem cell self-renewal and thereby significantly inhibits tumor growth, providing credence to the idea that targeting the unique aberrant microenvironment of cancer stem cells may be a critical aspect of effective cancer therapy. Other studies of the hematopoietic stem cell niche suggest that both signaling molecules and extracellular matrix components in the niche microenvironment can promote cell survival signals in acute myeloid leukemia, providing resistance to chemotherapeutic treatments *(8)*.

We recently reported that in basal cell carcinoma of the skin and in diverse other solid tumors, fibroblasts that comprise the tumor cell niche are, indeed, molecularly distinct from those that comprise the normal stroma *(9)*. In particular, genomic profiling revealed that unlike their normal counterparts, tumor-associated dermal fibroblasts express high levels of the secreted BMP antagonist *GREMLIN 1*. In contrast, the basal cell carcinoma cells themselves express BMP2 and BMP4. Gremlin 1 protein supports the basal cell carcinoma cells in a less differentiated, more expansive state *ex vivo*, suggesting that expression of secreted BMP antagonists by tumor-associated stromal cells may promote self-renewal of tumor cells *in vivo (9)*.

This chapter describes methods for identifying and evaluating the molecular signals that derive from fibroblasts in human tumors. Isolation of fibroblasts from human tissue allows comparison of normal- and tumor-derived cells from the niche. Candidate factors expressed by fibroblasts can then be confirmed *in vivo* by detecting RNA expression in human tissue using nonradioactive RNA *in situ* hybridization. Finally, to evaluate the functional consequence of proteins produced by niche fibroblasts, primary cancer cells can be isolated from human tissue and propagated *ex vivo* for studies examining the effects of exogenously added proteins and other factors.

2. Materials

2.1. Cell Culture

1. *Transport medium for tissue.* Dulbecco's Modified Eagle's Medium (DMEM, Invitrogen, Carlsbad, CA) supplemented with penicillin–streptomycin (Invitrogen) and Fungizone

antimycotic (Invitrogen) at final concentrations of 50 Units/L and 2.5 μg/mL, respectively. Sterile filter with 0.45-μm filter unit and store aliquots of 5 mL in individual 15-mL conical tubes at 4°C.

2. *Fibroblast growth medium.* DMEM supplemented with 10% fetal bovine serum (FBS, HyClone, Logan, UT), 50 Units/L penicillin streptomycin, and 2 mM L-glutamine (Invitrogen). Sterile filter as earlier and store at 4°C.
3. 70% Ethanol made with MilliQ water.
4. *Phosphate-buffered saline (PBS), calcium and magnesium free.* Prepare 10× stock with 9.2 g Na2HPO4, 2.0 g KH2PO4, 2.0 g KCl, 98.4 g NaCl (adjust to pH 7.4 with HCl if necessary) and autoclave before storage at room temperature. Prepare working solution by dilution of one part with nine parts MilliQ water.
5. 6-well flat-bottomed tissue culture dishes, as well as T75 and T175 tissue culture flasks by BD Falcon (Fisher Scientific).
6. Sterile scalpels, size no. 10 by Bard-Parker (Fisher).
7. Sterile curved dissecting forceps, serrated tips, 117 mm (Ted Pella, Inc., Redding, CA).
8. Solution of trypsin, 0.25% (Invitrogen).
9. *Freeze-down medium.* DMEM supplemented with 20% FBS, penicillin–streptomycin, and L-glutamine, plus 10% dimethyl sulfoxide (DMSO).
10. Nalgene "Mr. Frosty" freezing container (Fisher Scientific), at room temperature and filled with isopropyl alcohol per instructions (*see* **Note 1**).
11. Cryogenic vials by Corning (Fisher).

2.2. Nonradioactive In Situ *Hybridization*

1. It is imperative that RNase-free reagents be used during this procedure. Failure to do so can result in degradation of the RNA probe and complete loss of signal. Consequently, it is recommended that all equipment (pipettors, microcentrifuge tube racks, etc.) as well as benchtop surfaces be cleaned and then wiped with an RNase decontamination solution such as RNase Zap (Ambion) or equivalent.

The following pertains to the first day of the procedure:

2. Obtain paraffin-embedded sections containing tissue that is to be analyzed (*see* **Note 2**).
3. Digoxigenin (DIG)-labeled RNA probe complementary to RNA of interest (*see* **Note 3**).
4. Several liters of RNase-free water (Ambion) (*see* **Note 4**).
5. Xylene, 250 mL into each of two green solvent resistant dishes (Tissue Tek) (*see* **Note 5**).

6. Graded series of ethanol (2 × 250 mL 100%, 250 mL 95%, 250 mL 70%) made with RNase-free water and placed in white polypropylene staining dishes.
7. Slide holder with long handle (Tissue-Tek).
8. Four 50-mL conical tubes each containing 30 mL RNase-free water.
9. Two 50-mL conical tubes each containing 30 mL 1× Tris-buffered saline, made with 27 mL RNase-free water and 3 mL 10× Tris-buffered saline (TBS, BioRad).
10. One 50-mL conical tube containing 30 mL 1% hydrogen peroxide (H_2O_2), made with 20 mL RNase-free water and 10 mL 3% H_2O_2.
11. One 50-mL conical tube containing 40 mL of formamide (deionized, from American Bioanalytical)/sodium citrate-sodium chloride (SSC) made with 20 mL formamide, 10 mL 20× SSC (Ambion), and 10 mL RNase-free water.
12. Proteinase K. Dissolve one bottle of 100 mg Proteinase K powder (Promega) in 5 mL of RNase-free dH_2O. Freeze small aliquots (e.g., 250 μL) at –20°C. When ready to use, thaw aliquot and dilute to optimal concentration in 50 mL tubes with 37°C prewarmed TBS (*see* **Note 6**.)
13. 300 μL hybridization solution (DAKO) per one slide.
14. Microscope slide staining tray that can accommodate multiple slides in a horizontal position. Must also include lid.
15. Coverslips made by cutting pieces of Parafilm to size such that the slide is appropriately covered.
16. Sterile, RNase-free forceps.

The following pertains to the second day of the procedure:

17. *Blocking buffer*. 43.5 mL DEPC water, 5 mL 1 M Tris, 1.5 mL 5 M NaCl, 0.5 g casein (DIG Nucleic Acid Detection Kit, DAKO) (*see* **Note** 7.).
18. Two 50-mL conical tubes each containing 30 mL of 2× SSC (27 mL RNase-free water, 3 mL 20× SSC).
19. Remove 972 μL of this 2× SSC from the 50-mL conical tube and place into a 1.5-mL microcentrifuge tube.
20. Two 50-mL conical tubes each containing 50% formamide/ 2× SSC (13.5 mL RNase-free water, 15 mL formamide, 1.5 mL 20× SSC).
21. One 50-mL conical tube containing 0.08× SSC (30 mL RNase-free water, 48 μL 50× SSC from DAKO Genpoint Kit).
22. Eleven 50-mL conical tubes each containing 1× Tris-buffered saline/Tween (TBST) made from 27 mL RNase-free water, 3 mL 10× TBST (DAKO).

23. Four 50-mL conical tubes each containing 1× TBS (27 mL RNase-free water, 3 mL 10× TBS).
24. RNase cocktail (Ambion), stored at –20°C.
25. Parafilm cover slips, cut to size.
26. Dedicated slide holder for RNase treatment.
27. Dedicated forceps for RNase treatment.
28. Avidin and Biotin solutions from DAKO Biotin Blocking System.
29. Rabbit Ig (DAKO).
30. HRP-anti DIG (DAKO).
31. Biotinyl tyramide solution (DAKO, GenPoint kit).
32. Pap pen or diamond scribe.
33. DAB chromogen (300 μl DAB diluent + 6 μl DAB concentrate, GenPoint kit).
34. Three containers of 250 mL distilled water in polypropylene staining dishes.
35. 250 mL Mayers Hematoxylin in staining dish.
36. Xylene-compatible mounting medium such as Permount (COMPANY).
37. Cover glass of appropriate size to cover section.

2.3. Isolation and Treatment of Primary Human Skin Cancer Cells Ex Vivo

The following materials have been optimized for skin (basal cell carcinoma) and should be adjusted to include media that is appropriate for the particular tissue and cells of interest. Also, keep in mind that media requirements for mouse tissue can be different from those for human tissue.

1. *Transport medium for Tissue*: Keratinocyte Serum-Free Medium (SFM, Gibco) without supplement added (basal medium only), plus antibiotic-antimycotic (Invitrogen) at a final concentration 100 units of penicillin, 100 μg of streptomycin, and 0.25 μg of amphotericin B per mL. Sterile filter with a 0.45-μm filter unit. Aliquots of 5 mL in 15-mL sterile conical tubes should be stored in the dark at 4°C.
2. *Keratinocyte growth medium*: Keratinocyte Serum-Free Medium supplemented with supplied bovine pituitary extract (BPE) and Epidermal Growth Factor (EGF). Penicillin–streptomycin should be added at a final concentration of 50 Units/L. Sterile filter and store as earlier. Shelf life after supplementation is approximately 2 weeks. Use beyond this time is not recommended.
3. 70% Ethanol made with MilliQ water.
4. Hank's Balanced Salt Solution (HBSS, Invitrogen).

5. Dispase (Invitrogen), prepared at 5 mg/mL in HBSS and sterile filtered as earlier.
6. 6-well flat-bottomed tissue culture dishes.
7. 24-well plates and 100-mm dishes, both coated with rat tail collagen I (Fisher).
8. Sterile scalpels, size no. 10.
9. Sterile forceps (blunt ended) and sterile curved dissecting forceps, serrated tips, 117 mm (*see* **Note 8**).
10. 10% buffered formalin in prefilled containers (Fisher).
11. 0.05% trypsin–EDTA (Invitrogen).
12. HBSS with 15% FBS.
13. *Freeze-down medium.* Keratinocyte-SFM with supplied BPE and EGF, plus 10% dimethyl sulfoxide (DMSO) (*see* **Note 9**).
14. Nalgene "Mr. Frosty" freezing container, at room temperature and filled with isopropyl alcohol per manufacturer's instructions.
15. Cryogenic vials.

3. Methods

3.1. Isolating Niche Fibroblasts from Human Tissue

1. Fresh human tissue obtained from surgical specimens or other sources should be placed in transport medium as quickly as possible (*see* **Note 10**). Typically, 5 mL of medium in a sterile 15-mL conical tube is a sufficient volume to completely submerge samples of tissue up to 1 cm^3. Samples are kept on ice during transport to the laboratory.
2. All remaining steps are performed in a sterile tissue culture hood environment. First, prepare the necessary equipment, including sterile scalpel and forceps, one six-well tissue culture plate, and 70% ethanol. In addition, fibroblast growth medium and PBS should be prewarmed at 37°C. Take care to maintain sterility of forceps and scalpel at all times (they can be rinsed in 70% ethanol and carefully flamed to re-establish sterility if necessary).
3. Carefully remove tissue sample from the conical tube of transport medium using scalpel and/or forceps. Transfer tissue to a clean well of a 6-well tissue culture plate.
4. Rinse tissue liberally with 70% ethanol. Wash in prewarmed PBS.
5. While holding tissue firmly with forceps, use scalpel to remove the outer layer of tissue in all directions. This will

help reduce the possibility of contamination by eliminating the outer part of the tissue that may have encountered non-sterile surfaces during or after surgery.

6. Mince remaining tissue into approximately 1-mm^3 cubes.
7. Place minced cubes into the well of a fresh 6-well tissue culture dish.
8. Place the tissue culture dish containing the tissue into a tissue culture incubator at 37°C, 5% CO_2 for approximately 10–15 min *without any medium covering the tissue* (*see* **Note 11**).
9. When tissue has had time to adhere to the surface of the dish, remove dish from incubator and add approximately 2 mL of prewarmed fibroblast growth medium to the well. Medium should be added gently and slowly so as not to disturb the adhesion of the tissue to the plate. Gingerly return plate to tissue culture incubator, again taking care not to dislodge the tissue from the bottom of the well. Samples are maintained at 37°C and 5% CO_2.
10. Fresh fibroblast growth medium should be replaced approximately every 2 days. Fibroblasts will spontaneously grow out from the tissue anywhere from 2 to 14 days after the tissue is placed in culture (*see* **Note 12** and **Fig. 1**).
11. Once the well is confluent, cells should be trypsinized and expanded to flasks.
12. A portion of the cells should be frozen down and archived at as early a passage as possible. To freeze cells, trypsinize a confluent T175 flask with 5 mL trypsin, then once cells have detached, add 10 mL fresh media and spin down cells at 200 × *g* for 5 min. Remove supernatant and add 2 mL

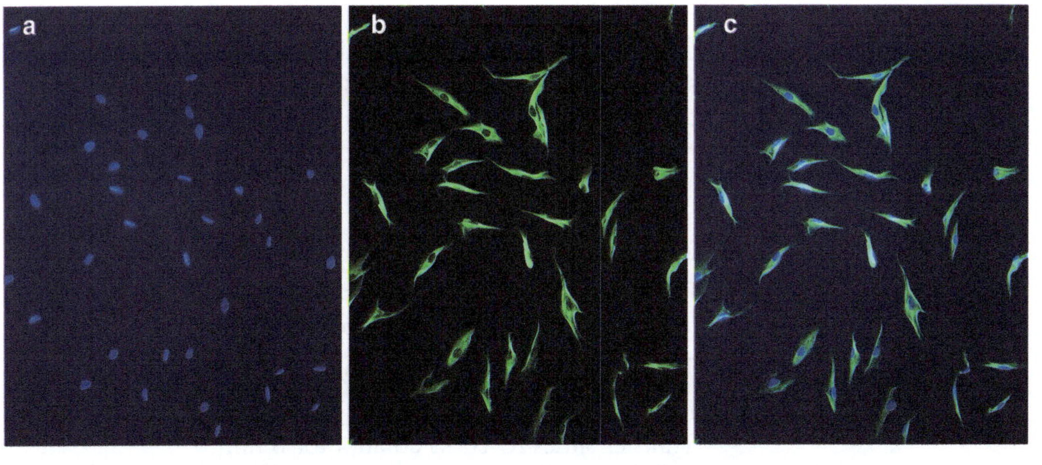

Fig. 1. Primary niche fibroblasts cultured from human skin tissue. (**a**) DAPI staining to illustrate nucleus in blue. (**b**) Immunofluorescence to illustrate vimentin in green. (**c**) Merge of (**a**) and (**b**).

freeze-down medium. Cells should be frozen at a concentration of approximately 5×10^6 cells/mL to 1×10^7 cells/mL. Place cell suspension in 2-mL cryogenic vials and then place at –80°C in a "Mr. Frosty" cryogenic container. After 24 h, transfer containers to liquid nitrogen. Be sure to note position of vials in the liquid nitrogen tank for future retrieval.

13. Remaining cells can be used for molecular analysis, such as global gene expression profiling by microarrays, quantitative RT-PCR analysis, Western blotting, and immunofluorescence, in order to identify candidate factors in the cancer stem cell microenvironment produced by tumor-associated fibroblasts.

3.2. Detection of RNA Expression in Human Tissue In Vivo *Using Nonradioactive RNA* In Situ *Hybridization*

This procedure is a modification of previous methods for detecting RNA in paraffin-embedded sections using nonradioactive *in situ* hybridization *(10)* and is a 2-day process.

Day One:

1. On the first day of the procedure, prepare reagents ahead of time as detailed in **Subheading 2.2**. Be vigilant about avoiding RNase contamination; use gloves at all time during the procedure.
2. Set a water bath with a 50-mL conical tube rack to 37°C. Preheat the two tubes of 1× TBS in this water bath.
3. Set a hybridization oven to 55°C.
4. Set a heat block that accommodates 1.5-mL microcentrifuge tubes to 65°C.
5. Arrange xylene and graded ethanol series in a staining rack in a chemical hood.
6. Label slides with paraffin-embedded sections according to the RNA probe to which they will be hybridized. Use of a pencil (writing on the white label of the slide) or a diamond scriber (etching on the reverse side of the glass slide such that glass shards will not compromise the tissue) is recommended, as pen can smear during the procedure. Be sure to always keep track of the side of the slide that contains the tissue!
7. Arrange slides to be stained in a slide rack with long handle. Deparaffin slides by placing rack (with slides facing down into the solution and label at top) in xylene for 3 min.
8. Transfer slides in rack to second xylene and incubate 3 min.
9. Rehydrate slides by incubating in a graded ethanol series: Transfer slides to 100% ethanol for 3 min.
10. Transfer slides to second 100% ethanol for 3 min.

11. Transfer slides to 95% ethanol for 3 min.
12. Transfer slides to 70% ethanol for 3 min.
13. Rinse slides by placing them in 50-mL conical tube of RNase-free water. Slides should be facing in opposite directions, with their backs touching (the surface that contains the tissue must be facing outward toward the solution!). In this manner, two slides can easily fit in a single conical tube.
14. Rinse again in the second tube of RNase water.
15. Incubate slides in tube with 1% H_2O_2 for 10 min.
16. Rinse twice in RNase-free water.
17. Place slides in prewarmed 1× TBS for 5 min at RT. During this incubation time, defrost Proteinase K and add 30 μL to the second tube of prewarmed TBS and keep at 37°C (*see* **Note 6**).
18. Incubate slides in diluted proteinase K at 37°C for 30 min.
19. Rinse slides twice in TBST for 3 min each.
20. During incubation with TBST, prepare 0.6 μL DIG-labeled probe diluted in 300 μL hybridization solution in a 1.5-mL RNase-free microcentrifuge tube. This is prepared for each slide according to the probe to be hybridized to that slide.
21. Mix the probe solutions by vortexing gently. Place at 65°C for 2 min in prewarmed heat block to denature RNA secondary structure. Immediately transfer tubes to ice.
22. Prepare the microscope slide tray for overnight hybridization by placing several pieces of paper towels on the bottom of the tray (underneath where the slides will be laid down). Soak the paper towels with 40 mL of the formamide/SSC solution. This will help keep slides moist during the overnight incubation period. Be sure towels are flat and will not interfere with horizontal positioning of slides.
23. Place slides in the slide tray in a horizontal position (be sure they are flat). Add 300 μL of the diluted probe solution directly onto the section.
24. Using RNase-free forceps, lay a precut piece of parafilm onto the section. Once all probes are added to all slides, place lid on slide tray.
25. Place slide tray in hybridization oven at 55° and incubate overnight or approximately 8–12 h.

Day Two:

26. On the second day of the procedure, prepare fresh all reagents necessary as detailed in **Subheading 2.2**. Again, take caution to avoid RNase contamination and wear gloves at all times.

27. Prewarm a water bath to 45°C and place the two tubes of 2× SSC, two tubes of SSC/Formamide, and one tube of 0.08× SSC into the bath in a 50-mL conical tube rack.
28. Remove slide tray from hybridization oven. Immediately lower oven temperature to 37°C.
29. Remove lid from slide tray and with RNase-free forceps, remove and discard parafilm cover slips. Tap off excess probe onto a Kimwipe or paper towel.
30. Transfer slides to 2× SSC solution in the 45°C water bath for 5 min. Slides are again positioned such that two slides occupy a single 50-mL conical tube, with the backs of the slide touching one another and the section side facing outward into the solution.
31. Incubate slides in prewarmed 2× SSC at RT for 5 min. Meanwhile, adjust water bath to 55°C and prewarm the SSC/formamide tubes and 0.08× SSC tube.
32. During the incubations in 2× SSC, prepare a work space for use with RNase. Use of a diaper or equivalent is recommended. In addition, a separate microscope slide staining tray dedicated for use with RNase is recommended to avoid contamination with other procedures that require RNase-free work.
33. Mix 972 μL of 2 × SSC with 28.5 μl RNase cocktail in a 1.5-mL microcentrifuge tube. Lay slides horizontally in the dedicated RNase slide tray, and place diluted cocktail on slides. Using RNase-dedicated forceps, lay parafilm cut to size on top of cocktail. Incubate slides in slide tray in hybridization oven at 37°C for 30 min to eliminate excess probe that has not found its cognate RNA.
34. Remove parafilm cover slip with RNase-dedicated forceps. Place slides, inverse-sandwich style, in prewarmed SSC/formamide for 30 min at 55°C. Discard RNase work area materials.
35. Perform a second stringent wash by placing slides in remaining SSC/formamide for 20 min, again at 55°C. Meanwhile, remove Biotin Blocking System reagents to thaw to RT.
36. Wash slides in prewarmed 0.08× for 20 min at 55°C.
37. Wash slides twice in 1× TBST for 3 min each at RT.
38. At this point, tissue sections are circumscribed with a pap pen or diamond scriber in order to minimize the necessary volumes of expensive reagents. On a blank microscope slide, press the pap pen to the slide to let it bleed. Then draw a circle around the tissue sections such that the tissue is fully circumscribed. Place slide in horizontal position in RNase-free microscope slide tray.

39. Add five drops of Avidin to the sections and incubate for 10 min at RT.
40. Wash slide in tube of 1× TBS for 3 min.
41. Place slide horizontally in tray and add five drops of Biotin, then incubate 10 min at RT.
42. During the incubation, prepare 1.5 mL blocking buffer in 1.5-mL microcentrifuge tubes each. (This is enough for four slides). Also per two slides, put 100 μL Rabbit Ig into 1.5-mL tube. Place 20 μL HRP-anti DIG into 1.5-mL tube.
43. Spin these four tubes in microfuge at maximum speed for 5 min.
44. Wash slides 3 times in 1× TBS for 3 min each. Rock slides gently on an orbital shaker.
45. Mix 300 μL blocking buffer and 15 μL rabbit Ig per slide, add to slide, and incubate for 30 min at RT.
46. During the incubation, prepare 1 mL blocking buffer, 50 μL rabbit Ig, and 10 ul HRP-anti-DIG. Add to slide and incubate for 30 min at RT.
47. Wash 3 times in 1× TBST for 5 min each, with rocking.
48. Add three drops biotinyl tyramide and incubate 15 min at RT.
49. At this point in the procedure, RNase-free conditions are no longer necessary. Wash 3 times in 1× TBST for 5 min each, with rocking.
50. Add three drops secondary streptavidin and incubate 15 min at RT.
51. Wash 3 times in 1× TBST for 5 min each, with rocking.
52. Prepare 300 μL DAB diluent and 6 μl DAB concentrate per slide.
53. Incubate section with diluted DAB mixture. Time depends on several variables, but typically is between 2 and 5 min. To allow for comparison of slides with the same RNA probe, development time should be standardized for each RNA probe (*see* **Note 13**).
54. Rinse slides well in staining dish of distilled water. Repeat rinses until brown color is no longer seen.
55. Counterstain nuclei by incubating slides in Mayers Hematoxylin for 15–30 s. Immediately blot with paper towels and rinse in distilled water. Run tap water over the slides in the sink until blue is no longer seen in the water.
56. Dehydrate sections by incubating slides in 70% ethanol, 95% ethanol, 2 changes of 100% ethanol, and two changes of xylene, each for 3 min.

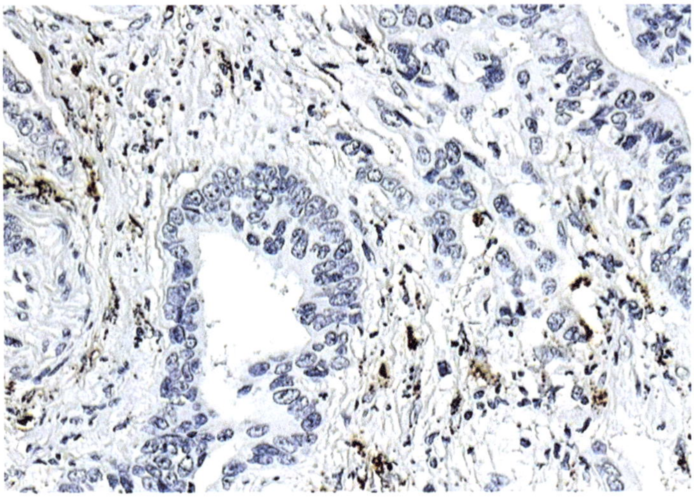

Fig. 2. In situ hybridization for *GREMLIN 1* RNA in human colon carcinoma. Expression of *GREMLIN 1* is indicated by clusters of punctate, brown color in tumor-associated stromal cells.

57. Place a few drops of mounting medium on the slide and cover with cover glass, being careful to avoid bubbles. Let slides dry in the chemical hood before visualization in the microscope. Positive signal appears as punctate, brown dots under the microscope and is considered *bona fide* if the corresponding sense RNA probe yields no staining (*see* **Fig. 2**).

3.3. Isolation and Treatment of Primary Human Skin Cancer Cells Ex Vivo

1. Fresh human skin tissue obtained from surgical specimens or other sources should be placed in transport medium as quickly as possible and kept on ice during transport to the laboratory.
2. All remaining steps are performed in a sterile tissue culture hood environment. Prepare sterile scalpel and forceps, one six-well tissue culture plate, one 12-well tissue culture plate, and 70% ethanol, HBSS, and dispase solutions. Take care to maintain sterility of forceps and scalpel at all times (they can be rinsed in 70% ethanol and carefully flamed to re-establish sterility if necessary).
3. Carefully remove tissue sample from the conical tube of transport medium using scalpel and/or forceps. Transfer tissue to a clean well of a 6-well tissue culture plate.
4. Rinse tissue liberally with 70% ethanol. Wash in prewarmed HBSS.
5. While holding tissue firmly with blunt-ended forceps, use scalpel to cut a small, cross-sectional piece of the sample. Drop sample in container of 10% buffered formalin and process and paraffin-embed it within a day for histological

confirmation. Alternatively, process and embed in OCT for frozen sections.

6. Trim fat off of remaining tissue and place in 3 mL of 5 mg/mL dispase solution at 4°C overnight (*see* **Note 14**).
7. The next day, separate epidermis from dermis. This can best be achieved by grasping tissue with blunt-ended forceps in one hand (epidermis side facing up) and very carefully peeling off the clear, top epidermal layer with fine dissecting forceps in the other hand (*see* **Note 15**).
8. Transfer epidermis to a fresh well of a 6-well dish. Add a small drop of HBSS to the tissue to keep it moist, as the epidermis is thin and can dry out quickly. Mince finely with sterile forceps and scalpel.
9. Incubate in 3 mL 0.05% trypsin–EDTA at 37°C for 15 min, with occasional mixing to disperse cells. Solution will often become somewhat cloudy when efficient dissociation has been achieved.
10. Neutralize with 5 mL 15% FBS in HBSS, spin down cells at 200 × *g* for 5 min., then resuspend in keratinocyte growth medium.
11. Plate cells onto 24-well collagen I-coated plates and incubate at 37°C in 5% CO_2. Media should be replaced every 2 days, and confluence of cells should be carefully monitored (*see* **Note 16**).
12. To assess the effects of recombinant proteins of interest, maintain cultured primary cells in Keratinocyte Growth Media (KGM) and incubate with the proteins at a range of concentrations and for the desired length of time (*see* **Fig. 3**). Assess cell number using triplicate counts with a

a b

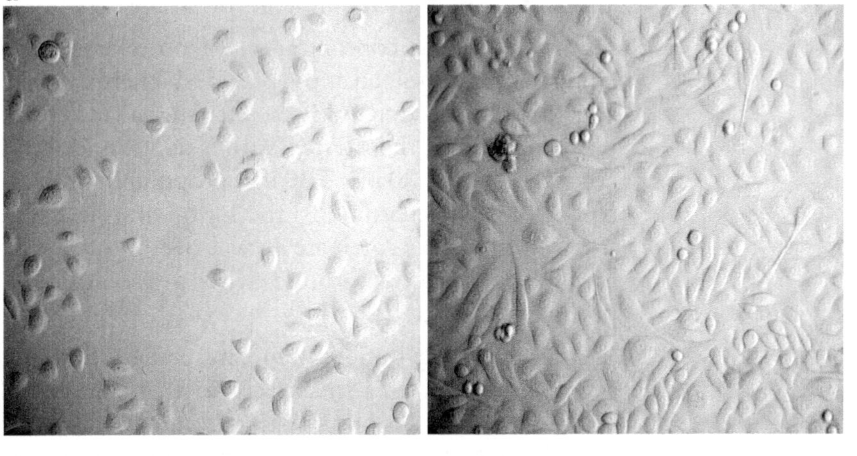

Fig. 3. Isolated primary human basal cell carcinoma cells *ex vivo*. Cells have been maintained in the absence (**a**) or presence (**b**) of 250 ng/mL recombinant Gremlin 1 protein for 4 days.

hemacytometer or other cell counting method to evaluate expansion of the cells. RNA can also be collected for later microarray analysis or Q-RT-PCR for markers of cell differentiation.

4. Notes

1. The Nalgene "Mr. Frosty" freezing container is recommended because it provides a controlled, 1°C/min cooling rate advantageous for successful cryopreservation of cells.
2. It is best to have sections of tissue cut onto microscope slides that are charged, such as poly-lysine coated slides. Cut sections can be stored in a clean, dry slide container until use. Some researchers recommend storage under vacuum.
3. The template used for making the DIG-labeled probe can be either a plasmid containing the gene of interest or a gel-purified PCR product. Remember to make an antisense RNA probe for detection of your RNA of interest. Labeling may be simplified by using the DIG RNA Labeling Kit (Roche Diagnostics).
4. If preferred, RNase-free water can be made instead of purchased, as DEPC-treated water is also suitable for this procedure.
5. Xylenes are hazardous if absorbed or inhaled. All work with xylene solutions should be performed in a fume hood, and gloves should used at all times. To avoid contact, it is recommended that the slide holders with long handles be used for steps with xylene, so that gloved hands do not need to touch the solution. Be sure to use Tissue-Tek's green solvent-resistant containers or the equivalent for xylene solutions.
6. Optimal working concentration of proteinase K should be empirically determined at the outset. As there can be lot-to-lot variability in the activity of the enzyme, a large batch should be prepared and then tested with a dilution series. Typically, around 30 µl per 30 mL TBS is appropriate, but this may need to be adjusted according to the lot of enzyme and the tissue used. Too much proteinase K will overdigest the tissue and lead to very high background, whereas too little can yield no positive staining.
7. Prepare blocking buffer by mixing and heating at temperatures up to 55°C in a small flask. Use of a dedicated stir bar (always used for this purpose) is recommended. Buffer can be stored at 4°C if used within a month, but solutions made fresh

are preferred to reduce background. Do not add sodium azide to the blocking buffer, as it can lead to inactivation of HRP.

8. Blunt-ended forceps are useful for obtaining a firm grasp on the tissue, while curved dissecting forceps are much more useful for fine tasks, such as removing epidermis from dermis.
9. MSDS recommends wearing safety glasses and a fume hood when working with DMSO, as it can cause eye damage and its vapor is an irritant. Thick rubber gloves are recommended, whereas nitrile gloves are not – they can dissolve upon exposure to DMSO.
10. Tissue sample collection should, of course, be performed in compliance with institutional review board (IRB) requirements.
11. This is a critical step in the procedure. Successful isolation of fibroblasts is often dependent on adherence of the tissue to the surface of the dish.
12. Success rate of cells growing out of the tissue has been approximately 60–70% in our hands. Once cells have grown out, their identity can be confirmed using immunofluorescence for cell lineage-specific markers.
13. For the first time with a given RNA probe, optimal development time can be determined by monitoring color under the microscope.
14. The temperature and length of incubation with dispase is a critical step that may need to be optimized for particular tissues. Higher temperatures or longer incubations can lead to loss of cell viability, whereas lower temperatures or shorter incubations can lead to incomplete separation of dermis and epidermis.
15. The epidermis is often quite thin and fragile, so its removal must be done in a gingerly manner. Because it is so thin, the epidermis should immediately be kept moist with HBSS so that it does not dry out during the dissection.
16. Confluence of cultured skin cells should be monitored daily, as they are particularly prone to contact inhibition and differentiation when too confluent. Cultures should be trypsinized and expanded before confluence exceeds 80%; otherwise, they will differentiate and not survive expansion. Note that if desired, potential contamination from fibroblasts or normal keratinocytes can be avoided by subjecting the culture to differential trypsinization and a transient increase in media calcium concentration, respectively *(11)*.

Acknowledgments

The author thanks Aaron D. Tward for assistance in preparing figures, Kelli Montgomery for aid with *in situ* hybridization protocol development, and Patrick O. Brown in whose laboratory all of this work was performed.

References

1. Moore, K. A. & Lemischka, I. R. (2006) Stem cells and their niches. *Science* 311, 1880–5.
2. Fuchs, E., Tumbar, T. & Guasch, G. (2004) Socializing with the neighbors: stem cells and their niche. *Cell* 116, 769–78.
3. Al-Hajj, M., Wicha, M. S., Benito-Hernandez, A., Morrison, S. J. & Clarke, M. F. (2003) Prospective identification of tumorigenic breast cancer cells. *Proc Natl Acad Sci U S A* 100, 3983–8.
4. Lapidot, T., Sirard, C., Vormoor, J., Murdoch, B., Hoang, T., Caceres-Cortes, J., Minden, M., Paterson, B., Caligiuri, M. A. & Dick, J. E. (1994) A cell initiating human acute myeloid leukaemia after transplantation into SCID mice. *Nature* 367, 645–8.
5. Singh, S. K., Hawkins, C., Clarke, I. D., Squire, J. A., Bayani, J., Hide, T., Henkelman, R. M., Cusimano, M. D. & Dirks, P. B. (2004) Identification of human brain tumour initiating cells. *Nature* 432, 396–401.
6. Sipkins, D. A., Wei, X., Wu, J. W., Runnels, J. M., Cote, D., Means, T. K., Luster, A. D., Scadden, D. T. & Lin, C. P. (2005) In vivo imaging of specialized bone marrow endothelial microdomains for tumour engraftment. *Nature* 435, 969–73.
7. Calabrese, C., Poppleton, H., Kocak, M., Hogg, T. L., Fuller, C., Hamner, B., Oh, E. Y., Gaber, M. W., Finklestein, D., Allen, M., Frank, A., Bayazitov, I. T., Zakharenko, S. S., Gajjar, A., Davidoff, A. & Gilbertson, R. J. (2007) A perivascular niche for brain tumor stem cells. *Cancer Cell* 11, 69–82.
8. De Toni, F., Racaud-Sultan, C., Chicanne, G., Mas, V. M., Cariven, C., Mesange, F., Salles, J. P., Demur, C., Allouche, M., Payrastre, B., Manenti, S. & Ysebaert, L. (2006) A crosstalk between the Wnt and the adhesion-dependent signaling pathways governs the chemosensitivity of acute myeloid leukemia. *Oncogene* 25, 3113–22.
9. Sneddon, J. B., Zhen, H. H., Montgomery, K., van de Rijn, M., Tward, A. D., West, R., Gladstone, H., Chang, H. Y., Morganroth, G. S., Oro, A. E. & Brown, P. O. (2006) Bone morphogenetic protein antagonist gremlin 1 is widely expressed by cancer-associated stromal cells and can promote tumor cell proliferation. *Proc Natl Acad Sci U S A* 103, 14842–7.
10. Iacobuzio-Donahue, C. A., Ryu, B., Hruban, R. H. & Kern, S. E. (2002) Exploring the host desmoplastic response to pancreatic carcinoma: gene expression of stromal and neoplastic cells at the site of primary invasion. *Am J Pathol* 160, 91–9.
11. Grando, S. A., Schofield, O. M., Skubitz, A. P., Kist, D. A., Zelickson, B. D. & Zachary, C. B. (1996) Nodular basal cell carcinoma in vivo vs in vitro. Establishment of pure cell cultures, cytomorphologic characteristics, ultrastructure, immunophenotype, biosynthetic activities, and generation of antisera. *Arch Dermatol* 132, 1185–93.

Chapter 15

Dendritic Cell Vaccines for Cancer Stem Cells

Serena Pellegatta and Gaetano Finocchiaro

Summary

Accumulating evidence suggests that only a fraction of neoplastic cells, defined as cancer stem cells (CSC), are responsible for tumor perpetuation. Recent data suggest that neurospheres (NS) from glioblastoma multiforme (GBM) are enriched in CSC. The characterization of this subpopulation of brain tumor cells with a potent tumorigenic activity supports the cancer stem cell hypothesis in solid tumors and may imply that cancer cells are differentially targeted by treatments, including dendritic cell (DC) immunotherapy. To test therapeutic strategies, a good model mimicking the characteristics of GBM-NS and GBM-AC (Adherent Cells) was necessary. One of the most frequently used murine brain tumor models is the GL261 glioma cell line. To see whether GL261 cells could mimic the growth of human GBM-CSC we let them grow in EGF/bFGF without serum. After 5 days neurospheres were visible in the culture medium and were proliferating continuously. The characterization in vivo and in vitro demonstrates that GL261-NS satisfy criteria used to identify CSC and are more immunogenic than AC. DC loaded with GL261-NS lysates protect mice against tumors from both GL261-NS and GL261-AC. Our results suggest that only DC vaccination against neurospheres can restrain the growth of a highly infiltrating and aggressive model of glioma and may have implications for the design of novel, more effective immunotherapy trials for malignant glioma and possibly other malignancies.

Key words: Dendritic cells, Cancer stem cells, GL261 glioma cells, Neurospheres, Glioblastoma multiforme

1. Introduction

Glioblastoma Multiforme (GBM) is the most frequent and aggressive brain tumor, with a median survival after diagnosis of 12 months *(1)*. A growing number of pharmaceutical and biological approaches have been developed to ameliorate this prognosis *(2)*. Only recently, however, a large randomized clinical trial has been able to establish that surgery plus the addition

John S. Yu (ed.), *Cancer Stem Cells*, Methods in Molecular Biology, vol. 568
DOI 10.1007/978-1-59745-280-9_15, © Humana Press, a part of Springer Science+Business Media, LLC 2009

of chemotherapy, during and after radiotherapy, may bring the medial overall survival to 14.6 months *(3)*. The need for new treatments is therefore urgent. Among the new treatments currently being investigated for malignant glioma, immunotherapy is theoretically very attractive, since it offers the potential for high tumor-specific cytotoxicity. Recent acquisitions on the molecular and cellular biology of dendritic cells (DC) may increase the chances of preparing effective "vaccines" against these tumors *(4)*. Immunotherapy plays a central role in the search for new treatments for GBM. Three clinical trials, in particular, were based on the use of tumor-homogenate pulsed DC for patients with recurrent gliomas and provided encouraging results *(5–9)*.

Vaccination strategies imply that DC activation should occur in the presence of the entire population of cancer cells. Recent data may challenge this approach by proposing that only a minority of tumor cells have the capacity to reproduce the tumor *(10)*. These tumor-initiating cells have also the potential to differentiate into cells with neural phenotypes, another attribute making them similar to neural stem cells. This revolutionary concept of brain Cancer Stem Cells may have relevant therapeutic implications *(11)*.

To define the criteria and methodology for DC-based vaccines we have performed preclinical experiments on immune-competent mice injected intracerebrally with syngeneic GL261 glioma cells and treated subcutaneously *(12)*, with DC loaded with a GL261-AC and NS lysate.

2. Materials

2.1. Cell Culture

1. CSC medium (amount made for 50 ml): Dulbecco's modified eagle medium:nutrient Mix F-12 (1×) liquid, 1:1 (D-MEM/F-12 GIBCO-Invitrogen, Carlsbad, CA, USA), penicillin/streptomycin (1:100) 500 μ B27 (1:50) 1 ml, Fungizone 200 μ bFGF 20 ng/ml, EGF 20 ng/ml, Heparin 5 μ g/ml
2. Penicillin/streptomycin sulfate (100×, Euroclone, Milan, Italy).
3. L-Glutamine (100×, Euroclone, Milan, Italy), stored in aliquots at –20°C
4. B-27 (50×; GIBCO-Invitrogen, Carlsbad, CA, USA), stored in aliquots at –20°C
5. Recombinant human fibroblast growth factor (FGF-2, 50 μg Tebu Bio, SantaCruz; CA, USA). Working solutions are prepared by dilution in 250 μ microg/ml of sterile water and stored in aliquots at –20°C.

6. Recombinant human epidermal growth factor (EGF, 500 µg; Tebu Bio, SantaCruz, CA, USA). Working solutions are prepared by dilution in 250 µg/mL of sterile water and stored in aliquots at –20°C.
7. Heparin (5 µg/ml; Sigma Aldrich, St. Louis, MO, USA). Working solutions are prepared by dilution in 5 µg/ml of sterile water and stored at 4°C.
8. Dulbecco's modified Eagle's medium (DMEM) (Euroclone, Milan, Italy) supplemented with 20% fetal bovine serum
9. Fetal bovine serum (Gibco-Invitrogen, Carlsbad, CA, USA). Stored at –20°C. Do not refreeze.
10. Differentiating medium: NSA medium (Euroclone,Milan, Italy) supplemented 3% serum, hormone mix, L-glutamine, penicillin/streptomycin.
11. Hormone mixture for Neurosphere medium (10× concentration, total amount made per batch = 200 ml). To prepare the hormone mixture, begin by making up the following medium to which you will add the various additions.

Base medium: 20 ml of 5× DMEM/F12 media, 4 ml of 30% glucose, 3 ml of 7.5% NaHCO3, 1 ml of 1 M Hepes, 150 ml of sterile distilled water, 200 mg of transferrin (Sigma T-2252), 50 mg Insulin (Sigma I-5500), 19.23 mg putrescine (Sigma P-7505; dissolve 19.32 mg of putrescine in 20 ml of sterile distilled water. Add to the medium mixture, the final concentration is 6×10^{-3} M); 20 µl of 3×10^{-3} M Selenium (Sigma S-9133; add 1.93 ml of sterile distilled water to 1 mg vial of selenium. Mix and transfer to a small sterile tube. Once made, this can be stored and –20 and reused. Add 20 µl to the media mixture), 20 µl 2×10^{-3} M progesterone (Sigma P-6149; add 1.59 ml of 95% ETOH to the 1 mg vial, mix and transfer to a small sterile tube. Store at –20°C. Can be reused. Add 20 µl to the media mixture (*see* **Note 1**)).

12. Phosphate-buffered saline 10× (PBS, Euroclone, Milan, Italy)
13. 25-cm^2 Flasks vent cap (Corning, Lowell, MA USA).
14. 75-cm^2 Flasks vent cap (Corning, Lowell, MA USA).

2.2. DC Vaccine Preparation

1. Iscove's medium (Sigma Aldrich, St. Louis, MO) supplemented with 5% fetal bovine serum.
2. Recombinant murine GM-CSF and recombinant murine IL-4 (5 ng/ml, Li StarFISH, Milan Italy). Working solutions are prepared by dilution in 100 ng/ml of sterile water and stored in aliquots at –20°C.
3. OptiMem (GIBCO, Invitrogen, Carlsbad, CA, USA).
4. DOTAP (Roche Applied Science, Mannheim, Germany).

5. Polystyrene tubes (Becton Dickinson, San Jose, CA, USA).
6. 6-well plates (Corning, Lowell, MA, USA).

2.3. Histology and Immunohistochemistry

1. Specific biotin-labeled secondary antibody (goat antimouse/rabbit, 1:200, Dako Cytomation, Glostrup, Denmark; rabbit antirat mouse adsorbed, 1:200, Vector, Burlingame, CA, USA and goat antihamster, 1:100, Vector, Burlingame, CA, USA). Stored at –20°C.
2. Peroxidase-conjugated streptavidin. (1:20, BioGenex, S.Ramon, CA, USA).
3. Diaminobenzydine (DAB, BioGenex, S.Ramon, CA, USA). Stored at 4°C.
4. Hydrogen peroxide. Stored at 4°C.
5. Rat monoclonal IgG1 anti-CD3, anti-CD4, anti-CD8 (1:20, Novocastra Laboratories, Newcastle, UK). Stored in aliquots at –20°C. Do not refreeze.
6. Mouse monoclonal IgG1 anti-Ki67 protein (1:50, clone B56, Pharmingen, Becton&Dickinson, San Jose, CA, USA).
7. Primary antibodies (Chemicon, Hofheim, Germany): mouse monoclonal anti-GFAP (stored at 4°C), rabbit monoclonal anti-beta-III tubulin (aliquots stored at –20C), mouse monoclonal antinestin (stored at 4°C).
8. Rabbit polyclonal antigreen fluorescence protein (GFP) (1:100, Santa Cruz Biotechnology, Santa Cruz, CA, USA). Stored at 4°C.
9. Paraformaldehyde (Sigma Aldrich): prepare a 4% (w/v) solution in fresh PBS.
10. Anti-Ki67 protein (BD Pharmingen, San Jose, CA, USA). Stored at 4°C.
11. Goat antimouse/rabbit (1:200, Dako Cytomation, Denmark).
12. Rabbit antirat mouse absorbed, 1:200, Vector, Burlingame, CA, USA.
13. Goat antihamster, 1:100, Vector, Burlingame, CA, USA.
14. FITC-conjugated swine antirabbit, Dako Cytomation, Glostrup, Denmark.
15. Biotin-labeled goat antihamster, Vector, Burlingame, CA, USA.
16. Streptavidin Texas Red conjugated, Vector, Burlingame, CA, USA.

2.4. Flow Cytometry

1. Primary antibodies (Chemicon, Hofheim, Germany) were: GFAP (1:200), beta-III tubulin (1:400), nestin (1:50).

2. Secondary antibodies (1:300, Cy3 and Cy5-conjugated, Jackson Immunoresearch Laboratories, UK). Aliquots stored at −20°C.
3. Phycoerythrin PE-conjugated mAbs (Pharmingen, Becton&Dickinson, San Jose, CA, USA): CD11c, CD80, CD86, MHC class I, MHC class II.
4. CellQuest software.
5. FACSFlow (Becton&Dickinson, San Jose, CA, USA).
6. Polystyrene tubes (Becton Dickinson, San Jose, CA, USA).
7. Saponin (stabilizer, ICN, Munich, Germany).
8. Goat serum (blocking, Sigma St. Louis, MO, USA).
9. Bovine serum albumin (BSA, Sigma St. Louis, MO, USA).
10. Permeabilizing solution: PBS1X-BSA 1%–saponin 0.5%.
11. Blocking buffer: PBS1X-BSA 1%– saponin 0.5%–goat serum 2%.

2.5. Cytotoxicity Assay

1. RPMI 1640 (Euroclone, Milan, Italy).
2. 2-Mercapto ethanol.
3. Sodium pyruvate.
4. Nonessential amino acids.
5. Recombinant human IL-2.
6. [^{3}H] Thymidine.
7. Trilux Microbeta 1450.
8. Mitomycin C.
9. ACK lysis buffer: 0.15 M NH_4Cl, 1.0 mM $KHCO_3$, 0.1 mM Na_2EDTA, pH 7.4.
10. 96-well flat bottom plates.

3. Methods

The observation that culture conditions may unveil the existence of different subtype populations in GBM may imply that cell hierarchies exist and can be targeted therapeutically.

The GL261 tumor was originally induced by intracranial injection of the chemical carcinogen 3-methylcholantrene into C57BL6 mice and maintained by serial intracranial and subcutaneous transplantation of small tumor pieces on the same mouse strain *(13)*. A cell line was established later from the GL261 tumor that was particularly used for immunotherapy investigations by several groups *(14, 15)*. To explore this perspective, we studied a murine model for glioma cancer stem-like cells derived from GL261 cell adherent line as pulsing agents for DC. CSC have been obtained from the cell line previously grown in serum (*see* **Note 2**). The recent review by Vescovi et al *(16)* provides a relevant contribution to the definition of a brain tumor stem cells. To satisfy criteria for CSC identification we have shown that GL261-NS have

- Extensive self-renewal ability
- Long-term proliferation
- Clonogenic potential
- Multilineage differentiation capacity (when switched to differentiating conditions they express markers of the three neural lineages: neurons, astrocytes, and oligodendrocytes)
- Aberrant differentiation properties
- Cancer-initiating ability

Our results show that GL261-NS behave differently from GL261-AC: they infiltrate aggressively the brain parenchyma and they seem attracted by ependymal regions and ventricles. Overall these features appear as reminiscent of the CSC phenotype. As a follow up to these experiments we performed DC vaccination against tumor from NS, and we have shown the efficacy of immunotherapy against this highly infiltrating and aggressive model of glioma.

We have suggested that the GL261 model can be of interest for further studies on DC vaccination of glioma subpopulations with CSC properties.

3.1. Preparation and Characterization of Neurospheres from an Adherent Cell Line

1. Cell culturing and propagation: the murine glioma cell line GL261-AC grows in DMEM (EuroClone, UK), 20% fetal bovine serum.

To evaluate whether GL261 cells could mimic the growth of human GBM-NS we let them grow in EGF/bFGF without serum. After 5 days neurospheres (GL261-NS) are visible in the culture medium and are proliferating continuously.

Every 4–5 days, neurospheres are collected, mechanically dissociated, and replated in the same conditions (8,000 cells/cm^2) (*see* **Note 3**).

2. Stable growth, long-term proliferation: cell proliferation is evaluated by a colorimetric method based on the WST-1 reagent (Roche Applied Science, Mannheim, Germany) at two time points: 24 and 48 h. WST-1 is added to 1×10^{-3} cells cultured in microplates. During incubation (2 or 4 h) WST-1 is converted into a colored, soluble formazan salt by viable cells. The amount of formazan is quantified by an ELISA plate reader at 420–480 nm.

Self-renewal is evaluated by proliferation kinetic of GL261-NS compared to GL261-AC. Cells are plated at 8,000 cells/cm^2

and the resulting cultures are collected every 4 days. The total number of viable cells is assessed at each passage by Trypan Blue exclusion (Sigma, St Louis, MO, USA). The same number of cells is plated every passage and the generated cells are cryopreserved. This low-density system for neural stem cells selects away differentiated cells leaving stem cells to proliferate and expand exponentially (*see* **Fig. 1**).

3. Clonogenic potential: clonogenic assay is performed using "Cloning by limiting dilution." This is a procedure for

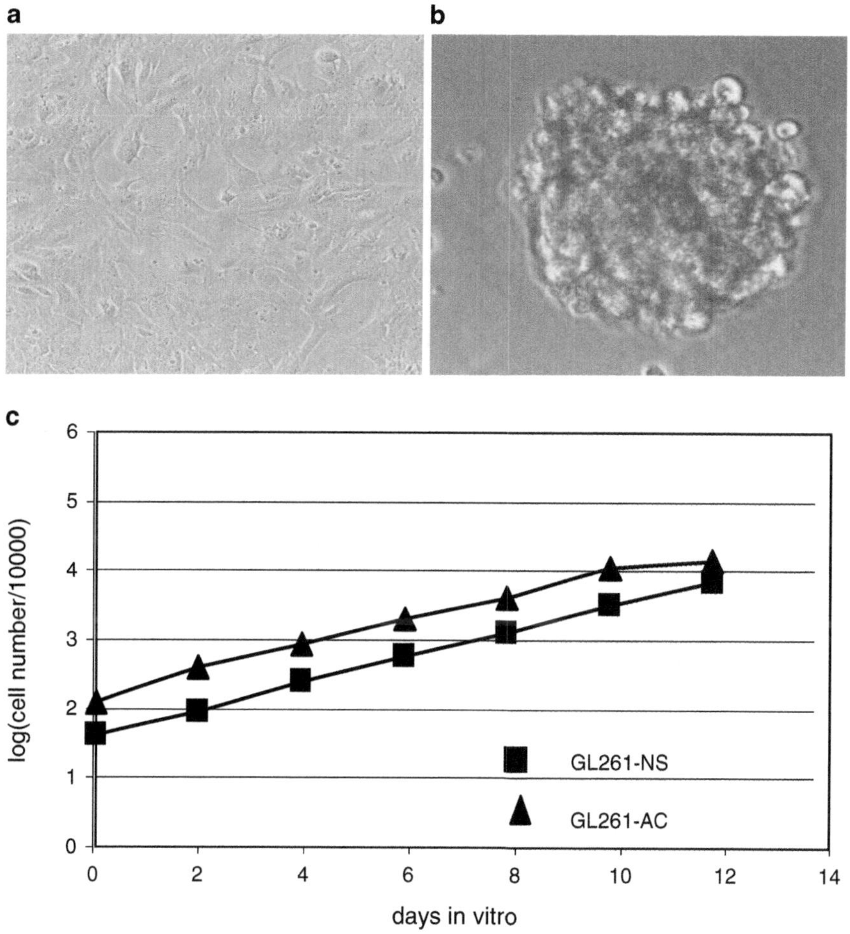

Fig. 1. Proliferation rate analysis: We culture GL261 in EGF/bFGF without serum. After 5 days neurospheres (NS) from GL261-AC (**a**) are visible in the culture medium and are proliferating steadily. Their growth increases exponentially regardless of passage number, while GL261-AC growth tends to plateauing. Cultures of GL261-NS (**b**) in EGF/bFGF are passaged every 3–4 days by dissociation into a single-cell suspension followed by replating in fresh growth medium. Proliferation kinetics (**c**) show that the growth of GL261-NS increases exponentially regardless of passage number, and the slope of this growth curve remains constant even after a significant number of passages. GL261-AC cultured in medium with serum show initial proliferation followed by a plateau phase.

separating cells based on the assumption that if a suspension of cells is diluted with enough culture medium, a concentration of cells will be produced such that an accurately measured volume of the diluted suspension will contain 1 cell. When this volume of the diluted suspension is placed into separated wells of a 96-well plate, each well should receive 1 cell/well. If this cell remains viable and proliferates, then an isolated cell clone will have been established in the well.

Cells derived from dissociation of neurospheres and GL261-AC are resuspended by standard subculture method. Viable cell count is performed and a cell suspension is prepared containing 100 cells/ml of medium to give a concentration of 10 cells/100 μl of medium. Another suspension is prepared that contains 10 cells/ml to give a concentration of 1 cell/100 μl. It is important to make dilutions that give at least 5–10 ml of each final cell suspension to obtain an entire 96-well plate.

The wells are inspected and then incubated al 37°C in 5% CO_2 for 5–7 days without changing the medium. Isolated clones of cells are observed for growth: an isolated clone that resulted from cell division of one isolated cell will usually have a uniform, circular border.

As a colony grows, it may be advisable to feed the wells every 4–7 days, typically, if growth is vigorous, 50% of the medium may be changed (*see* **Note 4**).

When colonies of a suitable size have formed, the cells are resuspended by trypsinization if adherent (GL261-AC), or careful pipetting if nonadherent (GL261-NS).

Each colony is subcultured into an individual well in a 24-well plate, which may contain conditioned medium if considered necessary. The total volume of medium per well should not exceed 1 ml. When sufficient cells are present subculture into 25-cm^2 flasks stocks of cloned cells should be frozen in liquid nitrogen as a sufficient number of cells is available.

4. Multilineage differentiation capacity: The expression of neural differentiation markers is assessed under four conditions:
   (a) GL261-NS cultured for 6 days with EGF/bFGF and no serum
   (b) GL261-NS dissociated in the presence of 3% FBS and without EGF-bFGF (differentiating conditions)
   (c) GL261-AC cultured in standard medium (DMEM, 20% FBS)
   (d) GL261-AC dissociated in the presence of 3% FBS and without EGF-bFGF (differentiating conditions). (*see* **Note 5**)

Cells are plated onto poly-L-lysine coated chamber slides, fixed in 4% paraformaldehyde, washed with PBS, permeabilized in 0.2%

NP-40, and blocked with goat serum. Cells are then incubated with primary antibodies followed by FITC-conjugated secondary antibody (1:300). Slides are counterstained with a mounting medium containing DAPI before examination by fluorescence microscopy. The primary antibodies (Chemicon, Germany) were: GFAP (1:200), beta-III tubulin (1:400), nestin (1:50). Analysis on adherent cells, dissociated neurospheres or differentiating cells was performed, using the same primary antibodies (antinestin, anti-GFAP e anti-beta-III-tub;1:200) and the intracellular staining protocol, by FACSCalibur flow cytometer (Becton Dickinson, CA, USA).

5. Evaluation of tumorigenicity by intracranial injection: tumorigenicity is determined by injecting 1×10^5 GL261 intracranially. GL261-AC are harvested by trypsinization and GL261-NS are manually dissociated, washed, and resuspended in PBS 1× in 3 μl/mouse and cell suspension are delivered into the left putamen. Intracranial tumor implantation is performed by stereotaxic injection through a 10-μl Hamilton syringe (coordinates with respect to bregma: 0.7 mm posterior, 3 mm left lateral, 3.5 mm deep into putamen).

3.2. DC Vaccination

1. Murine DC are prepared from the bone marrow (BM) of C57BL6N mice (H-2b). Under sterile conditions, BM is flushed from femurs, tibias, and humeri and depleted of red cells using ACK lysis buffer (ammonium chloride). Cells are incubated on ice for 3–5 minutes in the presence of cold ACK and then centrifugated at 300 × g for 5 minutes.
2. BM cells are plated in 6-well culture plates (2×10^6 cells/ 3 ml) in Iscove's medium, 5% FBS. After 2 hours of culture, floating cells are gently removed, and fresh medium supplemented with 5 ng/ml rmGM-CSF and 5 ng/ ml rmIL-4 is added. Medium is replaced on day 3. On day 7 of culture, adherent and nonadherent DC are removed with PBS 1X-EDTA 10 mM for FACS analysis and pulsing.

3.3. DC Loading with Tumor Lysate

1. The tumor lysate is obtained by sonication of GL261 cells in Opti-MEM medium (10^7 cells/500 μl) for 1 hour (*see* **Note** 6).
2. Tumor lysate (125 μg in 500 l Opti-MEM) and DOTAP are mixed in polystyrene tubes at room temperature for 20 min. DCs are washed twice in Opti-MEM medium and then resuspended at 5–10 × 10^6 cells/ml in 50-ml polypropylene tubes. The complex is added to DC and incubated at 37°C for 3–4 hours. After incubation, loaded DC are collected and centrifuged at 320 × g for 10 minutes and resuspended in sterile PBS1X (1×10^6 DC/150 μl/ mouse) for use as a vaccine.

3. For vaccinations, female C57BL6N mice (5 weeks) are treated with three subcutaneous injections, using an insulin syringe (U-100, 0.5 ml), spaced one week apart after tumor implantation.

3.4. Histology

1. For histology, brains and axillary lymph nodes are carefully removed, postfixed in 4% paraformaldehyde, and embedded in paraffin. Four-micron-thick sections are taken on a microtome. A representative number of brain and lymph nodes from each group are embedded in OCT and snap frozen in cold isopenthane in liquid nitrogen. Five-um-thick sections are then taken on a cryostat.
2. Sections for immunostaining were dewaxed and rehydratated through serial passages in xylene and alcohol. Endogenous peroxidase activity is blocked with 0.3% H_2O_2 in methanol; heating to induce epitope retrieval is obtained by microwave treatment or incubation in a water bath. Sections are incubated in blocking solution (5% normal serum from the secondary antibody species) prior to incubation with primary antibodies.
3. Negative controls are obtained by omitting primary or secondary antibodies, respectively. After washing, sections are incubated with the specific biotin-labeled secondary antibody (goat antimouse/rabbit, 1:200; rabbit antirat mouse absorbed, 1:200; goat antihamster, 1:100). Sections are then washed and incubated in peroxidase-conjugated streptavidin. (1:20) and the signal revealed with diaminobenzydine (DAB) and hydrogen peroxide (DAB, 1:50).
4. Slides are then counterstained with Hematoxylin. These primary antibodies are used: rat monoclonal IgG1 anti-CD3 anti-CD8 and anti-CD4 (1:20, clones CD3–12), mouse monoclonal IgG1 anti-Ki67 protein (1:50), mouse monoclonal IgG1 antinestin (1:500), rabbit polyclonal antigreen fluorescence protein (GFP, 1:100).
5. Detection of dendritic cells in the lymph node is obtained by direct detection of the green fluorescence protein or by using fluorescent-conjugated secondary antibodies (FITC-conjugated swine anti rabbit, 1:50; biotin-labeled goat antihamster, 1:100 and streptavidin Texas Red conjugated, 1:100). Images were taken on an Olympus BX60 fluorescent microscope with a DP 70 Olympus digital camera.

3.5. Flow Cytometry

1. Direct labeling: the monoclonal Antibodies (mAbs) are directly coupled to phycoerythrin (PE). One million cells are incubated with specific mAbs for 30 minutes at 4°C, washed twice with PBS1X (*see* **Note** 7), and resuspended in PBS 1X for analysis. The expression of surface epitopes is visualized

using a FACSCalibur flow cytometer and analyzed with the CELLQuest software.

2. Intracellular staining: cells are collected at 1×10^6/sample, fixed with 2% paraformaldehyde, and incubated for 15 minutes on ice.

Cells are washed once with permeabilizing solution for 5 minutes and centrifuged at 200 × g at room temperature.

Permeabilizing buffer (0.5 ml) is added and incubated for 10 mins at room temperature. Cells are washed once with the same buffer (0.5 ml) and after centrifugation are incubated in blocking buffer for 30 mins at room temperature.

Cells are washed twice with the same buffer and incubated with primary antibodies for 60 mins at room temperature.

After this, they are washed twice in 0.5 ml permeabilizing buffer and incubated with secondary antibodies for 30 minutes at room temperature.

Cells are washed twice in 0.5 ml of permeabilizing buffer, twice in 0.5 ml of PBS1X-BSA 1% and analyzed by flow cytometry.

3.6. *In Vitro* Cytotoxicity Assay

1. In vitro cytotoxicity assay is performed using the JAM test (*see* **Note 8**). In this assay, splenocytes obtained from vaccinated animals and controls are cocultured in vitro for 5 days with GL261 cells treated with Mitomycin C (Sigma Aldrich Co, St. Louis, MO).
2. Spleens are removed at different time points after first vaccination. Splenocytes are isolated by gently mashing the spleen using a 26-gauge needle in RPMI 5% serum. After red cell depletion using ACK, splenocyte are cultured in 25-cm^2 flasks (*see* **Note 9**) at 40/1 E:T (Effector/Target) ratio for 5 days in RPMI 1640 + 10% fetal bovine serum, penicillin/streptomycin, 2-mercaptoethanol, sodium pyruvate, nonessential amino acids, and recombinant human IL-2 (50 units/ml) in the presence of tumor cells treated with Mitomycin C. (*see* **Note 10**)
3. A GL261 glioma cell suspension is prepared at a concentration of 1×10^7 cells/ ml in PBS. Mitomycin C is added to a final concentration of 330 μg/ml in a 6-well plate draped in an aluminum foil and incubated for 1 hour at 37°C (*see* **Note 11** and **Fig. 2**).
4. GL261 target cells are labeled overnight with 1 μCi [^{3}H] thymidine (Amersham Biosciences, Milan, Italy). They are pelleted, washed once with culture medium, and distributed in triplicates (10^4 cells per well per 0.2 ml) into 96-well flat-bottomed microwell plates. After three washes with PBS, prestimulated cytotoxic effector cells are added to [^{3}H]-labeled GL261 (target cells), at different E:T ratios (1/100; 1/50; 1/25; 1/10), and incubated for 4–5 hours at 37°C 5%CO_2 in RPMI 1640 medium. Cultures were harvested onto

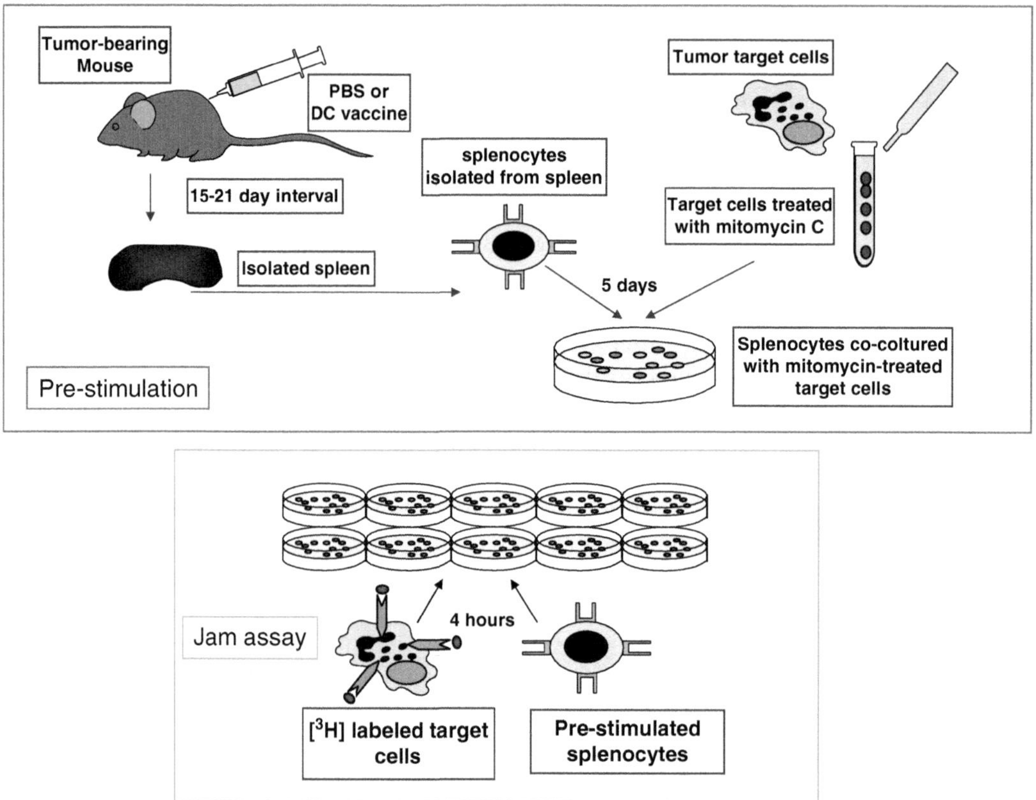

Fig. 2. Cytotoxicity is investigated in vitro using the JAM test. Splenocytes from spleen of vaccinated and control mice at different time points, are cocultured for 5 days with GL261 treated by Mitomycin C using a 40/1 E:T ratio, in the presence of recombinant human IL-2 (prestimulation). GL261-AC and GL261-NS labeled overnight with 1 μCi [^{3}H] thymidine are seeded with prestimulated splenocytes from control and vaccinated mice at different E:T ratios. This assay is important to evaluate the ability of splenocytes (including CD8 + T cells) to lyse autologous tumor cells in vitro. In particular our results suggest an induction of specific lytic activity against tumor cells by immunization with DC pulsed with lysates from NS. In the JAM assay splenocytes derived from each mouse are tested against GL261-AC and GL261-NS.

glass-fiber filters using a cell harvester. After washing and drying filters are placed in vials containing liquid scintillation fluid and counted using Trilux Microbeta 1450 (PerkinElmer Life Science, Boston, MA). The radioactivity measured corresponds to intact DNA, because DNA from dead cells is degraded into small fragments that pass through the filter. (*see* **Note 12**).

4. Notes

1. Once all the ingredients have been added and mixed, filter-sterilize using a 0.2-micron filter and divide into 5-ml aliquots in 15-ml sterile centrifuge tubes. Freeze at –20°C. It is important to avoid refreezing.

2. We have prepared neurospheres from other established cell lines present in our laboratory. They grow in vitro as aggregates, but they do not behave as CSC. The characteristics of GL261 cell cannot be generalized. The fact that CSC can still be obtained after long-term culturing of an established glioma cell line does imply a considerable level of plasticity.
3. We have found that GL261-NS grow better in 25-cm^2 Flasks. We use vent cap flasks to decrease contamination risks.
4. We have found that if growth is not vigorous, the frequency of medium changes should be decreased.
5. We have observed that GL261-AC do not grow under differentiating condition: they stop proliferation and fail to express neural marker.
6. Tumor lysate could be obtained by 3–4 cycles of freezing (–160°C) and thawing (37°C), or by sonication. We chose sonication and we found that 1 hour is required for complete lysis of GL261. Lysis is monitored by light microscopy. The absence of cell viability is evaluated by culturing for 24 hours: aliquots of lysate could be tested by subcutaneous injection into syngeneic or nude mice. The tumor lysate can be stored at –80°C.

This protocol can be adapted for many other cell culture systems. In particular for GL261 growing in suspension, cells can easily be dissociated and counted for the treatment protocol.

7. Two washes are important to avoid nonspecific labeling observed as positivity on flow cytometry analysis.
8. We chose the Jam assay because the standard 51chromium release assay has limited sensitivity, because of a high, nonspecific background release
9. The prestimulation of splenocytes is performed using vent cap 25-cm^2 Flasks. Usually we prefer to incubate flasks vertically in the incubator.
10. Since mitomycin C is very light sensitive, it is necessary to prepare a fresh stock solution each day for each experiment. Two washes are crucial, because any traces of mitomycin C left among the cells will reduce proliferative responses when splenocytes are added for coculture.
11. The choice of mitomycin C concentration depends on the cell line: the recommended use as a selection agent is at concentrations ranging from 100 to 800 μg/ml In cell culture, mitomycin C mitotically arrests cells. We have found this concentration (330μg/ml) of mitomycin to be optimal for our glioma cell line. An alternative treatment is irradiation: the appropriate dose should be determined empirically for each cell line; some cell lines require as much as

10,000–12,000 rad and may be more sensitive to mitomycin C treatment.

12. Percent-specific lysis is quantified using the following formula: %-specific lysis = (cpm spontaneous – cpm experimental)/cpm spontaneous × 100.

Acknowledgments

We acknowledge the support of AIRC, ARIN-Fondo di Gio, and Italian Minister of Health.

References

1. Holland, E. C. (2000) Glioblastoma multiforme: the terminator. *Proc Natl Acad Sci U S A*. **97**, 6242–6244.
2. Castro, M. G., Cowen, R., Williamson, I. K., David, A., Jimenez-Dalmaroni, M. J., Yuan, X., Bigliari, A., Williams, J. C., Hu, J., and Lowenstein, P. R. (2003) Current and future strategies for the treatment of malignant brain tumors. *Pharmacol Ther*. **98**, 71–108.
3. Stupp, R., Mason, W. P., Van Den Bent, M. J., Weller, M., Fisher, B., Taphoorn, M., Brandes, A. A., Cairncross, G., Lacombe, D., and Mirimanoff, R. O. Concomitant and adjuvant temozolomide (TMZ) and radiotherapy(RT) for newly diagnosed glioblastoma multiforme (GBM) (2004). Conclusive results of a randomized phase III trial by the EORTC Brain & RT Groups and NCIC Clinical Trials Group. *J Clin Oncol* (Meeting Abstracts). **22**, 2.
4. Ehtesham, M., Black, K. L., and Yu, J. S. (2004) Recent progress in immunotherapy for malignant glioma: treatment strategies and results from clinical trials. *Cancer Control* **11**, 192–207.
5. Yu, J. S., Wheeler, C. J., Zeltzer, P. M., Ying, H., Finger, D. N., Lee, P. K., Yong, W. H., Incardona, F., Thompson, R. C., Riedinger, M. S., Zhang, W., Prins, R. M., and Black, K. L. (2001) Vaccination of malignant glioma patients with peptide-pulsed dendritic cells elicits systemic cytotoxicity and intracranial T-cell infiltration. *Cancer Res*. **61**, 842–847.
6. Yamanaka, R., Abe, T., Yajima, N., Tsuchiya, N., Homma, J., Kobayashi, T., Narita, M., Takahashi, M., and Tanaka, R. (2003) Vaccination of recurrent glioma patients with tumour lysate-pulsed dendritic cells elicits immune responses: results of a clinical phase I/II trial. *Br J Cancer*. **89**, 1172–1179.
7. Rutkowski, S., De Vleeschouwer, S., Kaempgen, E., Wolff, J. E., Kuhl, J., Demaerel, P., Warmuth-Metz, M., Flamen, P., Van Calenbergh, F., Plets, C., Sorensen, N., Opitz, A., and Van Gool, S. W. (2004) Surgery and adjuvant dendritic cell-based tumour vaccination for patients with relapsed malignant glioma, a feasibility study. *Br J Cancer*. **91**, 1656–1662.
8. Yu, J. S., Liu, G., Ying, H., Yong, W. H., Black, K. L., and Wheeler, C. J. (2004) Vaccination with tumor lysate-pulsed dendritic cells elicits antigen-specific, cytotoxic T-cells in patients with malignant glioma. *Cancer Res*. **64**, 4973–4979.
9. Yamanaka, R., Homma, J., Yajima, N., Tsuchiya, N., Sano, M., Kobayashi, T., Yoshida, S., Abe, T., Narita, M., Takahashi, M., et al. (2005) Clinical evaluation of dendritic cell vaccination for patients with recurrent glioma: results of a clinical phase I/II trial. Clin *Cancer Res*. **11**, 4160–4167.
10. Clarke, M. F. (2004) Neurobiology: at the root of brain cancer. *Nature*. **432**, 281–282.
11. Al-Hajj, M., Becker, M. W., Wicha, M., Weissman, I., and Clarke, M. F. (2004) Therapeutic implications of cancer stem cells. *Curr Opin Genet Dev*. **14**, 43–47.
12. Pellegatta, S., Poliani, P.L., Corno, D., Menghi, F., Ghielmetti, F., Suarez-Merino, B., Caldera, V., Nava, S., Ravanini, M., Facchetti, F., Bruzzone, MG., Finocchiaroan G. (2006) Neurospheres enriched in cancer stem-like cells are highly effective in eliciting a dendritic cell-mediated immune response against malignant gliomas. *Cancer Res*. **66**, 10247–10252.

13. Ausman, J.I., Shapiro, W.R., & Rall, D.P. (1970). Studies on the chemotherapy of experimental brain tumors: development of an experimental model. *Cancer Res* **30**, 2394–2400.
14. Benedetti, S., Pirola, B., Pollo, B., Magrassi, L., Bruzzone, M. G., Rigamonti, D., Galli, R., Selleri, S., Di Meco, F., De Fraja, C., Vescovi, A., Cattaneo, E., and Finocchiaro, G. (2000) Gene therapy of experimental brain tumors using neural progenitor cells. *Nat Med* **6**, 447–450.
15. Plautz, G.E., Touhalisky, J.E., & Shu, S. (1997). Treatment of murine gliomas by adoptive transfer of ex vivo activated tumor-draining lymph node cells. *Cell Immunol.* **178**,101–107.
16. Vescovi, A.L., Galli, R., Reynolds, B.A. (2006) Brain tumor stem cells. *Nat.Rev. Canc.* **6**, 425–436.

Chapter 16

Maintaining and Engineering Neural Stem Cells for Delivery of Genetically Encoded Therapy to Brain Tumors

Jennifer Katz, Joseph Krueger, Brunhilde Felding-Habermann, and Evan Y. Snyder

Summary

Despite advances for the treatment of cancer, the prognosis for patients suffering from malignant brain tumors remains dismal. High-grade neoplasms, such as gliomas, are highly invasive and spawn widely disseminated microsatellites that have limited the efficacy of surgical and adjunctive therapies. The cancer stem cell hypothesis suggests that conventional chemotherapeutic treatments kill differentiated and differentiating cells which often form the bulk of the tumor. One major concern is that the cells which give rise to the tumor, the cancer stem cells, remain untouched and may be responsible for a relapse of the disease. Therefore, an adjunctive therapy to current cancer treatment is critical for the survivability of patients suffering from brain tumors. We have successfully engineered tumor-tropic neural stem cells to deliver antineoplastic gene products directly to the tumor-producing cells. This potential therapeutic strategy may safely eradicate tumor-producing cells in the brain while minimizing damage to normal, healthy cells.

Key words: Neural stem cells, Gene therapy, Brain tumors, Viral constructs, Transfection, Lentiviral transduction

1. Introduction

The neural stem cell began receiving attention once scientists recognized that it was a primordial self-renewing cell that can give rise to neurons, astrocytes, and oligodendroglia of the mature nervous system. Since neural stem cells can be isolated, expanded in culture and then reimplanted, the potential that neural stem cells hold as therapeutic agents for a variety of neuropathological conditions is promising *(1)*. Furthermore, it has previously

John S. Yu (ed.), *Cancer Stem Cells*, Methods in Molecular Biology, vol. 568
DOI 10.1007/978-1-59745-280-9_16, © Humana Press, a part of Springer Science+Business Media, LLC 2009

been reported that neural stem cells administered intracranially, migrate to and infiltrate experimental orthotopic primary brain tumors and deliver therapeutic agents that successfully inhibit tumor growth *(2, 3)*.

The protocol outlined in this chapter describes techniques that modify the cellular genome of murine neural stem cells for gene delivery approaches in brain tumor therapy. It is important to note, however, that stem cells isolated from origins other than mouse or the neuroectoderm have also been genetically modified to produce molecules of therapeutic value *(4–6)*. The transduction techniques summarized later are not specific for neural stem cells and may be performed on a variety of mammalian cell lines. However, due to the capacity of neural stem cells to differentiate in vivo following transplantation, as well as their ability to home toward brain tumors and track migrating tumor cells, the neural stem cell ;may be the optimal vehicle for delivery of genetically encoded therapy in the CNS *(3)*. Neural stem cells have been successfully engineered to deliver a variety of therapeutically relevant molecules, such as prodrug-converting enzymes *(3, 6, 7)* (e.g., cytosine deaminase, carboxylesterase, HSV-TK) or bioactive genes, including interleukin-12 *(8, 9)* and TRAIL *(10)*. Still, it is up to the user to determine the optimal stem cell source and therapeutic gene for their study.

The final step in generating a therapeutically useful neural stem cell line is to create neural stem cells that express the therapeutic protein. Modern retroviral and lentiviral transduction has dramatically improved the efficiency and stability of vector-based expression over traditional expression vectors and transfection techniques. In this regard, lentivirus is far more efficient at transducing mammalian cells than retrovirus due to its ability to infect both proliferative and nonproliferative cells, as well as its ability to infect each cell multiple times (multiplicity of infection [MOI]). The HIV-1-derived self-inactivating lentiviral constructs have proven to be useful for high-efficiency transfection of a variety of cell types *(11)*. Nonconcentrated viral preparations often give a rate of transduction approaching 100%, with high MOI, allowing sufficient expression of a therapeutic protein of interest (POI) *(12)*. The expression of this protein can be controlled under constitutive (cytomegalovirus [CMV]), ligand-activated (tetracycline-on/off), strong or weak and lineage-specific promoters; giving the user the desired control overexpression of the POI. The use of lentivirus ensures homogeneity in the transduced cells, without the need for clonal selection techniques that may introduce artifacts into the system. Beyond this, good lentiviral transduction eliminates the need for an antibiotic selection step, considerably decreasing the amount of time and effort required to generate a useful neural stem cell line.

2. Materials

2.1. Maintenance of Neural Stem Cells

1. Neural stem cell medium: Dulbecco's Modified Eagle's Medium + GlutaMax (DMEM) containing 10% fetal bovine serum (FBS), 5% inactivated horse serum, and 1% penicillin–streptomycin; store at 4°C.
2. 1× Dulbecco's Phosphate-Buffered Saline without calcium & magnesium.
3. 0.05% trypsin–ethylenediamine tetraacetic acid (EDTA); store at −20°C.
4. Sterile DMSO.
5. Falcon tissue-culture-treated dishes.
6. Cryogenic vials.

2.2. Producing Lentivirus in HEK 293T Cells

1. 0.1× TE buffer: 1 mM tris of pH 8.0, 100 μM EDTA of pH 8.0. Make stock buffer as 1× TE (10 mM tris of pH 8.0, 1 mM EDTA of pH 8.0). Sterile filter through a 0.22-μM filter; store at room temperature.
2. 2.5 M $CaCl_2$ in H_2O. Sterile filter through a 0.22-μM filter; store at −20°C.
3. 2× HBS: 281 mM NaCl, 100 mM Hepes, 1.5 mM Na_2HPO_4, pH 7.12. Sterile filter through a 0.22-μM filter; store at −20°C.
4. 293T medium: DMEM containing 5% FBS, 4 mM L-Glutamine, 1 mM MEM Sodium Pyruvate, 0.1 mM MEM Nonessential Amino Acids, and 1% penicillin–streptomycin.

2.3. Titering Your Lentiviral Stock

1. 293T medium: DMEM containing 5% FBS, 4 mM L-Glutamine, 1 mM MEM Sodium Pyruvate, 0.1 mM MEM Nonessential Amino Acids, and 1% penicillin–streptomycin.
2. Your lentiviral supernatant.
3. 6 mg/mL Polybrene® in H_2O.

2.4. Transduction of Cells

1. Your lentiviral supernatant (store at −80°C until use).
2. Mammalian cell line of choice (e.g., neural stem cells).
3. Complete culture medium for your cell line.
4. 6 mg/mL Polybrene in H_2O.
5. Appropriately sized tissue culture plates for your application.
6. Selection agent (if selecting for stably transduced cells).

3. Methods

Maintenance of Neural Stem Cells. Neural stem cell cultures should be maintained in a standard humidified incubator (37°C, 5% CO_2). Once the lines are established they can be carried in 10-cm tissue-culture-treated dishes. It is important to split cell lines only once a week at no more than a 1:10 dilution. Neural stem cells are contact inhibited and therefore cultures can become confluent during normal growth periods. However, there are some general guidelines to consider before proceeding with transplantation or injection of neural stem cells. When working with cells that have been previously frozen, do not use those that have been passaged for more than 6 weeks after the initial thaw. Cultures should be no more than 90% confluent, or they will begin to produce an extracellular matrix causing clumping of cells and yielding poor results. If cells do not appear as a single-cell suspension then they should not be used for in vivo experiments. The proper confluence can be maintained by splitting cultures at a dilution of 1:10 72–96 h before transplantation or injection. It is important to plate cells during the last split for immunocytochemical analysis to confirm undifferentiated state at the time of use.

Considerations for Choice of Viral Constructs. The general backbone for HIV-1-derived lentiviral constructs contains flanking HIV-1 long terminal repeats, (delta U3) regions for self-inactivation, and a (R region) polyadenylation signal *(13)*. Commercially available and privately generated HIV-1-based lentiviral vectors are similar in this manner; however, the addition of several elements to the viral backbone can improve stability and expression in transduced cells. These include the addition of the IFN-beta scaffold attachment region to increase expression, the central polypurine tract to facilitate nuclear import, and the woodchuck hepatitis virus posttranslational regulatory element to stabilize the mRNA *(14)*. These additional elements may be integrated into the lentiviral vector you plan to use, and improve the expression of the POI and thus therapeutic potential when arming stem cells.

The choice of promoter should also be considered when choosing a lentiviral construct. Typically, the CMV promoter is used for its strong induction of transcription and usefulness in a wide variety of neural cell types *(15)*. However, this strong promoting activity can possibly lead to toxicity through an overabundance of expressed protein. The use of a weaker promoter, such as the myeloproliferative sarcoma virus long terminal repeat negative control region deleted (MND promoter), allows sufficient expression in some cell types at a level lower than the CMV promoter *(15)*. Alternatively, the use of the Ubiquitin promoter has been proven to be effective for use in both human and murine cells *(16)*. Other lineage-specific neural promoters may be used if appropriate for the experiment planned.

For in vitro and in vivo studies, the ability to efficiently track the stem cells during the studied processes is critical. Thus, the lentiviral vector must contain a genetic tag for following the cells in live animals or fixed tissues *(17)*. For this, an arsenal of proteins that are detected in live or fixed cells through direct, substrate-based, or immunohistochemical means are available. For in vivo and in vitro tracking of live cells, choose fluorescent proteins such as GFP, YFP, CFP, etc. and bioluminescent proteins such as Firefly and Renilla luciferase. In fixed tissues, enzymatic reporters such as B-galactosidase and Alkaline Phosphatase are sensitive means to detect cells. Epitope tags such as the myc tag (myc), hemagglutin (HA), or histidine (HIS) tag fused to the POI are a good alternative for fixed tissues and cells. Any of these proteins or tags listed can be detected through typical antibody approaches. The genetic tag can either be present on the same expression vector as your POI, or otherwise independently expressed. In the latter case, using a different virus stock to label the cells from the stock used to force expression of your POI is suitable. However, modified lentiviral vectors that incorporate the internal ribosome excision sequence (IRES) allow simultaneous expression of both POI and reporter. Here, the IRES is sandwiched between the POI gene and reporter gene; thus, transcription of both proteins is driven by the same promoter. In this way, the delivery of the therapeutic protein and detection of the stem cells can occur with only one transduction.

3.1. Producing Lentivirus in 293T Cells

This procedure is to generate viral supernatant. Generally, this is sufficient for transfection of most cells. This protocol uses $CaCl_2$ precipitation for transfection of two 10-cm plates. Transformed Human Embryonic Kidney (HEK293T; 293T) cells are used for viral packaging. The actual construction of your packaging and expression constructs varies, so specific amounts mentioned in this protocol may vary. However, this protocol is fairly robust and should work for most types of lentiviral systems.

3.1.1. Day 1: Plating

1. Treat 10-cm plates with poly-L-lysine for 15 min at room temperature (50 μg/mL final)
2. Plate out 2×10^6 cells in 10 mL complete DMEM.
3. Grow overnight in 5% CO_2.

3.1.2. Day 2: Transfection

Generation of the lentivirus must be performed in a biosafety level 3 (BSL-3) facility. Make sure to follow your institutional guidelines.

1. In a laminar flow hood, using sterile tubes (FACS tubes are ideal), place packaging plasmids (13 μg GAG/POL, 7 μg VSV-G, 5 μg REV) and 10 μg lentiviral expression vector in 900 μL 0.1× TE. Make sure to dilute plasmids in sterile H_2O to appropriate volume and add 1× TE to make 0.1× TE.
2. Add 100 μL $CaCl_2$ solution, and vortex.
3. Add 1 mL of 2× HBS dropwise to tube, mixing slightly during this process.
4. Immediately add this solution dropwise onto the plate, spreading it out over the plate, gently swirling the plate as you add the solution.
5. Place cells into 5% CO_2 overnight (16 h).

3.1.3. Days 3–4: Viral Production

1. The next morning, change the media (use 10 mL). Note that the VSV G glycoprotein can cause 293T cells to fuse. They can appear as large, multinucleated cells (syncytia). This does not affect production of the lentivirus.
2. Grow 30–36 h, and then collect the media. Sterile filter (0.22 μM) and aliquot the supernatant into 1-mL aliquots. This supernatant is ready to use. Store virus aliquots at −80°C.

(Alternatively)

Harvest the viral supernatants up to 48 h posttransfection (Day 4).

(Alternatively)

After first supernatant harvest at 30–36 h, add 10 mL fresh media and collect new viral supernatant at 72 h. The viral yield for this collection will be substantially lower.

Caution: Remember that you are working with a nonreplicating, yet infectious, virus when you handle the supernatant. Use the appropriate precautions (BSL-2 precautions are recommended – follow your institutional guidelines).

3.2. Titering Your Lentiviral Stock

To easily perform a viral titer, a marker is needed to identify cells which have virus. In general, serial dilutions of the virus are made and the cells transduced. The marker is analyzed and the percentage of cells which express the marker is used to determine the titer. For this purpose, green fluorescent protein (GFP) is highly recommended due to the ability to quantitatively determine transduction efficiency using FACS analysis.

This protocol serves two purposes: To find the viral titer, and find the right concentration of virus to use if you wish to obtain an MOI of 1.0. Too much virus will give greater MOIs, which is good for protein expression but may cause toxicity.

1. Prepare tenfold serial dilutions ranging from 1×10^2 to 1×10^6 in 1 mL of culture media. You may prepare a wider range of serial dilutions (up to 1×10^{10}) if desired. Mix each dilution gently by inversion.

2. Follow the normal transduction protocol described later, using the entire 1-mL dilution.
3. Assay the cells for transduction efficiency. The lowest dilution that efficiently transduces all the cells (less than 5% nontransduced) is used to calculate the titer; any higher titration will result in an MOI greater than 1.
4. The exact MOI of 1.0 can be found between the tenfold dilution that leaves significant (more than 5%) nontransduced cells and the tenfold dilution that transduces all the cells. If desired, you may narrow this range down further by performing twofold dilutions in this range to find the exact MOI of 1.0. Keep in mind that there is a range of efficiencies in each transfection, and a calculated MOI of about 1.2 should be used to a practical MOI of 1.0.

3.2.1. Troubleshooting

If no or poor transduction efficiency is observed, perform the following diagnostics in this order:

1. Verify that your expression construct is correct (i.e., the gene is present, in frame, and not truncated or mutated).
2. Verify that your packaging plasmids are correct.
3. Verify your viral titer using an established reporter virus. Problems with virus production is usually a result of problems transfecting the packaging cells. The $CaCl_2$ transfection method is robust, but can be faulty if the HBSS is added improper$_l$y. If desired, switch to a lipid-based transfection technique.
4. If all the above are correct, consider the design of your expression construct. The use of the CMV promoter is not always appropriate. For example, the CMV promoter does not result in significant transcription in many stem cell types and murine cells. Consider a different promoter such as the Ubiquitin promoter.
5. Concentrate your virus if everything else seems fine (see the following section).

3.3. Concentration of Virus

Concentrating of virus is usually not necessary for in vitro transduction. One simple way to increase the concentration of virus for in vitro transduction is to simply use one-half of the media (5 mL instead of 10 mL) during viral production. However, if you plan to use the virus for in vivo transduction, you most often need a very concentrated virus. To accomplish this, simply centrifuge your viral supernatant at 50,000×*g* for 90 min. Carefully decant the supernatant and resuspend the viral pellet in 100 μL-1 mL of plain media (without FBS, etc.) or PBS.

3.4. Thawing Neural Stem Cell Lines

1. Thaw a frozen vial by placing the vial in a water bath maintained at a temperature of 37°C for 1–2 min. Do not agitate cells.
2. Remove the vial from the water bath as soon as the cells have thawed and decontaminate by wiping the vial down with 70% ethanol (*see* **Note 1**).
3. Pipet the contents of the vial (approximately 1–2 × 10^6 cells/vial) equally into one 10-cm dish containing 8 mL of prewarmed feeding medium.
4. Rinse the vial two times with 1 mL feeding medium and transfer to the dish.
5. Gently swirl each dish to evenly distribute cells.
6. Place dishes into the incubator.
7. Change the medium once the cells have attached to the plate (approximately 8 h).
8. Once cells have become confluent they can be trypsinized and gradually expanded as described in the next step.

3.5. Maintaining Neural Stem Cell Lines

1. Remove and discard culture medium by aspiration (*see* **Note 2**).
2. Rinse cell layer twice with 1× PBS (5 mL/10 cm dish) to remove all traces of serum.
3. Add 2 mL of 0.05% Trypsin–EDTA solution to the dish and incubate at 37°C for 2–5 min, then observe cells under an inverted microscope until cell layer is dispersed. (The serum in the medium deactivates the Trypsin–EDTA, so this is a crucial step that should not be left out). The dish can be gently tapped to help the cells detach. Cells that are difficult to detach can be placed back in the incubator for an additional 1–2 min.
4. Add 3 mL of feeding medium and aspirate cells by gently pipetting. Be careful not to introduce air bubbles while triturating.
5. Transfer 1 mL of the total 5 mL in the 10-cm dish equally (0.5 mL/dish) to two new 10-cm dishes containing 10 mL of prewarmed feeding medium. This should be done in a dropwise manner to ensure that each dish receives 10% of the original population. Fresh media can be added to the original dish so that the cells will reattach and proliferate as a backup should the split be unsuccessful.

3.6. Transduction of Neural Stem Cells

1. Day 1: The day before transduction, detach and count the cells, plating them in a six-well plate so they will be 40–60% confluent at the time of transduction. This is typically 2 × 10^5 cells per well. Make sure there is 1.5 mL of media in each well. Incubate cells per their normal growth conditions overnight.

2. Day 2: On the day of transduction, add 1 μL of 6 mg/mL (final concentration of 6 μ;g/mL) Polybrene to your cells. Swirl gently to mix. For high transduction efficiency and high MOI, simply add 500 μL of your thawed lentiviral supernatant to the media. Swirl gently to mix. Incubate cells per their normal growth conditions overnight (*see* **Note 3**). If you wish to more precisely transduce cells with low MOI, you will have to perform a viral titer (see prior section).
3. Day 3: The following day remove the media containing virus and replace it with 2 mL of complete culture medium. Aspirate this media and replace it two more times to remove any excess virus. Incubate cells per their normal growth conditions overnight, then passage and expand or harvest cells for expression analysis (*see* **Note 4**).

3.7. Preparing Neural Stem Cells for Transplantation or Injection

1. Remove and discard culture medium from dish.
2. Rinse cell layer twice with 1× PBS (5 mL/10-cm dish) to remove all traces of serum.
3. Add 2 mL of 0.05% Trypsin–EDTA solution to the dish and incubate at 37°C for 2–5 min, then observe cells under an inverted microscope until cell layer is dispersed.
4. Add 3 mL of feeding medium and aspirate cells by gently pipetting. Be careful not to introduce air bubbles while triturating.
5. Pipet cells into a 15-mL tube and centrifuge for 1 min at 1,000×*g*.
6. Remove supernatant by aspiration being careful to not disturb the cellular pellet.
7. Wash the cells by resuspending them in 10 mL of 1× PBS and gently triturate to form a single-cell suspension.
8. Centrifuge for 1 min at 1,000×*g*.
9. Repeat **steps** 7 and **8**.
10. After the last spin, remove the supernatant by aspiration and resuspend the pellet in 10 mL of 1× PBS and gently triturate cells. Take 10 μL of cell solution and count the number of cells.
11. Centrifuge the cell solution at 1,000×*g* and resuspend the pellet in the appropriate volume of 1× PBS at a concentration of 4×10^4 cells/μL.
12. Place cells on ice until ready to use (*see* **Note 5**).

3.8. Freezing Neural Stem Cells

1. Make freezing medium (e.g., 67% feeding medium + 20% FBS and filter; add 13% sterile DMSO).
2. Remove and discard culture medium from dish.

3. Rinse cell layer twice with 1× PBS (5 mL/10-cm dish) to remove all traces of serum.
4. Add 2 mL of 0.05% Trypsin–EDTA solution to the dish and incubate at 37°C for 2–5 min, then observe cells under an inverted microscope until cell layer is dispersed. Cells that are difficult to detach can be placed back in the incubator for an additional 1–2 min.
5. Add 3 mL of freezing medium and aspirate cells by gently pipetting. Be careful not to introduce air bubbles while triturating.
6. Pipet an equal volume of cells into cryogenic vials (approximately 1–2 × 10^6 cells per 1 mL).
7. Slow freeze cells by first placing vials of cells at −80°C for 24 h. Transfer the vials to −140°C for long-term storage.

4. Notes

1. All protocols involving neural stem cells should be carried out under aseptic conditions in the hood.
2. Volumes used in this protocol for neural stem cell culture are intended to be used with 10-cm dishes. Be sure to reduce or increase proportionally the amount of solutions for culture dishes of other sizes.
3. Transduction rates approaching 100% are typical of this protocol. However, depending on your cells, the POI expressed, and the viral titer some variability may occur. If you have a low yield of transduction, follow the troubleshooting steps outlined in the "titering your lentiviral stock" section.
4. If you are transducing cells with untitered viral supernatant and are concerned about possible toxicity, it is possible to incubate cells for as little as 6 h prior to changing the medium. However, typically little or no toxicity is observed.
5. Prior to transplantation or injection, cells can be maintained on ice for 2–3 h.

Acknowledgments

This work was supported by a Susan G. Komen Breast Cancer Foundation grant PDF0403205 and a CBCRP grant12IB-0122 to B. Felding-Habermann in consortium with EY Snyder.

References

1. Snyder, EY., Deichter, DL., Walsh, C., Arnold-Aldea, S., Hartweig, EA., and Cepko, CL. Multipotent neural cell lines can engraft and participate in development of mouse cerebellum. Cell, 68: 33–51, 1992.
2. Yip, S., Aboody, KS., Burns, M., Imitola, J., Boockvar, JA., Allport, J., Park, KI., Teng, YD., Lachyankar, M., McIntosh, T., O'Rourke, DM., Khoury, S., Weissleder, R., Black, PM., Weiss, W., and Snyder, EY. Neural stem cell biology may be well suited for improving brain tumor therapies. Cancer J., 9: 189–204, 2003.
3. Aboody, KS., Brown, A., Rainov, NG., Bower, KA., Liu, S., Yang, W., Small, JE., Herrlinger, U., Ourednik, V., Black, PM., Breakefield, XO., and Snyder, EY. Neural stem cells display extensive tropism for pathology in adult brain: evidence from intracranial gliomas. Proc. Natl. Acad. Sci. USA., 97: 12846–12851, 2000.
4. Elzaouk, L., Moelling, K., and Pavlovic, J. Anti-tumor activity of mesenchymal stem cells producing IL-12 in a mouse melanoma model. Exp. Dermatology, 15: 865–874, 2006.
5. Song, K., Benhaga N., Anderson, RL., and Khosravi-Far, R. Transduction of tumor-necrosis factor-related apoptosis-inducing ligand into hematopoietic cells leads to inhibition of syngeneic tumor growth in vivo. Cancer Res., 66(12): 6304–6311, 2006.
6. Kim, SK., Kim, SU., Park, IH., Bang, JH., Aboody, KS., Wang, KC., Cho, BK., Kim, M., Menon, LG., Black, PM., and Carroll, RS. Human neural stem cells target experimental intracranial medulloblastoma and deliver a therapeutic gene leading to tumor regression. Clin Cancer Res., 12(18): 5550–5556, 2006.
7. Aboody, KS., Najbauer, J., Schmidt, NO., Yang, W., Wu, JK., Zhuge, Y., Przylecki, W., Carroll, R., Black, PM., and Perides, G. Targeting of melanoma brain metastases using engineered neural stem/progenitor cells. Neuro-Oncol., 8(2): 119–126, 2005.
8. Ehtesham, M., Kabos, P., Kabosova, A., Neuman, T., Black, KL., and Yu, JS. The use of interleukin 12-secreting neural stem cells for the treatment of intracranial glioma. Cancer Res., 62(20): 5657–5663, 2002.
9. Yang, YS., Hiu, L. and Zhang, JN. Gene therapy of rat malignant gliomas using neural stem cells expressing IL-12. DNA Cell Biol., 23(6): 381–389, 2004.
10. Ehtesham, M., Kabos, P., Guitierrez, MA., Chung, NH., Griffith, TS., Black, KL., and Yu, JS. Induction of glioblastoma apoptosis using neural stem cell-mediated delivery of tumor necrosis factor-related apoptosis-inducing ligand. Cancer Res., 62(24): 7170–7174, 2002.
11. Logan, AC., Haas, DL., Kafri, T., and Kohn, DB. Integrated self-inactivating lentiviral vectors produce full-length genomic transcripts competent for encapsidation and integration. J. Virol., 78: 8421–8436, 2004.
12. Logan, AC., Nightingale, SJ., Haas, DL., Cho, GJ., Pepper, KA., and Kohn, DB. Factors influencing the titer and infectivity of lentiviral vectors. Hum. Gene Ther., 15: 976–988, 2004.
13. Miyoshi, H., Smith, KA., Mosier, DE., Verma, IM., and Torbett, BE. Transduction of human CD34+ cells that mediate long-term engraftment of NOD/SCID mice by HIV vectors. Science, 283: 682–686, 1999.
14. Logan, AC., Lutzko, C., and Kohn, DB. Advances in lentiviral vector design for gene-modification of hematopoietic stem cells. Curr. Opin. Biotechnol., 13: 429–436, 2002.
15. Van Damme, A., Thorrez, L., Ma, L., Vandenburgh, H., Eyckmans, J., Dell'Accio, F., De Bari, C., Luyten, F., Lillicrap, D., Collen, D., VandenDriessche, T., and Chuah, MK. Efficient lentiviral transduction and improved engraftment of human bone marrow mesenchymal cells. Stem Cells, 24: 896–907, 2006.
16. Schorpp, M., Jager, R., Schellander, K., Schenkel, J., Wagner, EF., Weiher, H., and Angel, P. The human ubiquitin C promoter directs high ubiquitous expression of transgenes in mice. Nucleic Acids Res., 24: 1787–1788, 1996.
17. Torbett, BE. Reporter genes: too much of a good thing? J. Gene Med., 4: 478–479, 2002.

Chapter 17

Measurement of Multiple Drug Resistance Transporter Activity in Putative Cancer Stem/Progenitor Cells

Vera S. Donnenberg, E. Michael Meyer, and Albert D. Donnenberg

Summary

Multiple drug resistance, mediated by the expression and activity of ABC-transporters, is a major obstacle to antineoplastic therapy. Normal tissue stem cells and their malignant counterparts share MDR transporter activity as a major mechanism of self-protection. Although MDR activity is upregulated in response to substrate chemotherapeutic agents, it is also constitutively expressed on both normal tissue stem cells and a subset of tumor cells prior to the initiation of therapy, representing a built-in obstacle to therapeutic ratio. Constitutive and induced MDR activity can be detected in cellular subsets of disaggregated tissues, using the fluorescent substrates Rhodamine 123 and Hoechst 33342 for ABCB1 (also known as P-gp and MDR1) and ABCG2 (BCRP1). In this chapter, we will describe the complete procedure for the detection of MDR activity, including: (1) Preparing single-cell suspensions from tumor and normal tissue specimens; (2) An efficient method to perform cell surface marker staining on large numbers of cells; (3) Flow cytometer setup and controls; (4) Simultaneous measurement of Hoechst 33342 and Rhodamine123 transport; and (5) Data acquisition and analysis.

Key words: Pre-existing multiple drug resistance, ABCB1 activity, ABCG2 activity, Hoechst 33342, Rhodamine 123, Epithelial tumors, Cancer stem cells, Flow cytometry

1. Introduction

1.1. MDR in Stem Cells

Multiple drug resistance (MDR) is mediated chiefly by a group of transmembrane transporter glycoproteins known as ATP-binding cassette (ABC) proteins. These energy-dependent cellular pumps are highly conserved in eukaryotes. Because they preferentially transport lipophilic cationic molecules, a single transporter can act on structurally dissimilar substrates. Biologically, they have been adapted to serve a wide variety of functions. In the central

John S. Yu (ed.), *Cancer Stem Cells,* Methods in Molecular Biology, vol. 568
DOI 10.1007/978-1-59745-280-9_17, © Humana Press, a part of Springer Science+Business Media, LLC 2009

nervous system they are expressed by capillary endothelial cells, forming the blood brain barrier *(1)*. In the gut they are expressed by villus tip enterocytes of the small intestine *(2)*, limiting the bioavailability of substrate drugs. In stem cells, MDR transporters, principally ABCG2 (also known as breast cancer resistance protein 1 (BCRP1)) and ABCB1 (also known as P-glycoprotein (P-gp) and multiple drug resistance protein 1 (MDR1)), operate at the single cell level, rather than at the tissue level, excluding xenobiotics and other toxins and making stem cells among the most resilient cells in the body. ABC transporters may be constitutively expressed, as they are in some stem cell populations *(3)*, or may be induced by exposure to substrate, as they are in memory T lymphocytes *(4)*.

MDR activity can be detected and quantified by exclusion of fluorescent substrates. The association of constitutive MDR expression with normal tissue stem cells (sometimes called adult stem cells) was first demonstrated by Lansdorp, who showed that the most primitive hematopoietic cells could be isolated from the bone marrow (BM) by sorting for cells which excluded the lipophilic, cationic dye rhodamine 123 *(5)*. Years later, this phenomenon was rediscovered by Goodell while sorting BM cells on the basis of DNA content (cell cycle). After staining BM cells with the membrane permeant DNA stain Hoechst 33342 she noted, in addition to the usual cell cycle populations, a very small population of viable cells which excluded the Hoechst dye. In very careful experiments, she determined that this population was highly enriched for cells capable of reconstituting hematopoiesis in radiation-ablated mice *(3)*. She called these cells *Side Population* cells because of their location on the far left side of blue fluorescence vs. red fluorescence histograms. Today, Hoechst 33342 excluding cells are commonly referred to as side population cells. Such cells can be found in a wide variety of tissues, and the side population strategy has been useful for isolating cells with high regenerative capacity from hematopoietic *(3, 5, 6)*, airway *(7)*, pituitary *(8)*, small intestine *(9)*, and testicular *(10)* tissues. However, given the wide array of biological functions that ABC transporters serve, and the fact that they are sometimes induced in transporter-negative cells by substrate exposure, the caveat must be given that MDR expression does not unequivocally identify stem cells and conversely, its absence does not rule out the self-renewing capacity most characteristic of stemness.

1.2. MDR in Cancer

Multiple drug resistance transporters are so named because they were first discovered in the context of antineoplastic therapy *(11)*. Their discovery solved the conundrum posed by the observation that cancer cells which developed resistance to a particular chemotherapeutic agent became simultaneously resistant to a wide variety of unrelated agents, including drugs with entirely different mechanisms of action. Today we know that MDR is constitutively expressed by a subset (usually a small subset) of neoplastic cells prior to treatment with substrate drugs. Treatment results in selection for drug excluding MDR active cells by a number of mechanisms, including regional gene activation *(12)*, gene amplification *(13)*, and modification of histone acetylation at the ABCG2 locus *(14)*. Recently it has been suggested that MDR activity in some cancers is regulated by the hedgehog signaling pathway *(15, 16)*, a key pathway in embryonic morphogenesis *(17)*. Further, although the mechanism remains unclear, there are data linking MDR expression to radiation resistance *(18)*.

MDR activity has been investigated in multiple types of cancer as a possible means of identifying the cancer stem cell. The vast majority of work has been done in cell lines, which have undergone generations of selection for characteristics favorable to in vitro growth in the absence of host- and therapy-mediated selective pressures. Not all tumorigenic cancer cell lines exhibit a side population. Investigators working with SP + cell lines derived from ovarian cancer *(19)*, breast cancer *(20)*, glioma *(21)*, prostate *(22)*, and thyroid cancer *(23)* all found enhanced tumorigenicity or in vitro clonogenicity in sorted side population cells. Harris et al. *(21)* and Mitsutake et al. *(23)* found that non-SP cells could give rise to SP cells. In contrast, Lichtenauer et al. found neither growth nor survival advantage in SP cells sorted from an adrenocortical carcinoma cell line *(24)*. Our own data in primary breast cancer isolates support plasticity in MDR expression. Sorted CD44+ CD90+ ABCG2− breast cancer cells gave rise to heterogeneous tumors which included a subset of ABCG2+ cells when explanted to NOD/SCID mice *(25)*. Therefore, the caveat given for normal stem cells, that MDR activity and stemness are not one in the same, holds for neoplastic cells as well *(26)*. Although MDR activity is upregulated in response to substrate chemotherapeutic agents, it is also constitutively expressed on both normal tissue stem cells and a subset of tumor cells prior to the initiation of therapy *(19, 25–27)*, representing a built-in obstacle to therapeutic ratio *(28)*. The take-home message is that a cancer cell which is both self-renewing (i.e., tumorigenic) and protected by MDR transporters constitutes a very difficult therapeutic target, having much in common with normal tissue stem cells. Thus, detection and isolation of MDR active cells by simultaneous measurement of rhodamine 123 and Hoechst 33342 transport *(29)* represents an important tool for investigation of those cancer cells capable of therapy resistance, dormancy, and reactivation.

2. Materials

1. Tissue requirement 1 g minimum.
2. 60-mm Petri dishes Falcon (PL-056).
3. BD Bard-Parker protected disposal scalpel, blade#10 (Fisher 02-688-78).
4. Collagenase/DNAase solution in PBS with calcium and magnesium.
   (a) Collagenase type I (Sigma C0130), 0.4% final concentration.
   (b) DNAse (Sigma D-5025-750KU), 350 kU/mL final concentration.
5. 50-mL Polypropylene conical tube, Falcon (Fisher 14-959-49A).
6. DMEM 1×, high glucose, SuperCase (Invitrogen 11965-118).
7. Cell strainer, 70-μm nylon, 26-mm diameter (Becton Dickinson 352350).
8. Cellector (TM) tissue sieve kit, 130 mL with pan, glass pestle, and screens (Bellco 205020).
9. Newborn calf serum (Atlanta Biologicals S11210).
10. Phosphate-buffered saline without calcium or magnesium (PBS-A, Sigma D5652).
11. PBS with $CaCl_2$ and $MgCl_2$ (PBS, Sigma D8662).
12. Ammonium chloride lysing solution 10× (500 mL).
    (a) 41.5 g Ammonium chloride.
    (b) 5.0 g Potassium bicarbonate.
    (c) 400 mL Glass distilled water.
    (d) pH to 7.2–7.4 with 1 M HCl or NaOH as needed.
    (e) QS to 500 mL with distilled water.
    (f) Dilute with distilled water to 1× prior to use.
13. Hemacytometer (VWR Scientific 48312-002).
14. Trypan Blue (Sigma T8154).
15. Histopaque (Sigma 10771).
16. 10-mL Pipettes, Falcon (Fisher 13-675-20).
17. 15-mL polypropylene conical, Falcon (Fisher 14-959-70C).
18. Mouse serum (Sigma M5905).
19. Microcentrifuge tube, 1.5 mL (Eppendorf Fisher 05-402-25).
20. L-Glutamine Gibco 25030-081.
21. 2-ME (Fisher BP176-100).

22. HEPES buffer (Sigma H-3375).
23. Rhodamine 123(Calbiochem 001004).
24. Hoechst 33342 (Molecular Probes H-3570).
25. Verapamil (Sigma V-4629).
26. Cyclosporin A (Sigma C1832).
27. Fumitremorgin C (Alexis Biochemicals 350-127-C250).
28. Vincristine (Sigma V8879).
29. Propidium Iodide (Sigma P-4170).
30. 70-μm Filter cap tubes, 12 × 75 (Fisher 08-771-23).
31. CompBead Antimouse Ig, κ (BD 552843).
32. SpectrAlign beads (DAKO, Cat. No. KO111).
33. 8-Peak Rainbow Calibration Particles (Spherotech, Libertyville, IL, Cat. No. RCP-30-5A).
34. Calibrite beads (PE) (BD 349502).
35. Calibrite (APC) (BD 340487).

3. Methods

3.1. Tissue Collection and Digestion

1. Place tissue into a tared 60-mm Petri dish.
2. If desired remove a small piece for histology (paraffin and OCT embedding).
3. Record the weight of remaining tissue.
4. Finely mince using sterile paired scalpels. Add a few drops of collagenase/DNase solution while mincing to prevent tissue from drying **(Fig. 1)**.
5. Place a 70-μm cell strainer in the mouth of a 50-mL polypropylene conical tube labeled "Filtrate."
6. Using a scalpel, a 10-mL pipette and PBS-A plus 2% calf serum (PBS-A-CS), transfer all tissue into the cell strainer. Repeat until all tissue fragments have been removed from the Petri dish, pipetting PBS-A-CS vigorously against the cell strainer to release individual cells (*see* **Notes 1** and **2**).
7. Bring volume of Filtrate tube to 50 mL with PBS-A-CS, centrifuge at 400 × *g* for 10 min, and discard supernatant. Resuspend in 2 mL PBS-A-CS and hold on ice.
8. Using a scalpel, transfer minced tissue in the strainer to an additional 50-mL polypropylene conical tube containing 10 mL collagenase/DNase solution. Label "Digest."

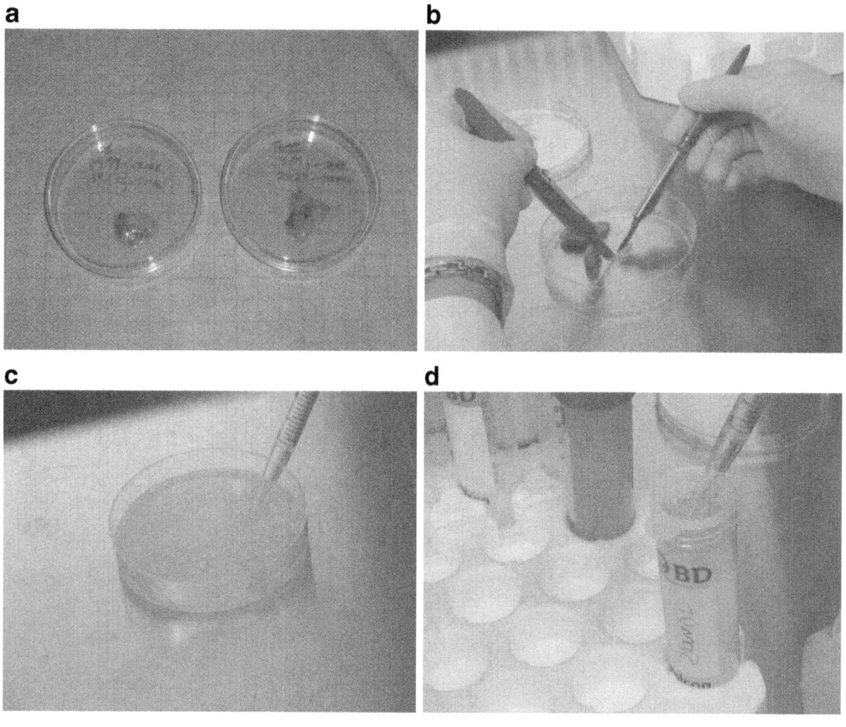

Fig. 1. Preparation of normal breast tissue and breast tumor for digestion. Freshly excised normal breast tissue (**a, left**) and breast tumor (**a, right**) was transported in ice-cold tissue culture medium and transferred to 60-mm Petri dishes. The tissue was minced into fine fragments using paired scalpels (**b**). A few drops of collagenase/DNase were added to the minced tissue to prevent drying and clumping. PBS-A is added to the Petri dish (**c**) and carefully pipetted through a cell strainer into a 50-mL conical tube (**d**, Filtrate). Tissue remaining in the strainer is digested with collagenase/DNase and the process is repeated as necessary.

9. Place Digest tube in shaking waterbath (e.g., Bellydancer, Stovall Life Science) at 37°C for 30 min, maximum agitation setting.
10. Place a new 70-μm cell strainer in the mouth of the conical tube labeled Filtrate.
11. Using a 10-mL pipette, transfer material from the Digest tube into the cell strainer on the Filtrate tube.
12. Add 10 mL PBS-A-CS to the Digest tube and transfer any remaining tissue to the cell strainer. Repeat if necessary.
13. Using a scalpel, transfer the undigested material from the cell strainer back into the Digest tube.
14. Add 10 mL collagenase/DNase solution and return Digest tube to the shaking 37°C waterbath for 30 min.
15. Bring volume of Filtrate tube to 50 mL with PBS-A-CS, centrifuge at 400 × *g* for 10 min, and discard supernatant. Resuspend in 2 mL PBS-A-CS and hold on ice.
16. Repeat **steps 10** and **11**.

17. Discard strainer from Filtrate tube.
18. Bring volume of Filtrate tube to 50 mL with PBS-A-CS, centrifuge at 400 × *g* for 10 min, and discard supernatant.
19. Add 45 mL of 1× NH_4Cl lysing solution and mix (*see* **Note 3**).
20. Centrifuge at 400 × *g* for 10 min at room temperature.
21. Pour off the supernatant, and loosen cell pellet.
22. Resuspend in 40 mL PBS-A-CS.
23. Centrifuge at 400 × *g* for 10 min at room temperature.
24. Resuspend in 2 mL PBS *with calcium and magnesium* plus 2% calf serum and hold on ice (*see* **Note 4**).
25. Count cells on a hemacytometer (Tuerk's solution to eliminate RBC, Trypan blue for viability) *(30)*.

3.2. Extracellular Staining

1. Pellet cells at 400 × *g* for 10 min at 4°C. Discard supernatant.
2. Resuspend cell pellet in 5 μL neat decomplemented (56°C, 30 min) mouse serum (*see* **Note 5**).
3. Pellet cells (400 × *g*, 10 min) and aspirate residual supernatant (*see* **Note 6**).
4. Stain the dry pellet for surface markers by the addition of 2 μL of each monoclonal antibody.
5. Add 2μL antibodies in the following order (*see* **Note** 7):
   (a) EpCAM-APC (Miltenyi Biotech, Bergisch Gladbach, Germany, Cat. No. 12000420)
   (b) CD90-PE (Beckman Coulter, Cat. No. IM1840)
   (c) CD45-APCCy7 (BD, Cat. No. 557833)
   (d) (*see* **Note 8**).
6. Incubate for 30 min on ice in the dark.
7. Dilute surface stained cell pellets in DMEM medium containing 10% calf serum, L-glutamine (2 mM), and 2-beta mercaptoethanol (50 μM) to a concentration of 2 × 10^6 cells/mL (*see* **Note 9**).

3.3. Simultaneous Measurement of Hoechst 33342 and Rhodamine 123 Dye Transport (Fig. 2)

3.3.1. Dye Loading in the Presence and Absence of Inhibitors

1. Load surface stained cells for 90 min at 37°C with 0.13 μM R123 and 8.12 μM Ho33342 in the absence and presence of inhibitors (**Table 1**, *see* **Notes 10–12**).
2. When the incubation is complete, wash the cells once (400 × *g*, 10 min) with ice-cold PBS-A-CS and resuspend to 5–10 × 10^6 cells/mL in ice cold PBS-A 10% CS (*see* **Note 13**).
3. Add Propidium Iodide, 10 μg/mL final concentration, as a viability dye (*see* **Note 14**).
4. Strain cells immediately prior to flow cytometric acquisition through a 70-μm cap filter.

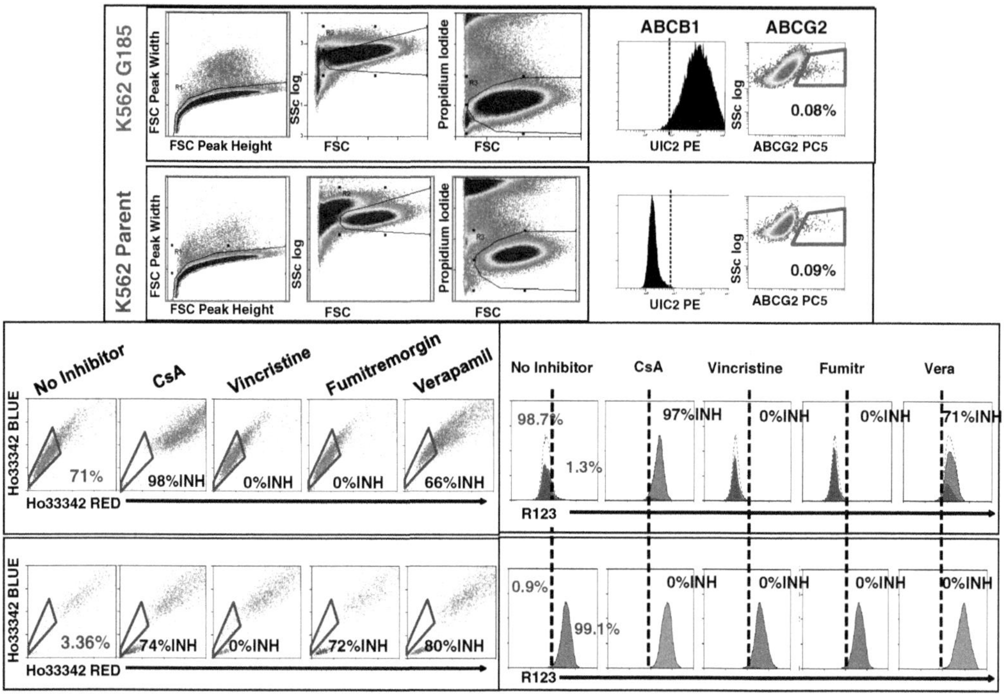

Fig. 2. Simultaneous detection of Ho33342 and R123 transport. Flow cytometry was performed on the parental cell line K562 (*bottom panels*, human erythroleukemia) and the MDR1 transfectant K562-G185 (*top panels*). A gating strategy was used to analyze only singlet viable cells. Monoclonal antibody staining on separate aliquots revealed that parental K562 cells were ABCB1 (MDR1) negative and expressed a very small subpopulation of ABCG2+, low side scatter cells (0.08% of viable cells, *top right*). In contrast, the transfectant line K562-G158, which is maintained in the presence of 100 ng/mL vincristine, is uniformly positive for ABCB1, and expresses a small subpopulation of ABCG2+ cells identical to that of the parent cell line. Both cell lines were coincubated with Ho33342 and R123 in the absence and presence of MDR inhibitors. In the absence of inhibitors, only a minor subpopulation within the parental K562 exhibited the SP phenotype (3.36%), whereas almost all K562-G185 (ABCB1 transfected) cells transported Ho33342 (71% SP). Addition of 5 μM cyclosporin A (CsA) abrogated transport in both cell lines. Vincristine had no effect on either cell line, whereas the ABCG2 specific inhibitor, fumitremorgin (1 μM), only inhibited Ho33342 transport in parental K562 (native ABCG2 mediated transport) and had no effect on Ho33342 transport through ABCB1. Verapamil (50 μM, a MDR nonspecific inhibitor) blocked SP phenotype in both cell lines. R123 transport was limited to cells with ABCB1 expression (K562-G185 only), the parental K562 had no constitutive ABCB1 activity or expression. The R123 dull phenotype was reversed by the addition of either CsA or verapamil but was not blocked by the addition of ABCG2-specific inhibitor fumitremorgin.

3.3.2. Instrument Setup and MDR Standards

IgG Capture Bead Staining

1. Vortex CompBeads thoroughly before use (*see* **Note 15**).
2. Label a separate 1.5-mL Eppendorf tube for each mouse monoclonal antibody conjugated to a *tandem dye* (e.g., ECD, PE-Cy5, PE-Cy7, and APC-Cy7).
3. Add one full drop (approximately 60 μL) of antimouse Ig CompBeads to each Eppendorf tube.
4. Centrifuge for 10 min at 400 × *g*. Carefully aspirate supernatant to ensure a "dry pellet" (*see* **Note 16**).
5. Sonicate each tube for 10 s in a water bath sonicator (*see* **Note 17**).

Table 1
Characteristics of common MDR inhibitors

Inhibitor	Vehicle	[Final concentration]	Notes
Vehicle controls (EtOH and DMSO)	Not applicable	1/1,000	
Verapamil	PBS-A	50 μM	Competitive MDR inhibitor
CsA	EtOH	5 μM	Potent dead end MDR inhibitor
Fumitremorgin	DMSO	10 μM	Highly specific ABCG2 inhibitor
Vincristine	PBS-A	1 μg/mL	Used to maintain ABCB1 expression in transfected cells

6. Add 2 μL of each antibody directly to beads (one antibody per tube) and gently reflux.
7. Incubate for 15 min at room temperature in the dark (*see* **Note 18**).
8. Add 1 μL of mouse serum; incubate 5 min at room temperature (*see* **Note 19**).
9. Add 100 μL of staining buffer and reflux.
10. Sonicate each tube for 10 s.
11. Add 1 mL of staining buffer. For manual compensation, add one drop of negative CompBeads to each test tube that contains antibody stained beads (*see* **Note 20**).
12. Centrifuge beads for 10 min at 400 × *g*; decant and carefully blot to remove residual supernatant (*see* **Note 21**).
13. Resuspend washed beads in 0.5 mL of staining buffer.
14. Transfer to 12 × 75 mm snap cap tubes for flow cytometry.
15. Sonicate for 10 s prior to acquisition on the flow cytometer.

Instrument Setup and Sample Acquisition

1. We use an eight-color MoFlo cell sorter equipped with ultraviolet, 488 and 633 nm lasers and an automated sample station capable of holding the sample at 4°C (Beckman Coulter, Fort Collins, CO). Other similarly equipped cytometers may be used. The filters used for each fluorescence channel are shown in **Fig. 3** (*see* **Note 22**).
2. The cytometer is calibrated to predetermined photomultiplier target channels prior to each use using SpectrAlign beads (DAKO, Cat. No. KO111) and 8-peak Rainbow Calibration Particles (Spherotech, Libertyville, IL, Cat. No. RCP-30-5A) (*see* **Note 23**).

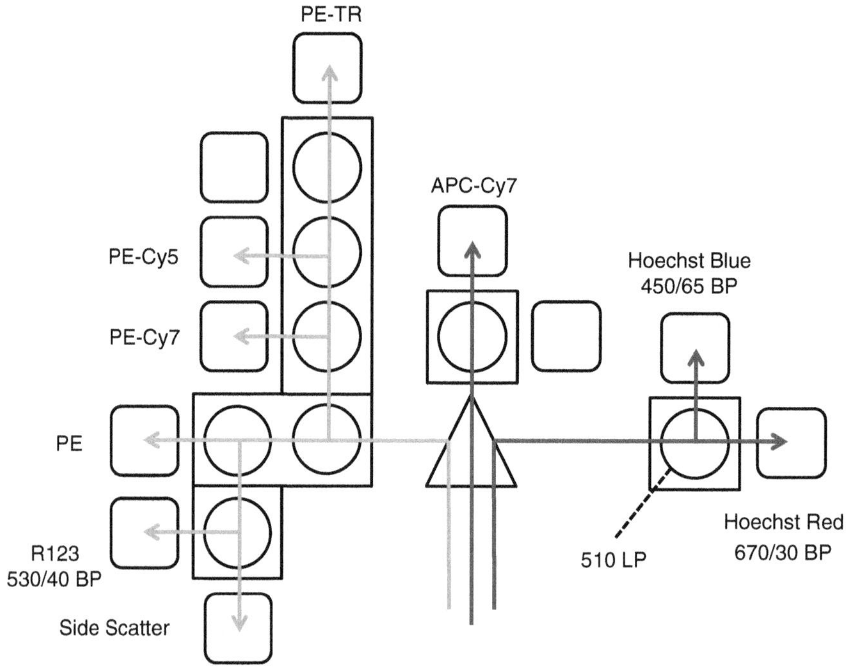

Fig. 3. Three-laser filter configuration to measure side population plus six additional fluorochromes. Routine analysis and sorting of SP cells is performed on the MoFlo cell sorter by using the 488-nm laser line as a primary exciter of FITC, PE, and PE tandem dyes. The 488 line is also used to provide forward (FSC) and side scatter (SSC) signals. Multiline UV is used to excite Hoechst 33342 for SP detection. An additional 633-nm laser is used to excite APC and APC tandem dyes. The MoFlo allows the operator to direct individual lines toward specific detector arrays on the Detection Table, as illustrated in the figure (*upper left*). The convergence of the three lasers on the sample stream (*top view*) and the paths of forward and side scattered light are shown in the *lower right*.

3. All fluorescence parameters are collected in the logarithmic mode, with the exception Hoechst 33342 emissions (blue and red) which are collected in the linear mode (*see* **Note 24**).
4. PMT settings for the two Hoechst channels must be fine tuned empirically for each sample because data are acquired in a linear mode and because fluorescence intensity varies with the total DNA per sample. For initial settings it is a great help to calibrate with cells having a known side population (*see* **Notes 10** and **22**), adjusting PMT gain until the G1/G0 peak is comfortably placed in the upper right third of the Hoechst Red by Hoechst Blue histogram (**Fig. 3**). For the side population itself, blue fluorescence intensity should be slightly greater than red fluorescence intensity, displacing the side population slightly upward on the histogram *(31)*.
5. Acquire unstained cells or beads first, and then each single stained sample (bead or cells) from the shortest emission wavelength (R123 stained cells) to the longest (e.g., CD45 APCCy7).

6. Calculate the compensation matrix from single stained tubes using automated software (supplied with most new cytometer software and all offline analysis packages).
7. Prior to running Ho33342/R123 loaded cells, minimal logical gating is performed in order to eliminate cellular clumps (doublets) *(32)*, subcellular debris (forward scatter threshold), and PI+ dead and dying cells (**Figs. 4** and **5**) MDR transfected reference cells (*see* **Note 12**) are particularly useful for this purpose.
8. Acquire Ho33342/R123 loaded cells, modifying "Hoechst red 670/20" and "Hoechst blue 450/65" PMT voltage so that nontransporting/non-SP cells with 2N DNA content are visualized in the last third of the Hoechst red 670/20 vs. Hoechst blue 450/65 histogram.
9. Acquire tumor/tissue cell suspensions loaded in the absence and presence of inhibitors (**Table 1**, *see* **Note 25**).

3.4. Data Analysis

We perform data analysis offline, using VenturiOne software (Applied Cytometry) which has been designed to accommodate very large datafiles. We create an analysis template into which we load our spectrally compensated data. Our strategy usually proceeds in three steps:

1. First we eliminate sources of interference with logical gates. These include cell-cell doublets and clusters, subcellular debris, and dead cells (**Fig. 5**).
2. Next we decide on our *classifier* parameters. In this example CD45 expression (**rt. 5**) serves as a primary classifier, and Hoechst dye exclusion (**Fig. 6**) is a secondary (branching) classifier.

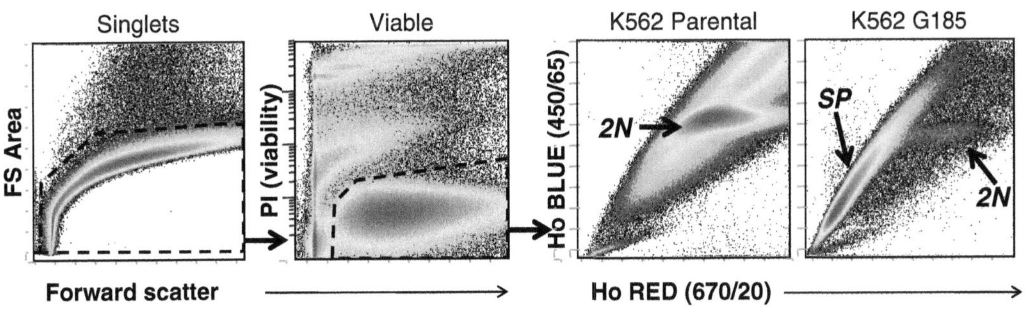

Fig. 4. Instrument set up for Hoechst 33342 detection using cell lines as process controls. Parental K562 and ABCB1-transfected K562-G185 cells were used as low-negative and positive controls, respectively. A gating strategy was used to restrict analysis to singlet viable (PI negative) cells. Approximately 650,000 viable singlet events were acquired for both samples. The majority of parental K562 had no Ho33342 transport. The majority of events had uniform Ho33342 fluorescence in both blue and red detection PMTs, representing cells with 2N DNA. Almost all ABCB1 transfected cells transported Ho433342 and therefore exhibited the *side population* (SP) phenotype, with greater blue than red fluorescence. Only a small minority of K562-G185 cells failed to transport Ho33342 and therefore had fluorescence intensities characteristic of 2N DNA cells.

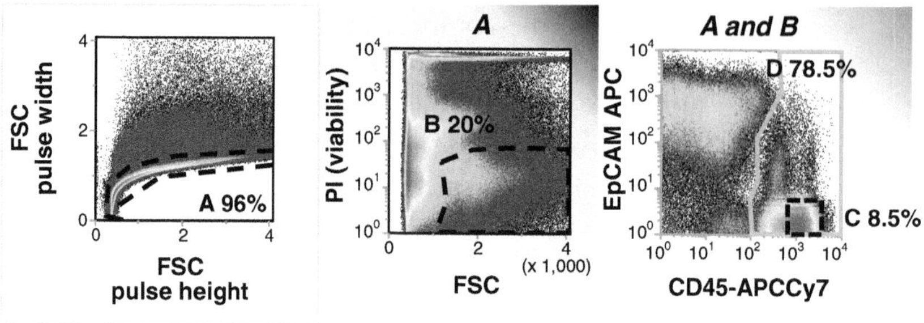

Fig. 5. Gating strategy to eliminate doublets, nonviable cells, and hematopoietic lineage positive events. Enzymatically and mechanically digested nonsmall-cell lung carcinoma was stained for the expression of surface CD45 and EpCAM (CD326); propidium iodide was added and cells were passed through a 70-μM filter immediately prior to acquisition. Singlet cells were identified on a 2-parameter histogram of forward scatter (FSC) pulse height vs. FSC pulse width (A). Viable cells which are impermeable to propidium iodide (PI) were identified (center histogram, B). Viable singlet cells were analyzed for the expression of CD45 (common leukocyte antigen, D) and the epithelial surface antigen EpCAM. CD45 bright/EpCAM-negative events were identified as lymphocytes (D) for use as a 2N DNA standard.

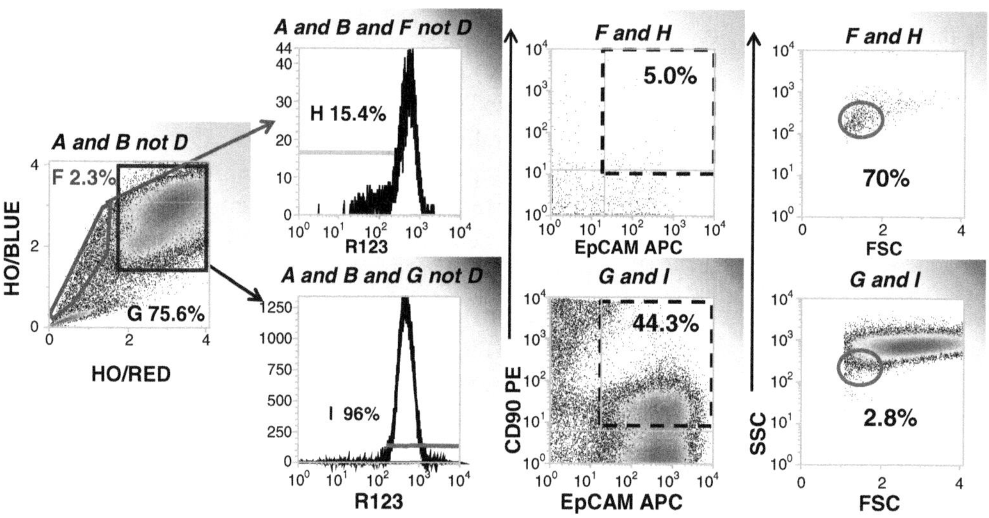

Fig. 6. Functional and phenotypic characterization of "side population" cells. Freshly isolated lung cancer cells (CD45− viable singlets) were divided into Ho33342 excluding (F) and retaining (G) populations. Rhodamine 123 (R123) exclusion and the surface expression of EpCAM, CD90 (Thy1) and light scatter properties (FSC, SSC) were evaluated. Side population cells (*top panels*) cotransported R123 (15.4%). Cells which cotransported Ho33342 and R123 were mostly EpCAM negative/CD90− and had low light scatter comparable to resting lymphocytes. Nonside population cells, which did not transport R123, were of mixed surface phenotype (EpCAM/CD90) and were predominantly (97.2%) large cells with high light scatter.

3. Next *outcomes*, parameters to be measured on each population defined by the classifiers, are determined. In this case **(Fig. 6)** the outcome variables are R123 dye efflux, CD90 and EpCAM expression, and light scatter (*see* **Note 26**).
4. Lymphocytes are often present in tumor specimens and are identifiable by bright CD45 expression and low light scatter **(Fig. 5)**. Since virtually none of these cells transports Hoechst 33342, the dye reaches the nucleus, where it binds to DNA

stoichiometrically. The great majority of lymphocytes have 2N DNA content, but a distinct population of cycling cells can also be seen **(Fig. 7)**. A population with very low blue and red fluorescence is often present as a source of artifact in digested tissue **(Fig. 8)**. These acellular events have sufficient light scatter to trigger the cytometer to acquire a pulse, but have no DNA content (which also prevents them from being eliminated by PI staining). Another potential source of artifact is cells with *laddered* DNA **(Fig. 8)**, the result of apoptosis-induced endonuclease release. The feature which distinguished them from *bona fide* side population cells is that cells with degraded DNA lie on the blue/red diagonal, whereas side population cells have greater blue than red fluorescence and are therefore slightly above the diagonal.

5. There is some confusion in the literature concerning the specificity of the side population, with early reports attributing the side population exclusively to cells expressing the ABCG2 transporter. **Figure 2** shows results obtained with K562 cells (parent) and cells transfected with an ABCB1 vector (K562 G185). Surface staining with anti-ABCB1 (UIC2) and anti-ABCG2 antibodies (*see* **Note 27**), and real-time PCR performed on these cells (not shown) confirmed that both cell lines had low ABCG2 expression but that only the transfectant had detectable ABCB1. The Instrument was set up using parental and ABCB1 transfected cell lines and presence of a battery of inhibitors. Two important points emerge from the analysis: (1) Both ABCG2 (parental and transfected) and ABCB1 (transfected only) mediate Hoechst 33342 transport and therefore confer the *side population* phenotype; (2) The transporters responsible for the side population and

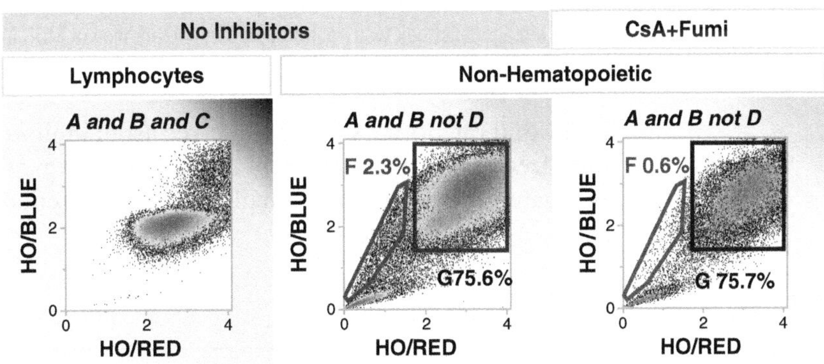

Fig. 7. Identification of "side population" cells by Ho33342 transport in freshly digested tumor tissue. Events were gated as shown in **Fig. 5**. Lymphocytes identified as CD45bright/EpCAM negative (**Fig. 5**) were largely homogeneous for Ho33342 uptake and DNA staining (2N, left histogram); no "side population" was detected. Nonhematopoietic (CD45–) cells evidenced a significant "side population" accounting for 2.3% of events (center histogram). Cells with 2N and >2N DNA content (non-Ho33342 transporting cells) comprised the majority of cells. Abrogation of the "side population" (right histogram) by the addition of MDR inhibitors cyclosporine and fumitremorgin confirms that Ho33342 elimination is mediated by MDR transporters.

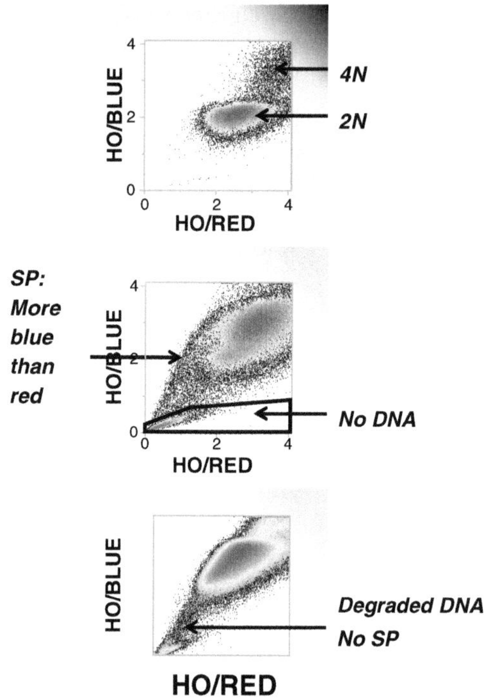

Fig. 8. Interpretation of Ho33342 exclusion histograms. Cells are permeant to the DNA-binding dye Ho33342. In the absence of constitutively active MDR transport, Ho33342 concentrates in the nucleus and binds to the DNA, resolving 2N and 4N populations (top histogram). DNA-bound Ho33342 autoquenches, shifting emission toward the red. Side population cells, which do not permit accumulation of Ho33342 in the nucleus, exhibit greater blue than red fluorescence, placing them above the diagonal (center histogram). Events in the lower left-hand corner are not Ho33342 transporting cells, but rather events that triggered on FSC but lack DNA. Another commonly seen artifact is illustrated in the bottom histogram. The diagonal streak exhibits equal red and blue fluorescence and represents cells with degraded DNA, as is seen in early apoptotic cells. In this example, the diagonal streak was not abrogated by the addition of MDR inhibitors.

R123 dull phenotypes may be inferred from inhibitor studies. Fumitremorgin, an exquisitely specific inhibitor of ABCG2, abrogated the side population phenotype in parental cells, but not in ABCB1 transfectants.

4. Notes

1. If material appears slimy this indicates DNA release. Add a few drops of DNase/collagenase solution and continue. DNAse activity is measured in Kunitz units. DNase is

necessary because cells that are killed during the disaggregation procedure release viscous DNA strands which entrap live cells.

2. Occasionally, tumors are too sclerotic for the minced pieces to be disaggregated through a nylon cell strainer. In these cases we substitute a stainless tissue sieve with a 10-mesh screen overlayed with a 100-mesh screen. Tissue fragments are forced through the screen with a glass pestle (*see* **Subheading 2**).
3. Treatment with ammonium chloride lyses red blood cells. This step may be omitted if sample is not visually contaminated with red blood cells.
4. DNase requires Ca^{2+} or Mg^{2+} for its activity.
5. Addition of mouse serum, which contains mouse immunoglobulins, blocks nonspecific binding of murine monoclonal antibodies.
6. We refer to this as a *dry* the pellet elsewhere in the protocol. Cells pelleted in a 15-mL conical tube and aspirated dry actually contain anywhere from 10 to 50 μL of residual liquid.
7. The choice of antibodies and fluorochromes is specific to the question being addressed and the available instrumentation, and can be modified at will. The order of antibody addition may influence staining as binding occurs very rapidly, and sequential addition of many antibodies progressively reduces the concentration of individual antibodies in the mixture *(26, 33)*.
8. We have found the following antibodies useful for subsetting MDR+ tumor cells: CD44-PECy7 (Abcam, Cat. No. AB46793), CD14-PECy5 (Beckman Coulter, Cat. No. IM2640U), CD33-PECy5 (Beckman Coulter, Cat. No. IM2647U), glycophorin A-PECy5 (Becton Dickinson, Cat. No. 559944). CD44 has been used by many investigators to identify tumorigenic cancer cells *(34)*. Anti-CD14, anti-CD33, and antiglycophorin A (hematopoietic markers) are all labeled with the same fluorochrome and collectively will serve as a *dump gate* during data analysis. This means that all positive events within the dump gate will be eliminated from the analysis.
9. Prewarm medium to 37°C. The medium formulation is not important and others may be substituted. Medium must contain serum, calcium, magnesium, and glucose and be buffered to an appropriate pH. Antimicrobials, especially and antifungal agents, should be avoided as some are MDR substrates.
10. Make all inhibitors at 1,000× final concentration to avoid adding an excessive amount of the vehicle (EtOH for cyclosporine).

11. Prepare tubes with inhibitors ahead of time and transfer cells immediately after the addition of R123 and Ho33342.
12. It is useful, especially when first establishing this assay, to use reference cell lines known to express MDR activity. We use the chronic myelogenous leukemia cell line K562 (low MDR activity) and K562 cells transfected with the ABCB1 vector G185 (very high MDR activity) *(35)* as process controls **(Fig. 4)**, incubated in the absence and presence of inhibitors **(Fig. 2)**. The G185 transfectants were kindly provided by Dr. Suresh Ambudkar.
13. Keep on ice under aluminum foil until acquisition; maximum wait time before acquisition is 1 h. Holding cells longer will result in loss of side population cells.
14. Viable cells exclude PI. PI stains dead and dying cells very brightly and facilitates their removal from the analysis. Since it is broadly fluorescent, PI-positive cells must be removed from the analysis with a logical gate **(Fig. 4)**.
15. Single stained compensation standards are essential for correct spectral compensation. We find it best to use peripheral blood mononuclear cells incubated with rhodamine 123 (per earlier protocol) for the FL1 standard, integrally stained Calibrite beads for PE and APC, and Ig capture beads (CompBeads) for antibodies conjugated to tandem dyes. Hoechst 33342 staining does not require spectral compensation. It should be noted that R123, which is acquired with the same filter as is used for FITC (530/40 BP), has more spillover into the PE channel than FITC and therefore requires more spectral compensation.
16. Beads do not pack as tightly as cells, so care must be taken not to aspirate the beads.
17. Sonication is not essential but is very useful because it disaggregates bead clumps better than vortexing or refluxing.
18. Tandem dyes are easily degraded by exposure to light. Reagents, stained cells, and stained beads must be carefully protected from ambient light. We perform staining in an unilluminated biological safety cabinet, and cover stained cells and beads with aluminum foil to minimize light exposure.
19. Addition of mouse serum prevents clumping of beads due to antibody crosslinking.
20. Negative beads are not required for some automated compensation algorithms, such as those implemented on Beckman-Coulter instruments or in VenturiOne software.
21. Decanting and blotting must be done in one smooth motion to prevent loss of beads.

22. The sample station is prechilled and kept at 4°C during the entire acquisition.
23. Determining balanced PMT settings is an art in itself and has been addressed by us elsewhere *(32)*. Once these settings have been established, and bead target channels have been determined, they can be reproduced from experiment to experiment by adjusting PMT voltage to place the brightest peak of the Rainbow particles in the predetermined target channel. The intensity of Hoechst 33342 staining will vary from specimen to specimen, because it is measured on a linear scale and is highly dependent on the total amount of DNA in the sample. PMT voltages for the Hoechst Red and Blue channels are adjusted for each sample to place the median fluorescence of the 2N population at about channel 192 (of 255) (**Fig. 4**).
24. Our MoFlo is an analog model in which log or linear modes of amplification are user selected. Newer digital flow cytometers use high-resolution linear amplification exclusively and logarithmic transformations are performed mathematically.
25. CD45− Lin− CD44+ *side population* cells may be very rare events. We routinely acquire 10 million events per sample at rates not exceeding 10,000 events/second, to assure sufficient cells for subset analysis.
26. Which parameters are used to define classifiers, and which define outcomes is somewhat plastic and depends on the goal of the analysis. For example, one could classify cells into several populations on the basis of CD90 and EpCAM expression and then look at dye efflux as an outcome. The *classifier/outcome* strategy has the advantages of focusing the analysis and avoiding the *all possible permutations* problem inherent in multiparameter data files.
27. These antibodies interfere with transporter activity and cannot be used with dye transport assays.

Acknowledgments

The authors would like to thank Melanie Pfeifer and Amber McCauslin for technical assistance. This work was supported by grants BC032981 and BC044784 from the Department of Defense, the Hillman Foundation, and the Glimmer of Hope Foundation. Vera Donnenberg is a CDMRP Era of Hope Scholar.

References

1. Schinkel AH, Mol CA, Wagenaar E, van Deemter L, Smit JJ, Borst P. Multidrug resistance and the role of P-glycoprotein knockout mice. Eur J Cancer 1995;31A:1295–1298.
2. Benet LZ, Izumi T, Zhang Y, Silverman JA, Wacher VJ. Intestinal MDR transport proteins and P-450 enzymes as barriers to oral drug delivery. J Control Release 1999;62:25–31.
3. Goodell MA, Brose K, Paradis G, Conner AS, Mulligan RC. Isolation and functional properties of murine hematopoietic stem cells that are replicating in vivo. J Exp Med 1996;183:1797–1806.
4. Donnenberg VS, Burckart GJ, Zeevi A, et al. P-glycoprotein activity is decreased in CD4+ but not CD8+ lung allograft-infiltrating T cells during acute cellular rejection. Transplantation 2004;77:1699–1706.
5. Udomsakdi C, Eaves CJ, Sutherland HJ, Lansdorp PM. Separation of functionally distinct subpopulations of primitive human hematopoietic cells using rhodamine-123. Exp Hematol 1991;19:338–342.
6. Chaudhary PM, Roninson IB. Expression and activity of P-glycoprotein, a multidrug efflux pump, in human hematopoietic stem cells. Cell 1991;66:85–94.
7. Giangreco A, Shen H, Reynolds SD, Stripp BR. Molecular phenotype of airway side population cells. Am J Physiol Lung Cell Mol Physiol 2004;286:L624–L630.
8. Chen J, Hersmus N, Van Duppen V, Caesens P, Denef C, Vankelecom H. The adult pituitary contains a cell population displaying stem/progenitor cell and early embryonic characteristics. Endocrinology 2005;146:3985–3998.
9. He DN, Qin H, Liao L, et al. Small intestinal organoid-derived SP cells contribute to repair of irradiation-induced skin injury. Stem Cells Dev 2005;14:285–291.
10. Riou L, Bastos H, Lassalle B, et al. The telomerase activity of adult mouse testis resides in the spermatogonial alpha6-integrin-positive side population enriched in germinal stem cells. Endocrinology 2005;146:3926–3932.
11. Ling V, Thompson LH. Reduced permeability in CHO cells as a mechanism of resistance to colchicine. J Cell Physiol 1974;83:103–116.
12. Wang YC, Juric D, Francisco B, et al. Regional activation of chromosomal arm 7q with and without gene amplification in taxane-selected human ovarian cancer cell lines. Genes Chromosomes Cancer 2006;45:365–374.
13. Chen GK, Lacayo NJ, Duran GE, et al. Preferential expression of a mutant allele of the amplified MDR1 (ABCB1) gene in drug-resistant variants of a human sarcoma. Genes Chromosomes Cancer 2002;34:372–383.
14. Calcagno AM, Fostel JM, To KK, et al. Single-step doxorubicin-selected cancer cells overexpress the ABCG2 drug transporter through epigenetic changes. Br J Cancer 2008;98:1515–1524.
15. Sims-Mourtada J, Izzo JG, Ajani J, Chao KS. Sonic Hedgehog promotes multiple drug resistance by regulation of drug transport. Oncogene 2007;26:5674–5679.
16. Peacock CD, Wang Q, Gesell GS, et al. Hedgehog signaling maintains a tumor stem cell compartment in multiple myeloma. Proc Natl Acad Sci U S A 2007;104:4048–4053.
17. Lum L, Beachy PA. The Hedgehog response network: sensors, switches, and routers. Science 2004;304:1755–1759.
18. Johnstone RW, Ruefli AA, Tainton KM, Smyth MJ. A role for P-glycoprotein in regulating cell death. Leuk Lymphoma 2000;38:1–11.
19. Szotek PP, Pieretti-Vanmarcke R, Masiakos PT, et al. Ovarian cancer side population defines cells with stem cell-like characteristics and Mullerian Inhibiting Substance responsiveness. Proc Natl Acad Sci U S A 2006;103:11154–11159.
20. Patrawala L, Calhoun T, Schneider-Broussard R, Zhou J, Claypool K, Tang DG. Side population is enriched in tumorigenic, stem-like cancer cells, whereas ABCG2+ and ABCG2– cancer cells are similarly tumorigenic. Cancer Res 2005;65:6207–6219.
21. Harris MA, Yang H, Low BE, et al. Cancer stem cells are enriched in the side population cells in a mouse model of glioma. Cancer Res 2008;68:10051–10059.
22. Patrawala L, Calhoun-Davis T, Schneider-Broussard R, Tang DG. Hierarchical organization of prostate cancer cells in xenograft tumors: the CD44+ alpha2beta1+ cell population is enriched in tumor-initiating cells. Cancer Res 2007;67:6796–6805.
23. Mitsutake N, Iwao A, Nagai K, et al. Characterization of side population in thyroid cancer cell lines: cancer stem-like cells are enriched partly but not exclusively. Endocrinology 2007;148:1797–1803.
24. Lichtenauer UD, Shapiro I, Geiger K, et al. Side population does not define stem cell-like cancer cells in the adrenocortical carcinoma cell line NCI h295R. Endocrinology 2008;149:1314–1322.
25. Donnenberg VS, Luketich JD, Landreneau RJ, DeLoia JA, Basse P, Donnenberg AD. Tumorigenic epithelial stem cells and their

normal counterparts. Ernst Schering Found Symp Proc 2006;5:245–263.

26. Donnenberg VS, Landreneau RJ, Donnenberg AD. Tumorigenic stem and progenitor cells: implications for the therapeutic index of anti-cancer agents. J Control Release 2007;122:385–391.

27. Lou H, Dean M. Targeted therapy for cancer stem cells: the patched pathway and ABC transporters. Oncogene 2007;26:1357–1360.

28. Donnenberg VS, Donnenberg AD. Therapeutic index and the cancer stem cell paradigm. In: Bagley R, Teicher B, eds. *Stem Cells and Cancer Series: Cancer Drug Discovery and Development.* New York: Springer, Humana Press; 2009.

29. Bertoncello I, Williams B. Hematopoietic stem cell characterization by Hoechst 33342 and rhodamine 123 staining. Methods Mol Biol 2004;263:181–200.

30. Turgeon ML. Clinical hematology: theory and procedures (4th ed). Philadelphia, PA: Lippincott Williams & Wilkins; 2005.

31. Ibrahim SF, Diercks AH, Petersen TW, van den Engh G. Kinetic analyses as a critical parameter in defining the side population (SP) phenotype. Exp Cell Res 2007;313: 1921–1926.

32. Donnenberg AD, Donnenberg VS. Understanding clinical flow cytometry. In: O'Gorman MR, Donnenberg AD, eds. *Handbook of Human Immunology* (2nd ed). Boca Raton: CRC Press Taylor and Francis; 2008:181–220.

33. Donnenberg AD, Donnenberg VS. Rare-event analysis in flow cytometry. Clin Lab Med 2007;27:627–652.

34. Al-Hajj M, Wicha MS, Benito-Hernandez A, Morrison SJ, Clarke MF. Prospective identification of tumorigenic breast cancer cells. Proc Natl Acad Sci U S A 2003;100:3983–3988.

35. Kane S, Reinhard D, Fordis M, Pastan I, Gottesman M. A new vector using the multiple drug resistance gene as a selectable marker enables overexpression of foreign genes in eukaryotic cells. Gene 1989;84:439–446.

Index

T

V

X

Z

GPSR Compliance

The European Union's (EU) General Product Safety Regulation (GPSR) is a set of rules that requires consumer products to be safe and our obligations to ensure this.

If you have any concerns about our products, you can contact us on ProductSafety@springernature.com

In case Publisher is established outside the EU, the EU authorized representative is:

Springer Nature Customer Service Center GmbH
Europaplatz 3
69115 Heidelberg, Germany

Batch number: 10615113

Printed by Printforce, the Netherlands